CIE GUIDES TO INSECTS OF IMPORTANCE TO MAN

1 LEPIDOPTERA

J D Holloway, J D Bradley & D J Carter

Edited by
C R Betts

C·A·B International Institute of Entomology

British Museum Natural History

C·A·B International Institute of Entomology
56 Queen's Gate
London SW7 5JR
UK

Published by
C·A·B International
Wallingford
Oxon OX10 8DE
UK

Tel: Wallingford (0491) 32111
Telex: 847964 (COMAGG G)
Telecom Gold/International Dialcom: 84:CAU001
FAX: (0491) 33508

ISBN 0 85198 594 7
ISSN 0952-1461

Printed in the UK by The Cambrian News Ltd. Aberystwyth.

CONTENTS

FOREWORD

This manual has been developed for use by trainees attending the *International Course on Applied Taxonomy of Insects & Mites of Economic Importance* which has been held annually since 1979 by the CAB International Institute of Entomology (formerly Commonwealth Institute of Entomology) in London. It provides the basis for a practical introductory course for applied entomologists who are involved in the provision of taxonomic services for agriculture, horticulture, and forestry. It will also be of value to all entomologists who require an introduction to the taxonomy and biology of the Lepidoptera, and is the first of a series of practical training manuals which will eventually provide similar accounts of Coleoptera, Diptera, Hemiptera, Hymenoptera, and other relevant Orders of insects.

The manuals will be co-published with the British Museum (Natural History) enabling us to market them at an acceptable price. Both Institute and Museum staff will contribute to each work in the series and authorship will be acknowledged accordingly.

The main objectives of this manual are to facilitate identification of the most important families and subfamilies of Lepidoptera that are likely to be encountered in tropical, sub-tropical and temperate agriculture, horticulture and forestry. It is NOT intended as a pest species identification guide, though every effort has been made to give examples from genera of economic importance, and to refer in the bibliography to works that will help with identification to generic and specific levels.

In the keys prominence has been given wherever possible to characters that are readily appreciable and that do not involve dissection. Illustrations are provided as a 'running glossary' of morphological terms used, and to highlight key characters. Many of the higher taxonomic categories of the Lepidoptera are difficult to define in terms of clear-cut characters; indeed the higher classification is continually being modified by fresh research. Thus if our keys facilitate a success rate of over 80%, we will not be dissatisfied. Nevertheless, we would appreciate any constructive criticism and suggestions for improvement that can be noted for future revision of the work.

ACKNOWLEDGEMENTS

We would like to thank our many colleagues in Britain and overseas who have reviewed sections of the manual or provided information for it, particularly:
Mr. P.R.Ackery, Mr. D.S. Fletcher, Mr. D.T. Goodger, Mr. M.R. Honey, Dr. I.J. Kitching, Dr. G.S. Robinson, Dr. K. Sattler, Dr. M.J. Scoble, Mr. M. Shaffer, Mr. R.L. Smiles, Mr. K.R.C. Tuck, Mr. R.I. Vane-Wright, Mr. A. Watson (Department of Entomology, British Museum (Natural History)); Dr. J.S. Dugdale (DSIR Entomology Division, New Zealand); Dr. E.S. Nielsen (CSIRO Division of Entomology, Australia); Dr. R.W. Poole (USDA Systematic Entomology Laboratory, U.S.A.); Dr. J. Rawlins (Carnegie Museum, Pittsburgh, U.S.A.); Dr. S.E. Miller (Bishop Museum, Hawaii); Mr. M.E. Epstein (University of Minnesota, U.S.A.); Dr. G. Tarmann (Tiroler Landsmuseum, Innsbruck); Dr. J.W. Schoorl (Amsterdam); Mr. H.S. Barlow (Malaysia).

More than a hundred trainees have now used and tested earlier versions of the manual, pinpointing snags and anomalies and showing ways of improving the text. We thank them all for their patience, particularly those who used the earliest versions.

Much of the new artwork has been produced for the manual by the authors and also by Mr. Brian Hargreaves, Dr. Clive Betts and Mr. G.J. duHeaume (CIE); some of which was financed by a generous grant from Shell (U.K.) Ltd. The bulk of the keyboarding of text for the manual was done by Amitha Godage (CIE).

The final editing and production of camera-ready copy was carried out by Dr. Clive Betts who also designed the cover.

INTRODUCTION

The total number of known species of Lepidoptera makes up about 10 per cent of the Animal Kingdom and is roughly equivalent to the number of flowering plant species known. Many more await discovery and description and many so-called species will prove to be diverse complexes when examined more critically. The present total number of species approaches 200,000, most of which are moths (Heterocera) and only about 15,000 are butterflies (Rhopalocera). The moths are divided rather loosely into macrolepidoptera and microlepidoptera, a division that is largely artificial with a historical rather than taxonomic basis, but which is likely to remain for practical reasons in that methods of study differ somewhat and the literature is largely segregated.

Most of the species feed on green plants and consequently can be in direct competition with man, requiring counter-measures and control, but some are beneficial, many are aesthetic, and most, through their diversity and association with vegetation, may reflect the ecological stability of natural environments in the stability and persistence of their own populations. They are an integral part of most natural ecosystems: It has been calculated (D.Janzen, pers comm.) that in a forest in Costa Rica 80 per cent of defoliation was by caterpillars and, of those, 90 per cent were Lepidoptera.

Lepidoptera have a complete four-stage development or metamorphosis and are therefore classified as holometabolous. The four stages are: the egg or ovum (plural: ova), the embryonic stage; the caterpillar or larva (plural: larvae), the principal feeding and growing stage; the chrysalis or pupa (plural: chrysalids, pupae), a transition stage; finally the adult or imago (plural: imagos, imagines), the principal dispersive stage and sole reproductive one.

The adults are characterised by their broad-scaled and typically patterned wings and the presence, sometimes vestigial, of a tongue (haustellum or proboscis) which is often long and coiled, adapted for sucking liquid nourishment, particularly nectar, honeydew and the juices of fruits, but also used for obtaining dissolved salts from urine, ash, carrion and the lachrymal secretions of mammals. Some Trichoptera (caddis-flies) have thinly scaled wings but lack the tongue.

Variation in size is considerable, but gigantism is not necessarily indicative of economic importance, and most of the injurious species are small to medium sized. The largest moths in the world are the Atlas Moths, *Attacus atlas* and *Coscinocera hercules* (Saturniidae), of the Indo-Australian tropics and the rare Owlet Moth, *Thysania agrippina* (Noctuidae), of Brazil, which have wingspans around 300mm. The largest in body bulk are moths of which the genus *Xyleutes* (Cossidae) found in Australia, the females of which have abdomens up to 70mm long and 20mm or more in girth, with wingspans approaching 200mm. *Xyleutes* females sometimes lay more than 18,000 eggs. The largest butterflies are in the swallowtail family, Papilionidae, which includes the Birdwings (*Ornithoptera*, *Troides* and their allies) from the Indo-Australian tropics with wingspans of around 200mm. The smallest Lepidoptera are the pigmy moths (Nepticulidae), the smallest in the world being the European *Stigmella acetosae* with a wingspan of 2-3mm.

Lepidopterous larvae generally feed on the foliage and other parts of living plants, either externally as browsers and shelter-dwellers or internally as borers and miners. Some have adapted to a life on plant produce such as dried fruits, seeds, nuts, and flour. Some forage dead or decaying vegetation, others feed on fungus or keratinous material of animal origin such as hair, wool, horn or feathers. Some galleriine Pyralidae feed as larvae on wax in the nests of bees. Predation and parasitism are rare but isolated cases are known from a few families where the larvae prey on scale insects, leafhoppers and other Hemiptera. Some of the more important categories from the economic standpoint are elaborated below.

Many Lepidoptera coming into the pest category are restricted to a particular geographical region and often to a particular crop or host-plant, but others, perhaps the majority, are geographically widespread and/or have a wide range of host-plants. For

3

example, in the noctuid genus *Othreis* there are numerous species in the Indo-Australian tropics but it is the most widespread species, primarily *fullonia* and *materna*, that are most often recorded as pests piercing and damaging citrus fruit. *O.fullonia* also appears to have an expanded host range that may have contributed to its success: its relatives appear to be restricted to the plant family Menispermaceae but *fullonia* will feed as a larva on Leguminosae, particularly *Erythrina*. In the pyraustine genus *Crocidolomia* the species *binotalis* is a widespread pest of Cruciferae in the Old World tropics, yet all its congeners, as far as is known, are more localised and feed on the related family Capparidaceae. Again a host switch appears to have been agent in the development of a pest species within a genus of non-injurious species. Therefore the recording of life histories and host-plants of non-pest species may be a vital process in expanding our understanding of the biology of pests.

Inevitably a number of pest species have become established in new areas, usually as result of the activities and increased mobility of man and his biological economic resources, or because he has created areas of suitable habitat for them. Apart from the aerial mobility of the adults, their migratory or dispersive habits, many species are carried about at some stage in their life cycle by commerce to other parts of the world. Accidental introductions usually soon perish, but some species survive and flourish, often on a dramatic scale if their natural predators and parasites are absent. Therefore, when identifying suspected lepidopterous pests, the possibility of non-indigenous species must be considered. This is especially true for entomological services responsible for plant quarantine.

CHARACTERISATION OF THE LEPIDOPTERA

The name 'Lepidoptera' is derived from the Greek terms meaning 'scaly-winged'. This is the most obvious feature distinguishing Lepidoptera from other insects. The setae (macrotrichia) covering the wing membrane have become modified into flattened, overlapping scales. In the majority of Lepidoptera scales are also present on the head, thorax, legs, and abdomen. These scales, through pigment or structure, contribute colour and pattern to the wings, perform some aerodynamic function, and may provide a limited means of thermoregulation.

Many of the other features associated with the Lepidoptera, such as a maxillary tongue, are really only applicable to the more advanced representatives. Kristensen (1984b) has listed 26 features he considers as definitive apomorphies for the Lepidoptera, and also gives 21 features that serve to unite the Lepidoptera and Trichoptera as sister-groups, making up the Amphiesmenoptera.

The 26 Lepidopteran features given are mainly found in the adult as follows:

Head: loss of median ocellus; postmedian process on corporotentorium (for insertion of neck muscles); an intercalary sclerite laterally in membrane between antennal scape and pedicel; characteristics of flexure and lack of antagonistic muscles in the maxillary palp; presence of a slender craniostipital muscle; arched postlabium with long piliform scales; chemoreceptors in a depression on the terminal segment of the labial palp; salivarium without longitudinal dorsal muscle; two features of the nervous system.

Thorax: "hair-plate" near anterior apex of laterocervical sclerite; prothoracic endoskeleton with free arm arising from bridge between sternum and pleuron; tergopleural apodeme on mesothorax; "prescutal arm" on metathorax; epiphysis on inner surface of fore-tibia; wings with dense covering of broad scales; metathoracic spiracle with single anterior lip.

Abdomen: first tergite with paired lateral lobes articulating with second sternite;

valve of male genitalia primarily undivided; protractor muscles of aedeagus originating within the valves; cerci absent in both sexes.

Visceral: abdominal nerve cord with at most five ganglia and unpaired connections; mesothoracic aorta curving upwards to dorsum.

Larval head: elongated pleurostome; maxillary palp with less than five segments.

INJURIOUS LEPIDOPTERA

DEFOLIATORS

The majority of injurious species are those with larvae that feed on foliage. Macrolepidoptera particularly fall into this category: examples such as armyworm and cutworm (Noctuidae) outbreaks which occur in all warm regions spring immediately to mind. Amongst the microlepidoptera the Tortricidae are important: in northern temperate regions *Tortrix viridana* periodically defoliates oak, while in the Oriental tropics *Homona coffearia* and *Cydia leucostoma* are important defoliators of tea. *Plutella xylostella* (Yponomeutidae) is a pest of Cruciferae throughout the world and, although the larva is very small, it often occurs in such numbers that damage can be considerable. This migratory species is remarkable in that it can flourish in most climates and at altitudes up to 4,000m.

In severe outbreaks by lepidopterous defoliators, crops and trees can be completely stripped of foliage but mostly the effect may be more subtle and hard to quantify economically. Minor defoliation can affect yields of crops directly or even indirectly, for example through defoliation of shade trees for a cash crop such as coffee. Shade trees planted to improve the quality of human life may also be affected by defoliation, for example the periodic stripping of *Samanea* trees (Leguminosae) in South East Asia by species of the noctuid genus *Rhesala* or, with associated rashes and irritation caused by airborne larval hairs, the stripping of pines in Mediterranean areas by species of *Thaumetopoea* (Thaumetopoeidae). It may be very hard to predict which species will be serious pests when a new crop is planted in a new area, such as in Malaysia when young oil palms planted in areas recently cleared of forest were attacked by several Lepidoptera, mainly Limacodidae. The effects dwindled as the plantations matured. Some pest outbreaks may be more or less unique events such as the stripping of areas of *Shorea albida* (Dipterocarpaceae) swamp forest in Sarawak for a brief period by an unidentified lymantriid species.

It is impossible for lepidopterists to restrict their attention to a few well known pest genera within a few families; defoliating pests are found in most families, even small ones such as Thaumetopoeidae and Eupterotidae, and often only occur one to a genus. Add to this the families predominant in other categories of damage and the task of the economic lepidopterist is seen to be immense.

LEAF ROLLERS AND TIERS

A number of species compound the damage they do in defoliation by also cutting a large area of the leaf lamina to scroll into a protective case or by tying leaves together with silk. The hesperiid butterfly *Erionota thrax* scrolls the leaves of bananas in the Oriental tropics. Numerous species in the families Pyralidae and Tortricidae come into this category. One of the best known is the pyralid *Syllepte derogata* which rolls the leaves of cotton throughout the Old World tropics. A number of pyralids roll the leaves of rice including *Cnaphalocrocis medinalis* and species of the genus *Marasmia*.

LEAF MINERS

Many microlepidoptera, particularly in the families Nepticulidae, Incurvariidae, Gracillariidae and Lyonetiidae, have larvae small enough to feed in a mine or tunnel between the upper epidermis and lower epidermis of a leaf. Several are of major economic importance, such as the gracillariid *Conopomorpha cramerella* and the lyonetiid *Leucoptera coffeella*. The character of the mine and the deposits of frass within it is often of taxonomic value so the affected leaves should be preserved.

STEM BORERS

Species from several different families have larvae that bore in the stems of annual crops or shrubby ones. Cereal stem borers occur in a number of different families, but particularly the Noctuidae (e.g. *Busseola* and *Sesamia*), and the Pyralidae. In the latter, the subfamily Crambinae includes some of the most injurious stem borers of graminaceous crops. The best known is probably *Chilo partellus* the larva of which attacks maize and sorghum in the Old World tropics. In the Oriental Region several *Chilo* are found on rice, e.g. *C. polychrysa* whose larva hollows the stem and causes wilting or dead heart, and *C. auricilia* which eats in to the rice stem, causing it to snap. The subfamily Phycitinae contains *Maliarpha separatella*, a rice borer in the African and Oriental tropics, and *Emmalocera depressella*, widespread on the Indian subcontinent, which bores into the root-stock of sugarcane. The subfamily Schoenobiinae contains important rice and sugar cane borers in the genus *Scirpophaga*.

Cossidae such as *Zeuzera* species feed as larvae in the stems of shrub crops such as coffee, and Noctuidae such as *Penicillaria jocosatrix* bore in the stems of mango and other shrubs. Forest trees grown for commercial purposes are often subject to severe injury, for example by pyralids of the *Hypsipyla* generic complex in the Indo-Australian tropics.

TIMBER BORERS

Cossidae of the genera *Cossus* and *Xyleutes* and Hepialidae of the genus *Endoclita* can cause damage to timber crops in the Indo-Australian tropics.

ROOT AND TUBER FEEDERS

Among the most important tuber and fleshy stem feeders are gnorimoschemine microlepidoptera of the family Gelechiidae. These include *Phthorimaea operculella* which is a major pest of potato tubers and will also attack tomato and egg plant; all are Solanaceae. The Cossidae include root feeders such as *Azygophleps albovittata* which attacks groundnuts in Nigeria. The larvae of many Hepialidae damage the roots of pasture and other grasses, but mainly in temperate regions. Some noctuids such as *Luperina testacea* are also pasture root feeders.

FLOWER FEEDERS

Lepidoptera from several different families feed as larvae on the flowers of plants and may thus affect yields if the crop is a fruit or a seed. Larentiine Geometridae in the *Gymnoscelis-Chloroclystis* complex often feed on flowers as do Noctuidae in the subfamilies Nolinae and Acontiinae (*Eublemma*).

A serious pest of flower buds and bolls of cotton at all stages is *Pectinophora gossypiella* (Gelechiidae), which is found in most cotton growing regions of the world. The larva is difficult to detect because there is little external evidence of its presence.

6

SEED AND FRUIT FEEDERS

Relatively few macrolepidoptera feed on fruit and seeds, such as the sarrothripine noctuid genera *Characoma* and *Nanaguna* and a few lycaenid butterflies on the pods of Leguminosae, but this habit is much commoner amongst the microlepidoptera.

The phycitine pyralid *Etiella zinckenella* is highly injurious to the pods of legumes and has a circumtropical distribution, but can easily be confused with several related species. Numerous examples occur in the Tortricidae, especially the olethreutine genus *Cydia*. These include *C. pomonella* which bores in to the fruits of apple and pear in temperate regions, and *C. molesta* which attacks the fruit and shoots of peach in Europe, N.America and Australia. *Cryptophlebia leucotreta* is a ubiquitous pest of citrus and other cultivated fruit in Africa.

FRUIT PIERCERS

A number of large noctuids have, as adults, strengthened and modified tongues that enable them to pierce the skins of fruits and suck out the juices. In doing so they introduce fungi that cause the fruit to rot. The most renowned of such species is *Othreis fullonia* but several of its close relatives and species of *Calyptra*, *Anomis* and *Ophiusa* are also involved.

EYE FREQUENTERS AND BLOOD SUCKERS

Several noctuids and some geometrids, including relatives of species in the previous category, fly to the eyes of slow-moving mammals during the night to feed on their lachrymal secretions. They may introduce diseases in this way and thus be of veterinary importance. The problem is being studied in Thailand and Malaysia by Dr Hans Banziger. The species include catocaline and noctuids and also some geometrids such as ennomines of the genus *Semiothisa*. Many of the species involved are also attracted to carrion and may rely on such feeding habits to obtain protein, or salts of sodium or potassium necessary for muscle activity.

A very few noctuids such as *Calyptra eustrigata* have a tongue of the fruit piercing type adapted for penetrating mammalian skin and sucking blood.

STORED PRODUCT PESTS

A number of species that were presumably adapted to feeding on fallen seeds and detritus of both animal and vegetable origin have become pests of stored food products and domestic and textile pests. The most widespread and injurious of these belong to the families Tineidae, Oecophoridae, Gelechiidae and Pyralidae.

Sitotroga cerealella (Gelechiidae) is second only in importance to the grain weevils on stored grain, and also important on processed cereals are *Nemapogon granella* (Tineidae) and *Ephestia kuehniella*, *Plodia interpunctella* and *Pyralis farinalis* (Pyralidae). *Ephestia cautella* (Pyralidae) has in recent years become a major pest on a variety of processed vegetable products. *Hofmannophila pseudospretella* (Oecophoridae) is an almost cosmopolitan household pest, feeding on most dried vegetable or animal material.

Many tineines (Tineidae) feed on keratinous material of animal origin such as hair, wool, horn or feather, the 'clothes moths' of the genera *Tinea* and *Tineola* being well known examples.

BENEFICIAL LEPIDOPTERA

The activities of Lepidoptera are not always detrimental to man. They can also be beneficial: as pollinators, silk producers, in biocontrol as both defoliators and predators, through their aesthetic value and as indicators of environmental quality.

POLLINATORS

Adults of most Lepidoptera make some sort of contribution to the pollination of flowering plants dependent on insects. Many plants, usually recognisable through the correlation of their long corolla tubes with the exceptional length of the tongues of some Lepidoptera such as species of Sphingidae, are specifically moth-pollinated. Tobacco flowers are usually pollinated by Sphingidae and an extreme instance is seen in the pollination of the Madagascan orchid *Angraecum sesquipedale* by *Xanthopan morgani*.

Complex symbioses have evolved, such as that between the yucca and *Tegeticula yuccasella* (Prodoxidae) where there is total interdependence for cross-pollination and larval food resources.

SILK PRODUCERS

Many Lepidoptera larvae spin a silken cocoon within which to pupate. The cocoons of species in the superfamily Bombycoidea are especially well developed and several have come to be exploited by man for their silk. The classical commercial silkworm is *Bombyx mori* (Bombycidae) but several Saturniidae, such as species of the genus *Antheraea*, produce silk (termed wild silk) that is of commercial value though not as fine as that produced by *mori*.

AGENTS OF BIOLOGICAL CONTROL

The sometimes intense defoliation of plants by Lepidoptera larvae can, on occasions, be turned to advantage in controlling weed species introduced from one region of the world to another. Potential control agents have to undergo elaborate tests to ensure their specificity to the plants to be controlled or they might themselves become pests when introduced. A sound taxonomic investigation is an essential part of this process. The population dynamics of the control agent and its potential for adaptation to a new environment are important factors in its success or otherwise.

The classic example of success using Lepidoptera in biological control is the control of *Opuntia* (prickly pear) in Australia by the phycitine pyralid *Cactoblastis cactorum*.

PREDATORS

A few Lepidoptera are of some benefit to man as predators or ectoparasites of scale insects or other homopteran pests such as psyllids, fulgorids and cicadellids. The small family Epipyropidae is well known for this in the Indo-Australian tropics but there are also acontiine Noctuidae, batrachedrine Momphidae and phycitine Pyralidae that are of minor importance in this respect.

AESTHETIC VALUE

One of the reasons why the Lepidoptera, particularly the butterflies and bombycoid moths, have been more extensively collected and studied than other Orders is that many have great aesthetic quality in their wing patterns. This has, in some instances, given them a commercial value. In many parts of the world they are collected in large numbers from the wild for sale to collectors or to be made up into souvenir articles for tourists. The commoner species can withstand moderate harvesting of this kind in the same way as any other biologically renewable resource. But it is necessary to monitor such activities carefully and to regulate the commercial collection of rarer species by legislation (to the extent of prohibition in the case of endangered ones). If such an industry is commercially important it can lead to the conservation of natural habitats on which the species concerned depend.

In some parts of the world, such as Papua New Guinea, viable butterfly and moth farming projects have been set up.

Lepidoptera could be said to be of indirect commercial value in that they are favourite subjects for definitive issues of postage stamps, these often making a substantial contribution to the revenues of the issuing state. In Asia, Australasia and the Pacific sets have recently been issued by Australia, the Cocos-Keeling Is., Malaysia, New Caledonia, New Zealand, Norfolk I., Pitcairn I. and Tuvalu.

INDICATORS OF ENVIRONMENTAL QUALITY

Increases in human population combined with advances in technology have, in the second half of this century, subjected the ecosystems of the world to pressures with which they were never adapted to cope. The result has usually been dramatic change to those ecosystems, particularly when they are fragile as in the tropics, both humid and semi-arid. Such changes will, in the long term, affect the human populations dependent on those ecosystems: erosion is enhanced by forest clearance, fertility impaired by wasteful agricultural methods, and desertification is increased by overgrazing and harvesting of firewood.

There is thus a need to develop long term resource management policies for these ecosystems based on an understanding of the ecological processes involved in their maintenance, ensuring a sustained yield of agricultural or forest products for human benefit as well as reservoirs of natural habitat to maintain the biological diversity of world ecosystems and their untapped potential for human welfare. Associated with this is a need for techniques to monitor changes such as degradation and regeneration.

One such technique is to use indicator organisms. Insects, through their diversity, rapidity of generation turnover and therefore susceptibility to change, are prime candidates, particularly phytophagous insects where specificity to particular plant taxa and therefore to vegetation type is likely to be high. Lepidoptera may prove particularly suitable, being readily sampled with light-traps and relatively well collected and studied taxonomically.

THE IDENTIFICATION OF LEPIDOPTERA

In the Lepidoptera, identification is usually based on the imago because immature stages are insufficiently known taxonomically. Here it might be stressed that the field entomologist can make a valuable contribution by rearing a proportion of larvae through to the adult stage while at the same time preserving others and pupae (see notes on preservation of specimens) for museum storage and reference. In this way authentic collections can be formed for future research.

It is especially important to preserve voucher material of all stages in a recognised institution when a description of a newly discovered life history is published, even when of an apparently common and well known species. If its taxonomy later proves to be more complicated than envisaged originally, the material can be re-examined and the correct name associated with the life history description. For example, the name *alope*, applied indiscriminately to all noctuids of the genus *Lacera* in the Indo-Australian tropics was found by Holloway (1979) to embrace a complex of seven species, three of which range widely from India to Melanesia. It is therefore uncertain to which species belong life histories attributed to *alope* in the literature unless voucher material of adults has been preserved.

Identification of imagos is normally first attempted by examination of gross external characters such as size, wing shape, colour-pattern and markings, body vestiture, antennae, palpi, legs and other readily visible structures. Success can depend very much on the condition of the specimen(s), and identification is facilitated when these are pinned through the thorax and have the wings spread flat. It is essential to base descriptions of new species on good, well prepared material. Superficial visual diagnosis must usually be backed up by critical examination of structural characters.

Conventional keys based on the sort of morphological characters used for other orders of insects, such as on sclerite shape and sculpturing, distribution of setae, tarsal morphology and even wing venation, are much less practical for Lepidoptera as most of these structures are obscured by dense scaling, and their examination often requires elaborate preparatory work. A crucial part of the key presented later is the absence, presence and position of the auditory, or tympanal, organs of moths. These are often difficult to see, especially when a character such as the position of the first abdominal spiracle relative to the counter-tympanal hood is involved as in the distinction between Noctuidae and Arctiidae. Such characters are even more difficult to see in the smaller members of the families concerned and a beginner might be excused for considering them absent or making other mistakes. An experienced lepidopterist rarely uses such characters, preferring to rely on a general 'feel' for the groups that is very hard to put into words in a key.

Nevertheless, we hope that the keys to suborders, superfamilies and families, despite their limitations, will be useful to a student trying to acquire familiarity with the Lepidoptera.

There are very few satisfactory keys to genera in the Lepidoptera, mainly because most genera (and indeed many higher categories such as noctuid subfamilies) are defined on very weak characters and embrace species that prove on dissection and closer examination not to be related at all. Most identifications within a family are made with reference to plates of illustrations or a reliable insect collection, by means of matching wing patterns and other readily appreciable characters. This process has led many entomologists to consider that Lepidoptera identification is easy, but this is grossly misleading. It takes many years of experience of the literature or of a major reference collection to be able to narrow the field rapidly, i.e. to know where to look.

Even then, the matching of external characters probably only leads to the assignation of a specimen to be identified to a species complex where all members are externally very similar. The key characters of the genitalia are again obscured by the scaling: recourse to dissection is necessary.

Nomenclatural problems are also more acute in the Lepidoptera, again because the classification has been based in the past on weak external characters, but also because the group has been very popular, particularly with amateurs who, whilst often very proficient, sometimes do not appreciate in their enthusiasm the complexity of the job they are tackling or, for logistical reasons, do not have access to all relevant type material. It is not unusual to find that a species has been described more than once in different subfamilies of one family or even in different families. The weak definition of genera means that species are frequently changing their generic combination, which is very confusing for the field entomologist.

The superficiality with which Lepidoptera taxonomy has been pursued in the past has resulted in synonymies being proposed on flimsy evidence without examination of type material. The following scenario is unfortunately far from rare. Brown describes species A,B, and C. Smith decides they are conspecific and places B and C as synonyms of the oldest name, A. Fresh material collected from the same geographical area comes into the hands of Jones who realises there are in fact three species, decides one is A, accepting the published synonymy of Smith, and describes the other two as new, X and Y, without examining the type specimens of the names A,B, and C. It may transpire when this is in the literature that Jones' concept of A is wrong, that his A is C, his X is B and his Y is A. The genus *Ganisa* in the Eupterotidae contains three species that have been treated in this way.

The sort of nomenclatural confusion outlined above can only be unravelled by recourse to type material. Therefore it is essential that such type material be (a) well curated on a permanent basis and (b) available to bona fide scientists for study, preferably, for logistic reasons, in one place rather than scattered all over the globe.

COLLECTION AND PREPARATION OF SPECIMENS

The ideal Lepidoptera specimen is one that has been reared, then killed and spread whilst still in perfect condition. Such a specimen has maximum information content both biologically (full data on rearing, host-plant etc. should be kept) and taxonomically. Therefore care must be taken to inspect emergence cages regularly and collect any adults before they damage themselves in flight; however, the wings should be allowed to harden fully before killing.

Wild-caught material is usually obtained by netting or light-trapping. Again it is vital that maximum taxonomic information content be preserved by ensuring each specimen is collected, killed and spread in as close to perfect condition as possible. The scales on the wings of Lepidoptera are easily rubbed off and also easily wetted by killing agent. This is particularly so for the more fragile microlepidoptera.

If light-trapping is the method used, the best means of acquiring perfect specimens is to collect off a white sheet set behind the lamp. Larger specimens can be taken in a killing jar in small numbers, and transferred to a larger jar when quiescent. Smaller moths should be collected live individually in small glass tubes and killed either by placing the tube in a deep freeze or one by one immediately prior to spreading the following morning (see below).

KILLING, PINNING AND WING-SPREADING

KILLING

Micro-moths and very small macro-moths are best killed with either ammonia (0.880) or ethyl acetate since these leave the specimen in a relaxed condition ready for pinning and spreading. A plug of absorbent cotton wool dampened with the killing agent is dropped to the bottom of a glass jar or vial and covered with a thin layer of non-absorbent cotton wool. Keep the sides of the container dry. Exposure to fumes for 10-15 mins. is usually sufficient. Chloroform, trichloroethylene, and carbon tetrachloride are also quick but tend to leave specimens in a stiffened condition. The conventional killing jar containing potassium cyanide (KCN) embedded under a layer of plaster of paris is more suitable for larger moths and butterflies but must be used *with caution*. Specimens which become too stiff for spreading should be left in the jar or in a relaxing box for 12-24 hrs.

PINNING

Micro-moths: stainless steel headless pins (minuten pins). Useful sizes are: diameters .010, .0124, .0179
Macro-moths and butterflies: stainless steel pins with nylon heads. Useful sizes

are: Nos. 2-5; length 1 1/2 inches (approx. 37mm). Nos 0 & 1 are too flexible and are not recommended.

Specimens should be pinned through the middle of the mesothorax, the head of the pin tilting very slightly forward. Micro-moths on minuten pins should be staged, preferably on a short length of polyporous, cork or suitable plastic material which holds the pin firmly, rather than on card, which tends to let the specimen spin. A No. 5 pin is used for staging.

SETTING or WING SPREADING

When setting boards are not available the rough-spreading of the wings is a simple and satisfactory compromise. It is important to spread the wings since this not only facilitates examination and identification but leaves the specimen in a good condition for relaxing and setting later.

Both macro and micro moths can be rough-spread by pinning into plastazote (fine plastic foam) and moving the wings gently up into the spread postion with a needle. The wings should stay in that position through friction with the foam or may be held in place by further minuten pins (try not to puncture the wing).

For macro-moths it is important that the insect be placed two-thirds up the pin from the point to allow plenty of space for data labels. Therefore a thick piece of plastazote will be needed for rough-spreading, or a thinner sheet set up on a frame.

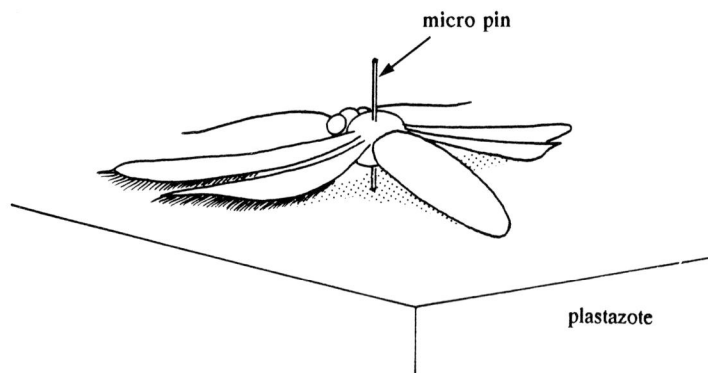

Wing spreading on plastazote

If setting boards are available, there are several points to remember:
1. The groove in the board should be slightly wider than the body and legs of the moth or butterfly; the wings should not extend beyond the edge of the board.
2. Select the pin to fit the insect, for example:
    small Geometridae - size 2
    medium Geometridae, small Noctuidae - size 3
    large Geometridae, medium Noctuidae - size 4
    large Noctuidae, most Sphingidae and Bombycoidea - size 5.
3. Pin the insect down in the groove so the wings are flush (level) with the surface of the board; if too high or too low, the wings get bent at the base.

12

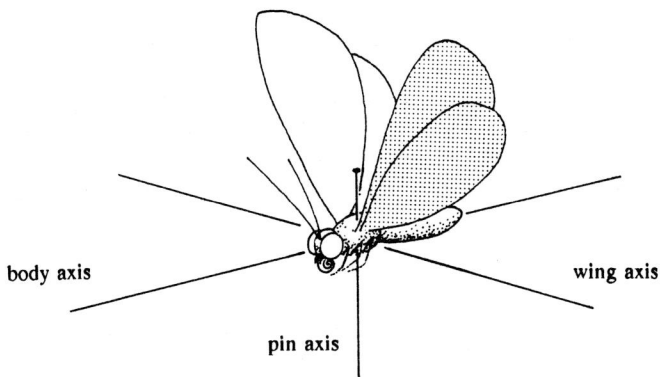

Three axes involved during setting

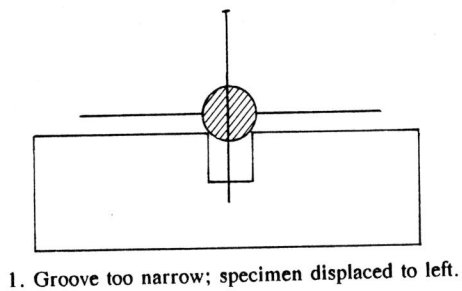

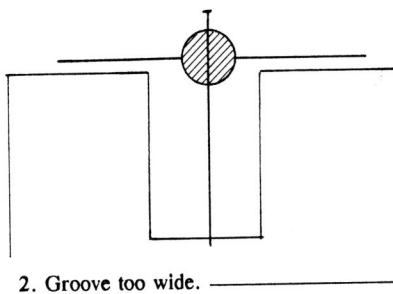

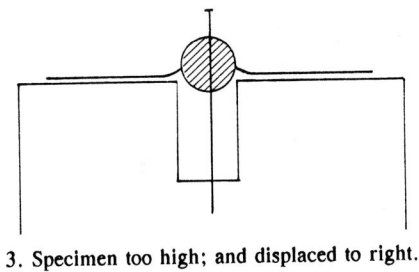

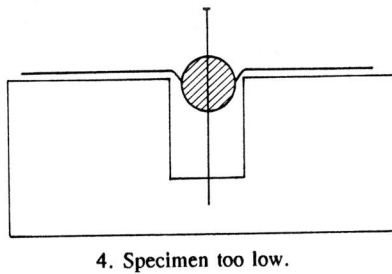

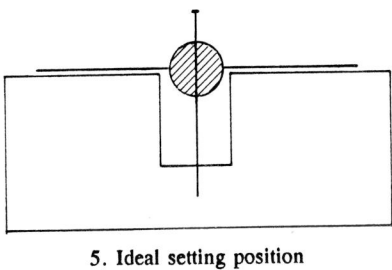

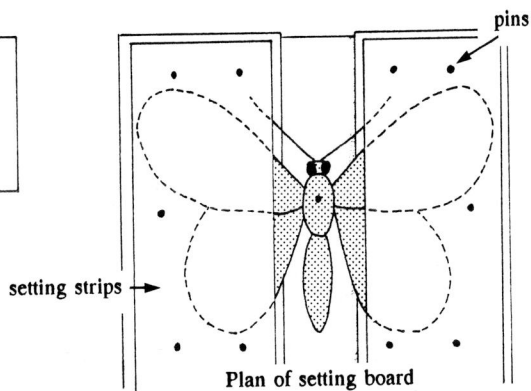

1. Groove too narrow; specimen displaced to left.

2. Groove too wide.

3. Specimen too high; and displaced to right.

4. Specimen too low.

5. Ideal setting position

Plan of setting board

Using a setting board: 1-5 sections through wing axis of body.

13

4. Correct display involves right-angles: of the pin to the body axis; of the forewing dorsum to the body axis.
5. Anchor setting paper strips to the head of the board (away from you) with pins. Then, whilst the 'major' hand moves the wings forward with a needle or pin by pressure behind the main veins, the 'minor' hand can, with one finger, tense or relax the strip allowing the wings to be held or moved.
6. Don't place the edge of the setting strip too close to the edge of the groove.
7. Keep moving the hindwings up with the forewings or they may become uncoupled and flip up.
8. When the wings are in the correct position, anchor the strips with pins to maintain tension.
9. Keep the boards in a dry atmosphere or *gentle* oven for one to two weeks before unpinning. In the tropics a cupboard illuminated by a normal electric light bulb can provide the right conditions. Beware of depredation by ants.
10. Add label data as soon as possible after unpinning.

PRESERVING LARVAE AND PUPAE

For study of morphology and chaetotaxy larvae and pupae should be preserved in fluid. A simple but satisfactory method is to drop the larva or pupa into a vial containing about 9 parts 80% alcohol and 1 part glacial acetic acid. The addition of acetic acid helps prevent discoloration. After 48 hours transfer specimens to 80% alcohol to which a little glycerine is added.

An alternative method is to kill the larva by dropping it into water that has just come off the boil, cooling it in cold water after a few minutes. It can then be transferred to the alcohol and glycerine mixture.

GENITALIA DISSECTION

CURSORY EXAMINATION and TEMPORARY SLIDE PREPARATION

Equipment

stereo microscope with linear magnification up to at least x100
dissecting needles (dental canal-finders)
camel-hair or sable hair brushes, sizes 1 - 2
fine-pointed forceps (watchmaker's)
test tubes and glass beaker (250ml - 500ml)
glass dishes or watchglasses (flat-bottomed) or artists' palettes
30% - 50% alcohol
10% solution of potassium hydroxide (KOH) - caustic potash
microscope slides and coverslips
glycerine (in Canada balsam bottle with glass dropper-rod)

Useful additional items are a small hypodermic syringe (2ml) with a fine needle (sizes 25G - 30G, about 13mm. long) and acetic acid.

Procedure

Detach abdomen: if the insect is pinned and already dried-out apply upward pressure with the tips of the fine forceps placed under the tail-end; if the insect is freshly killed and not dried-out (but see below) the abdomen will need to be severed by gripping with the fine forceps immediately behind the metathorax. Immerse abdomen in test tube containing about 5ml of 10% KOH. Either, a) leave abdomen to soak for 12-24 hrs. in KOH at room temperature, or b) heat test tube in a water bath (e.g. beaker half filled with water)

14

to near boiling point and leave to simmer for 5-10 mins. The abdomen should become less opaque and sink to the bottom of the test tube. A Dri-Block electric heater is an alternative to the water bath. Maceration of fresh material is difficult and rarely gives satisfactory results so should be avoided.

Remove the softened (macerated) abdomen from KOH with dissecting needle or forceps and transfer to watchglass or palette of 30-50% alcohol. Descale and cleanse, teasing out body contents by pressing with needle and brush, working from the genitalia towards the anterior of the abdomen. After initial cleaning transfer to fresh alcohol for examination. The genitalia can now be separated from the body wall for easier examination, those of the female being dissected out by using the needles or fine forceps to cut or gently tear the membranous tissue between the 7th and 8th segments, care being taken not to sever the ductus bursae. It is often useful to stain female genitalia lightly, after separating the 7th and 8th tergites, so as to reveal the ductus and bursa more clearly.

Preparations which are difficult to get properly clean in alcohol because of crystalline or oily deposits should be dissected instead in weak acetic acid (about 30%). After cleaning return to alcohol before proceeding.

In some groups, e.g. Noctuidae, distention of the vesica of the male's aedeagus can aid examination of cornuti and scobination. This is achieved by inserting the hypodermic needle into the base of the aedeagus via the entry hole of the ductus ejaculatorius and inflating the vesica with alcohol.

If required, a temporary slide preparation can be made by taking the dissection direct from alcohol and mounting in glycerin.

After examination the abdomen and genitalia can be glued on thin card and pinned with the specimen. Only a minute quantity of water-soluble glue should be used, in order that the dissection can be easily floated off in water or KOH.

PERMANENT SLIDE PREPARATIONS

Additional equipment
        stain: 1% aqueous solution of Mercurochrome (i.e. 1gm in 100ml distilled water),
        or 1% (approx.) solution of Chlorazol Black E in 20%-30% alcohol (i.e. a few
        crystals of Chlorazol Black in 100ml of alcohol).
    drop bottle(s) for stock solution of stain
    95% alcohol
    mountant medium: a proprietary brand, e.g. Euparal
    Canada balsam bottle (loose top and rod)
    mountant solvent: e.g. Euparal Essence

Procedure
Staining usually follows after cleaning of the abdomen and dissection of the genitalia as already described. However, staining during dissection can help reveal transparent structures such as the female bursa copulatrix, and thus make dissection easier. If cleaned in acetic acid the dissection should be thoroughly washed in 30% alcohol before staining.

Staining: a drop or two of stock solution is added to water or weak alcohol in a watchglass and the dissection immersed for 2-10 mins., depending on the strength of the stain. Avoid overstaining; staining should be sufficiently heavy to show details of fine membranous structures, such as the ductus bursae of the female, and weakly sclerotized structures. It is preferable to use a weak stain and repeat the process rather than stain too heavily. If overstaining occurs, return the preparation to acetic acid for clearing. After staining and final cleaning and dissection, transfer the preparation to 95% alcohol for at least 10-15 mins.

The preparation should now be sufficiently dehydrated and hardened for mounting. First put a drop or two of the solvent in a palette or onto a microscope slide and lift abdomen and genitalia from the alcohol into it with a needle. If the specimen is large, excess alcohol can be removed by lightly dabbing onto filter paper before placing into the solvent. A blob of mountant is then dropped onto the middle of a slide (cleaned with alcohol) and should be sufficient to spread and prevent undue pressure when the coverslip is applied. The dissection is then lifted from the solvent to the mountant with a needle, and the genitalia and abdomen are manipulated into position (see below) and the coverslip applied with fine forceps.

The slide preparation is kept flat and allowed to harden in a covered tray for several days at room temperature, or in a slide-drying oven at 40-50°c for 48 hrs.

ARRANGEMENT OF GENITALIA and LABELLING OF SLIDE

Method of treatment varies according to the family or species-group, but generally the genitalia in both sexes are dissected from the abdomen and mounted alongside it on the slide in ventral aspect. Whenever possible the valvae of the male genitalia are opened out and pressed flat on the slide, a sufficient quantity of mountant being used to prevent undue distortion through weight of the coverslip. When the valvae cannot be spread apart and remain closed, e.g. as in some Gelechiidae, the genitalia are usually mounted in a lateral position. The aedeagus of the male is dissected out, but if this cannot be done without causing structural damage it should be left in situ.

Slide preparations should be numbered and labelled and a separate label with the slide number should be pinned with the specimen.

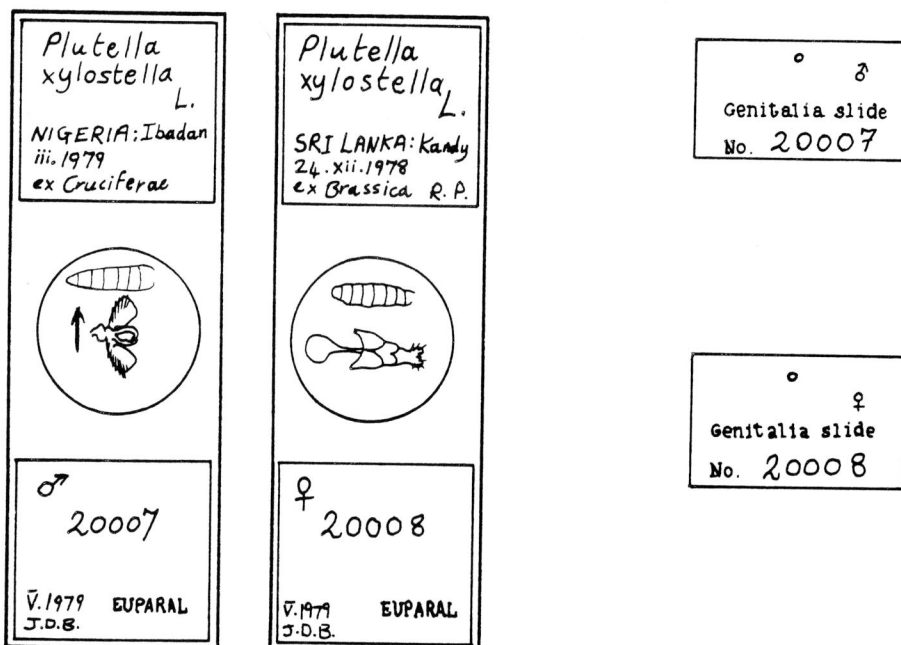

CHECKLIST OF HIGHER TAXA OF THE LEPIDOPTERA

Following the scheme proposed by Nielsen (1985) the order Lepidoptera may be broadly classified into four groups or "suborders": Zeugloptera, Aglossata, Heterobathmiina and Glossata. The Glossata may be further divided into Dacnonypha, Neopseustina, Exoporia and Heteroneura, and the Heteroneura into Monotrysia and Ditrysia. The Glossata: Heteroneura contains 99 per cent of all known species of Lepidoptera, many of which are pests.

Much research is currently being done on the higher classification; the linear arrangement below may not be fully up to date but reflects the general concepts of relationships of the commonly recognised superfamilies and families. An asterisk * indicates families of economic importance; those marked in the Ditrysia are keyed. An estimate of the number of *described* species is given for each family, and an indication of the geographical distribution of the family.

SUBORDERS, INFRAORDERS and their divisions are printed in **bold** capitals, SUPERFAMILIES in capitals and Families in upper and lower case. A more refined classification and review of the primary evolutionary lineages in the Lepidoptera are given by Kristensen (1984b). The most recent classificatory scheme for the Lepidoptera is by Minet (1986).

	Species	Geographical distribution
ZEUGLOPTERA		
MICROPTERIGOIDEA		
Micropterigidae	100	Temperate regions of both hemispheres
HETEROBATHMIINA		
HETEROBATHMIOIDEA		
Heterobathmiidae	2	Temperate South America
AGLOSSATA		
AGATHIPHAGOIDEA		
Agathiphagidae*	2	Eastern Australasia

GLOSSATA: DACNONYPHA (sensu Nielsen, 1985)

ERIOCRANIOIDEA (sensu Nielsen, 1985)

Eriocraniidae	20	Holarctic
Acanthopteroctetidae	3	Western U.S.A.
Lophocoronidae	2	Southern Australia

GLOSSATA: NEOPSEUSTINA

NEOPSEUSTOIDEA

Neopseustidae	10	Oriental tropics (montane), South America (Chile)

GLOSSATA: EXOPORIA

MNESARCHAEOIDEA

Mnesarchaeidae	6	New Zealand

HEPIALOIDEA

Hepialidae*	500	Worldwide, especially Australia and New Zealand
Prototheoridae	7	South Africa
Neotheoridae	1	Brazil
Anomosetidae	1	Australia
Palaeosetidae	3	Australasia

GLOSSATA: HETERONEURA

MONOTRYSIA

NEPTICULOIDEA

Nepticulidae*	600	Worldwide
Opostegidae	50	Worldwide
Palaephatidae	30	Mostly Austral S. America; Australia

INCURVARIOIDEA

Heliozelidae*	100	Worldwide, excluding New Zealand
Incurvariidae*	300	Worldwide

Adelidae	100	Worldwide
Cecidosidae	10	Austral S. America, S. Africa
Prodoxidae*	50	Mostly Holarctic; one sp. from S. America

TISCHERIOIDEA

| Tischeriidae | 50 | Holarctic to tropics |

DITRYSIA

TINEOIDEA

Psychidae*	1,700	Worldwide
Eriocottidae	250	Old World
Arrhenophanidae	10	Neotropical
Tineidae*	3,500	Worldwide
Lyonetiidae*	500	Worldwide
Roeslerstammiidae	20	Indo-Australian (Amphitheridae) tropics, S. America
Gracillariidae*	1,700	Worldwide
Douglasiidae	20	Holarctic, Australia

YPONOMEUTOIDEA

Yponomeutidae* (including Acrolepiidae, Argyresthiidae, Plutellidae)	1,700	Worldwide
Heliodinidae (including Schreckensteiniidae; family delimitation uncertain)	300	Most regions, mainly tropical
Ochsenheimeriidae*	20	Palaearctic to Himalayas; 1 sp. introduced in N. America
Epermeniidae*	70	Worldwide
Glyphipterigidae*	250	Worldwide

SESIOIDEA

Sesiidae*	1,000	Worldwide
Choreutidae*	400	Widespread, mainly Old World tropics
Brachodidae*	100	All regions except Nearctic

IMMOIDEA

| Immidae* | 250 | Pantropical |

GELECHIOIDEA

Oecophoridae*	7,000	Worldwide
Elachistidae*	300	Worldwide, mainly Holarctic
Blastobasidae*	450	Worldwide
Pterolonchidae	10	Mainly Mediterranean, 1 in S. Africa, 1 in America
Coleophoridae*	1,200	Mainly Holarctic; some introductions in Australia and New Zealand
Agonoxenidae*	4	Eastern Australasia
Momphidae*	300	Worldwide
Cosmopterigidae*	1,200	Worldwide
Scythrididae	300	Mainly Holarctic - Eurasia
Gelechiidae*	5,000	Worldwide

ALUCITOIDEA

Alucitidae*	80	Mainly temperate and warm regions
Copromorphidae*	60	Mainly Indo-Australian
Carposinidae*	200	Widespread, mainly warm regions

PYRALOIDEA

Pyralidae*	20,000	Worldwide
Thyrididae*	600	Mainly Old and New World tropics
Hyblaeidae*	20	Mainly Indo-Australia; Neotropics

PTEROPHOROIDEA

Pterophoridae*	500	Worldwide
Tineodidae	10	Indo-Australia - mainly tropical
Oxychirotidae	3	Indo-Australia - mainly tropical
Lathrotelidae	1	South Pacific (Rapa I.)

TORTRICOIDEA

Tortricidae*	5,500	Worldwide

COSSOIDEA

Cossidae*	1,000	Worldwide
Metarbelidae*	100	Old World tropics (Arbelidae)
Ratardidae*	11	Oriental tropics
Dudgeoneidae*	5	Indo-Australian, South Africa, South America (Chile)

ZYGAENOIDEA (after Common, 1970)

Chrysopolomidae*	25	African

Dalceridae*	60	New World, mainly tropical
Megalopygidae*	220	New World tropics
Limacodidae*	850	Worldwide, mainly tropical
Epipyropidae*	32	Indo-Australian tropics mainly, though worldwide
Heterogynidae*	10	W. Europe, S. Africa
Zygaenidae*	300	Widespread in Old World; Nearctic
(including as subfamilies Himantropteridae, Anomoeotidae)		
Cyclotornidae	5	Australia

BOMBYCOIDEA

Endromidae*	1	Palaearctic
Mirinidae	1	Palaearctic
Lasiocampidae*	2200	Worldwide
Anthelidae*	100	Australasia
Eupterotidae*	300	Old World tropics
Bombycidae*	60	Old World tropics
Lemoniidae*	20	Mediterranean, S. Palaearctic, Oriental
Carthaeidae*	1	Australia
Oxytenidae*	60	New World tropics
Cercophanidae	30	New World tropics
Apatelodidae*	250	Mainly New World tropics; Nearctic
Brahmaeidae*	20	Old World
Saturniidae*	1,300	Worldwide
Sphingidae*	1,000	Worldwide
(sometimes set apart in SPHINGOIDEA)		

MIMALLONOIDEA (recently separated from BOMBYCOIDEA)

| Mimallonidae* (=Lacosomidae) | 250 | New World tropics mainly; Nearctic |

NOCTUOIDEA (first four families sometimes separated as NOTODONTOIDEA).

Notodontidae*	3,200	Worldwide
Thaumetopoeidae	50	Africa; Mediterranean to N. India; Australasia
Dioptidae	500	Mainly New World tropics; Nearctic
Thyretidae	150	African tropics
Noctuidae*	21,000	Worldwide
(Nolidae, Cocytiidae, Agaristidae now placed as subfamilies; Dilobidae also included here).		
Arctiidae*	11,000	Worldwide
(Ctenuchidae now placed as subfamily; Hypsidae now split between Aganainae and Arctiinae)		
Lymantriidae*	2,160	Worldwide

GEOMETROIDEA

| Drepanidae* | 800 | Worldwide excluding New World tropics |
| Cyclidiidae* | 13 | Oriental and eastern Palaearctic |

Thyatiridae*	200	Worldwide but mainly north temperate
Axiidae*	5	Mediterranean
Geometridae*	20,000	Worldwide
Uraniidae*	113	Worldwide, mainly tropical
(including Microniidae as subfamily)		
Epiplemidae*	500	Worldwide, mainly tropical
Sematuridae*	35	New World tropics
(= Apoprogonidae	1	Southern Africa)
Epicopeiidae	5	Oriental
(sometimes placed in the Uraniidae)		

CALLIDULOIDEA (sometimes subordinated to GEOMETROIDEA)

Callidulidae*	100	Indo-Australian tropics
Pterothysaniidae	12	Oriental; Malagasy; African tropics

CASTNIOIDEA (subordinated to SESIOIDEA by Minet, 1986)

Castniidae*	150	New World tropics; Oriental; Australia

PAPILIONOIDEA (senso lato; mainly Rhopalocera - butterflies)

Hedylidae	40	Neotropical
Hesperiidae*	3,100	Worldwide
(including Megathymidae; sometimes separated as HESPERIOIDEA)		
Lycaenidae*	4,000	Worldwide
(including Riodinidae)		
Nymphalidae*	8,200	Worldwide
(sensu lato, including Danaidae, Satyridae etc.)		
Libytheidae*	10	Worldwide excluding high latitudes
Pieridae*	2,000	Worldwide
Papilionidae*	550	Worldwide

CHARACTERISATION OF SUBORDERS, INFRAORDERS & DIVISIONS
(JDB)

SUBORDER ZEUGLOPTERA

Small day-flying moths, the most archaic of the Lepidoptera possessing functional mandibles (jaws). Tongue (haustellum, proboscis) not developed; maxillary palpi long, 5-segmented; labial palpi short, 2 to 4-segmented; fore- and hindwings with similar venation (homoneurous); wing-coupling jugate (a small lobe called the jugum or fibula at the base of the forewing projects over the hindwing); hindwing without frenulum; female genitalia with a single genital opening (cloaca) on sterna 9-10, functioning as combined copulatory aperture, egg-pore (oviporus) and anus (monotrysian-type genitalia).

Pupae are decticous-exarate (with functional mandibles and free appendages); abdominal segments 1-7 movable, 8-10 fused; subterranean, in a silken cocoon.

Larvae with adfrontal sutures absent; antennae long, 3-segmented; ocelli conjoined; thoracic legs simple with coxa, trochanter and femur fused; abdomen with claviform setae, segments 1-8 each with a pair of non-muscular, 3-segmented, conical prolegs; exophagous (free-living).

MICROPTERIGOIDEA

MICROPTERIGIDAE

Non-economic. Micropterigids are very small 'primitive' moths, (wingspan 7-15mm) and occur mainly in temperate regions. Adults are predominantly diurnal, with usually metallic bronze-golden or bronze-purple colouration. They are somewhat sedentary and tend to congregate on flower heads, often in moist situations in partial sunlight, and occasionally swarm. They differ from other Lepidoptera in having masticatory jaws for eating pollen or fern spores. The larvae live low down amongst surface litter or roots of grasses and other vegetation, browsing on tender green tissues or fungal hyphae.

Main genera: *Micropterix, Sabatinca.*

References: Heath, 1976; Kristensen, 1984b (with colour illustrations of primitive lepidopteran families); Nielsen, 1985; Zagulyaev, 1978.

SUBORDER HETEROBATHMIINA

HETEROBATHMIOIDEA

HETEROBATHMIIDAE

Non-economic. Placed in a separate suborder by Kristensen (1984b) and Minet (1986). Monogeneric (*Heterobathmia*) and restricted to temperate South America, comprising about ten known species, only two of which are described. Adults resemble eriocraniids externally, but have the mandibles developed, and have been observed on flowers of *Nothofagus*; the larvae mine in the leaves.

SUBORDER AGLOSSATA

AGATHIPHAGOIDEA

AGATHIPHAGIDAE

Of minor economic importance, and placed in separate suborder by Speidel (1978). Represented by two species: *Agathiphaga queenslandensis* Dumbleton, which is the larger with a wingspan of about 13 mm and occurs in Eastern Australia (Queensland), and the slightly larger *A. vitiensis* Dumbleton from Fiji, Solomon Is., Vanuatu, New Caledonia. Adults of both species are sombre-coloured, with brown markings on a grey-brown background, and have prominent mandibles and 5-segmented labial palpi. Their larvae develop in the cones of kauri pines (*Agathis*), feeding in the seeds and pupating within; the adults emerge through a hole pre-cut in the wall of the seed by the larva, the pupal exuviae protruding after eclosion.

References: Common, 1974; Kristensen, 1967, 1984a, b.

SUBORDER GLOSSATA

INFRAORDER DACNONYPHA

Small, primitive day-flying moths with mandibles present and functional or intermediate between those of the Zeugloptera and Ditrysia, the mandibles being obsolescent and instead a rudimentary tongue developed. Maxillary palpi 5-segmented, 4th segment longest; labial palpi 2- or 3-segmented; fore- and hindwings with similar venation (homoneurous); wing-coupling jugate; frenular bristles sometimes present in hindwing; female genitalia with a single genital opening on sterna 9-10 (monotrysian).

Pupae are decticous-exarate (mandibles long, curved and serrated, or atrophied; appendages free); abdominal segments 1-7 movable, 9-10 fused; subterranean or in larval feeding place; in a silken cocoon.

Larva has head with ecdysial lines joining anterior margin; apodous (without thoracic legs or abdominal prolegs); endophagous (living internally).

ERIOCRANIOIDEA

ERIOCRANIIDAE

Non-economic, and essentially Holarctic. Eriocraniid adults are similar to micropterigids in their generally iridescent metallic coloration, and are sometimes common in deciduous woods, flying in small swarms about their host-plants in sunshine and resting on the leaves and buds (not flowers) or in dull weather sitting passively on the twigs. Larvae are apodous leaf miners on Betulaceae, Corylaceae and Fagaceae, forming blotches (sometimes originating as a gallery).The mines may be recognised from those of other blotch-miners by the long, interwining threads of frass.

Main genus: *Eriocrania*

References: Davis, 1978; Heath, 1976; Kristensen, 1984b; Zagulyaev, 1978.

ACANTHOPTEROCTETIDAE and LOPHOCORONIDAE

Non-economic. Rare and specialised, these families are confined to the Nearctic region and Australia respectively.

Main genera: *Acanthopterocteta, Lophocorona.*

References: Common, 1973, 1974 (Lophocoronidae); Davis, 1978 (Acanthopteroctetidae); Kristensen, 1984b.

INFRAORDER NEOPSEUSTINA

NEOPSEUSTOIDEA

NEOPSEUSTIDAE

Non-economic. A rare and relict group of characteristically round-winged, medium-sized (wingspans up to 30 mm), somewhat lacewing-like moths with long antennae. About 10 species are known, at least half of them from the forested regions of the Himalayas to China and Taiwan, the others from the *Nothofagus* forests of southern South America (Chile).

Main genera: *Apoplania, Neopseustis.*

References: Davis, 1975; Davis & Nielsen, 1980; Kristensen, 1984b; Nielsen, 1985.

INFRAORDER EXOPORIA

Large to very large, mainly day-flying and crepuscular moths, generally regarded as primitive because of their homoneurous venation, but showing considerable specialisation in the intermediate ditrysian form of the female genitalia. Antennae short; mandibles absent; tongue rudimentary or absent; maxillary palpi vestigial or absent; labial palpi 2- or 3-segmented; fore- and hindwings with similar venation, forewing with humeral vein; wing-coupling jugate, the jugum (or fibula) strongly developed and projecting; frenulum absent; female genitalia with separate copulatory (ostium) and ovo-anal (egg-pore + anus) openings on sterna 9-10 either connected by seminal groove (usually external) or contiguous (ditrysian-exoporian).

Pupae are adecticous-obtect (with vestigial non-functional mandibles; appendages fused to body); abdominal segments 1-7 free in male, 2-6 free in female, with segmental whorls of spines or ridged dorsally; usually free in larval tunnel, protruded at eclosion.

Larva has head with 6 ocelli; thoracic legs and abdominal prolegs developed, thoracic legs 5-segmented, abdominal prolegs on segments 3-6 and 10 with crochets multiserial, in an ellipse; internal wood and pith feeders, often in roots, or subterranean.

HEPIALOIDEA

HEPIALIDAE

Economic. Worldwide in distribution, but with the majority of the 500 or more known species occurring in the southern hemisphere, the largest and most strikingly colourful in Australia, New Zealand and South Africa, with wingspans over 150 mm. Hepialid moths are strong fliers, and many habitually fly low down over grasses and other vegetation at dusk, those of some of the smaller species hovering in a characteristic 'ghostly' manner, hence the common names of swift and ghost moths. Larvae tend to be polyphagous or oligophagous. Those of the larger species which tunnel in the living wood of trees and shrubs, or on the bark regrowth round the entrance to the tunnel, may take 3 or 4 years to reach maturity and can severely harm the plant. Although the females of some of these giants can lay enormous numbers of eggs - up to 18,000 eggs has been estimated for the Australian *Abantiades magnificus* (Lucas) - relatively few survive to the adult stage. Some species live in vertical tunnels in the ground and either feed on the surface of roots or emerge at night to feed on foliage or litter near the entrance to the tunnel. In most regions some of the smaller species are occasionally sufficiently abundant to damage pasture and grassland and less frequently to affect vegetable root crops, vines and cultivated trees and shrubs.

Main genera: *Abantiades, Aenetus* (= *Charagia*), *Endoclita, Gorgopis, Hepialus, Leto, Thitarodes, Trictena, Oncopera, Oxycanus.*

Pest species: *Endoclita hosei* (Tindale), Malaysia to Borneo, on cocoa (*Theobroma*) and various other shrub trees; *Endoclita sericeus* (Swinhoe) - cocoa ring bark borer - on a wide range of cultivated trees and shrubs, including coffee and tea; *Gorgopis libania* (Stoll) - gorgopis lawn moth - Central and southern Africa, on Gramineae; *Hepialus humuli* (Linnaeus) - ghost swift - Palaearctic, on most Gramineae, *Cannabis, Cynara, Humulus, Pisum*, etc; *Hepialus lupulinus* (Linnaeus) - garden swift - Palaearctic, as *H. humuli.*

References: Carter, 1984; Common, 1970; Pinhey, 1975 (South Africa); Heath, 1976 (British Is.); Sukhareva, 1978 (Palaearctic); Tindale, 1932-64 (Australia); Nielsen & Robinson, 1983 (southern S. America).

ANOMOSETIDAE, NEOTHEORIDAE, PALAEOSETIDAE and PROTOTHEORIDAE

Non-economic. Rare and jointly represented by only 12 known species found only in the southern hemisphere.

References: Common, 1970; Kristensen, 1978.

MNESARCHAEOIDEA

MNESARCHAEIDAE

Non-economic. A relict group represented by six known species restricted to New Zealand.

Reference: Kristensen, 1968; Gibbs, 1979.

INFRAORDER HETERONEURA

MONOTRYSIA

Minute to small day-flying micro-moths, primitive (plesiomorphic) in having the female genitalia monotrysian but more advanced (apomorphic) than the preceding suborders in having the fore- and hindwing venation dissimilar and in some cases greatly reduced. Usually with short tongue; maxillary palpi 5-segmented or vestigial; labial palpi 2- or 3-segmented; wing-coupling frenate (forewing with retinaculum, hindwing with frenulum of one (male) or more (female) bristles); female genitalia with single genital opening (cloaca) on sterna 9-10 (monotrysian).

Pupae are adecticous-obtect (with mandibles; appendages loosely fused to body); abdominal segments free and movable except 8-10; in a tough silken cocoon, protruded at ecdysis.

Larvae are virtually apodal, the thoracic legs and abdominal prolegs being reduced to fleshy protuberances; endophagous (mining in leaves and bark), sometimes exophagous in later instars, constructing portable cases and feeding externally.

28

NEPTICULOIDEA

NEPTICULIDAE

Of minor economic importance. The 600 or so described species - about half of them in the genus *Stigmella* - comprising this family are distributed in most parts of the world, including the highest mountains. All are diminutive micro-moths with wingspans around 3-6 and include one of the smallest known lepidopterans, *Stigmella acetosae* (Stainton) from Europe. The forewings are typically heavily fringed, giving them a somewhat rounded appearance, and are often spotted or banded with glittering gold or silver markings, but sometimes are dull-coloured. The antennal eye-caps and erect hair-like scales of the head are characteristic. The moths are diurnal and fly rather erratically, keeping close to the foodplant, and after settling will run rapidly over the leaves or bark. The eggs are comparatively large and are laid singly at the feeding site, so that the larva can eat directly through the base into the substrate. The larvae are mostly miners in leaves or bark, and the form of the mine and, in the case of leaf mines, the position of the remains of the egg chorion on the leaf, can aid in identification. Frass is usually retained in the mine and is deposited in a central track and not scattered as in the mines of Diptera and Hymenoptera. Pupation usually takes place outside the mine, usually in an oval cocoon of silk attached to a leaf or spun amongst detritus on the ground.

Main genera: *Ectoedemia, Stigmella, Trifurcula.*

References: Carter, 1984; Emmet, 1976 (British Isles); Hering, 1951, 1953; Wilkinson & Scoble, 1979; Zagulyaev, 1978.

Pest species: *Ectoedemia atricollis* (Stainton), Palaearctic, on apple (*Malus*); *Stigmella ipomoeella* Gustafsson, Sri Lanka, on sweet potato (*Ipomoea*); *Stigmella malella* (Stainton) - apple pigmy moth - Palaearctic, on apple (*Malus*); *Stigmella pomella* (Vaughan), Palaearctic, on apple (*Malus*).

PALAEPHATIDAE

Non-economic. Restricted largely to temperate forests of southern Argentina and Chile (28 species), with one genus reported from Australia.

References: Davis, 1986; Nielsen, 1987.

OPOSTEGIDAE

Mostly non-economic. A family with few genera of micro-moths widely distributed but nowhere common. The forewings are typically white with a dark pattern, and the adults are further characterised by their extreme venational specialisation and the large size of the eye-cap which is proportionately much larger than in the Nepticulidae and is covered with regularly imbricated scales and ribbed on the underside. The head has a tuft of hair-like scales between the antennae but otherwise is smooth-scaled. Larvae are mostly miners in bark or leaves of low-growing plants; the pupa is formed in an oval cocoon outside the mine.

Pest species: *Opostega* sp. in S. America boring in the cambium of *Nothofagus dombeyi* (Fagaceae); one Japanese species mines the trunk of *Betula* (Betulaceae).

References: Carey et al, 1978; Zimmerman, 1978.

INCURVARIOIDEA

Mostly non-economic. A comparatively numerous and diverse assemblage of relatively large primitive micro-moths currently divided into six families: Heliozelidae, Adelidae, Crinopterygidae, Incurvariidae, Cecidosidae and Prodoxidae. Adelidae and Incurvariidae contain the majority of the species and are represented in most regions, while the Prodoxidae are a small, specialised complex of about 20 known species widely distributed in the northern hemisphere, but the greatest number of genera and species occurring in North America.

Adults are mostly diurnal and visit flowers; some adelids congregate in small swarms. Adelids and incurvariids commonly have metallic coloration and a fasciate forewing pattern, but male adelids are distinguished by their exceptionally long antennae, sometimes several times the length of the forewing and proportionately longer than in any other group of Lepidoptera. Members of other families have antennae less than the length of the forewing. Adult Heliozelidae can be recognised by their reduced mandibles and the absence of M-CuA in the hindwing. Heliozelids also only fly in sunshine and are very small in size. The American yucca moths (Prodoxidae) are white or greyish-white, and non-metallic in coloration.

Larvae of Adelidae and Incurvariidae mine leaves, stems, flower heads, fruit, etc.; some later cut out oval lenticular (flattened) cases which they use as portable shelters. Larvae of Prodoxidae are mostly pod borers of Agavaceae (*Agave*, *Yucca*), but also tunnel in stems, twigs, buds and flower heads. Cecidosid larvae are gall-makers. In the prodoxid genus *Tegeticula* the maxillary palpi are modified to form a tentacle used to pollinate yucca flowers in which the eggs are laid - without this pollination the seeds in which the larvae must feed will not develop. Heliozelid larvae mine in leaves, petioles and twigs of shrubs and trees: when mature, the larva excises an oval case and lines it with silk in which to pupate; the pupa is protruded at eclosion. Pupation in the other groups usually occurs within the larval feeding place, but sometimes elsewhere.

Families and main genera: Heliozelidae (*Antispila*, *Coptodisca*, *Heliozela*); Incurvariidae (*Incurvaria*); Prodoxidae (*Lampronia*, *Nemophora*, *Prodoxena*, *Tegeticula*).

Pest species: *Paraclemensia acerifoliella* Fitch (Incurvariidae) a forest pest in N. America; *Lampronia capitella* (Clerck) (Prodoxidae) - the currant shoot borer - Palaearctic, North America (?introduced), on red, white and black currant, gooseberry (*Ribes*), at first in the fruit, later boring in the buds and shoots; *Lampronia rubiella* (Bjerkander) (? = *corticella* Linnaeus) - raspberry moth - Palaearctic, North America (?introduced), on raspberry, loganberry and blackberry (*Rubus*), first in the fruit, later in the developing buds; *Coptodisca splendoriferella* (Clemens) (Heliozelidae), North America, on apple (*Malus*); some members of the genus *Antispila* Hubner (Heliozelidae) are pests of grape vines.

References: Carter, 1984; Common, 1970, 1974; Davis, 1967; Emmet, 1976; Nielsen & Davis, 1985; Nielsen, 1985; Zagulyaev, 1978.

INFRAORDER HETERONEURA

DITRYSIA

Advanced haustellate moths and butterflies, comprising 99 per cent of known Lepidoptera, very small to very large, diurnal, crepuscular or nocturnal. Mandibles absent; tongue (haustellum) usually well developed, often long and coiled, sometimes secondarily degenerate; maxillary palpi 1- to 5-segmented or vestigial; labial palpi usually 3-segmented and well developed; fore- and hindwing venation dissimilar (hetero- neurous); wing-coupling frenate (forewing with retinaculum, without jugum or fibula, hindwing with frenulum of one or more bristles), or amplexiform (frenulum and retinaculum absent, humeral area of hindwing expanded to underlap forewing); female genitalia with separate copulatory aperture (ostium) on sternum 8 and ovo-anal (egg-pore + anus) aperture (oviporus) on sterna 9-10 (ditrysian-exoporian). Auditory (tympanal) organs sometimes developed on the meta-thorax (Noctuoidea) or on the 1st or 2nd abdominal segment (Geometroidea and many Pyraloidea). Butterflies (Papilionoidea) may be distinguished from most moths by their clubbed antennae and amplexiform wing-coupling. In moths the antennae are rarely clubbed and the wing-coupling is usually frenate, but if amplexiform (e.g. Bombycoidea) the antennae are not clubbed.

(eg. Papilionoidea)

Pupae are adecticous, obtect (without mandibles; appendages fused at least partially to body; abdominal segments fused and immovable except 4th, 5th and 6th, and in advanced families males with 7th segment fused to 8th, and sometimes all segments immovable (e.g. Papilionoidea)); in cocoon or cell and protruded at ecdysis, or exposed and attached to sub- strate by a cremaster (e.g. Papilionoidea); specialised adaptations in structure and coloration often occur in Papilionoidea.

Larva nearly always with thoracic legs well developed (5-segmented) and with two or more pairs of crochet-bearing abdominal prolegs; when legs and abdominal prolegs are developed, prolegs claspers are always present on 10th abdominal segment.

NOTE: Minet's classification recognises the four suborders: Zeugloptera, Aglossata, Heterobathmiina and Glossata indicated above, but combines Tischerioidea, Incurvarioidea and Ditrysia in the infraorder Eulepidoptera (of Glossata).

KEYS TO HIGHER TAXA

33

KEY TO SUBORDERS, INFRAORDERS AND DIVISIONS

1. Venation of fore- and hindwings dissimilar, hindwing with not more than 6 veins arising from cell with either frenulum present (most moths) or an enlarged humeral area (butterflies, some moths); forewing with jugum absent......(see over page).....

   2

-- Venation of fore- and hindwings similar, hindwing with more than 6 veins arising from cell and frenulum absent or weak; forewing with jugum present 3

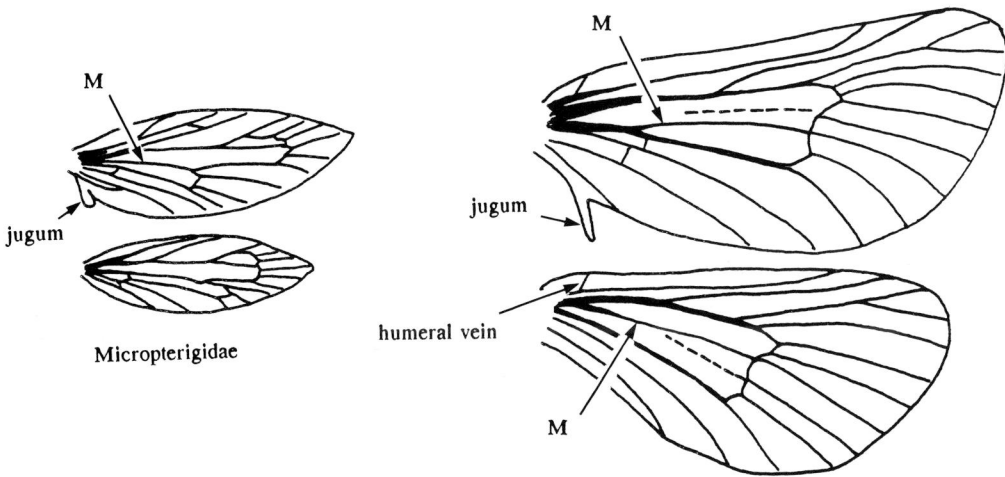

Micropterigidae

Hepialidae: *Hepialus*

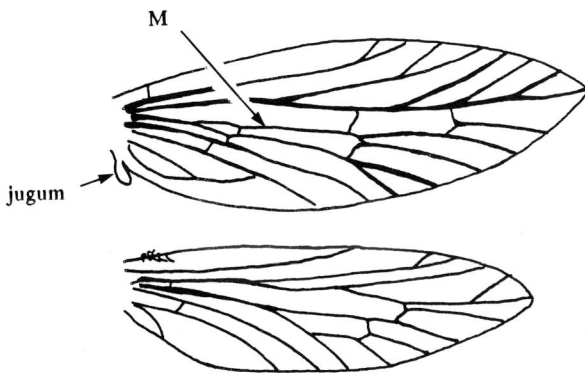

Eriocraniidae

Homoneurous venation: Zeugloptera, ~nonypha, and Exoporia

35

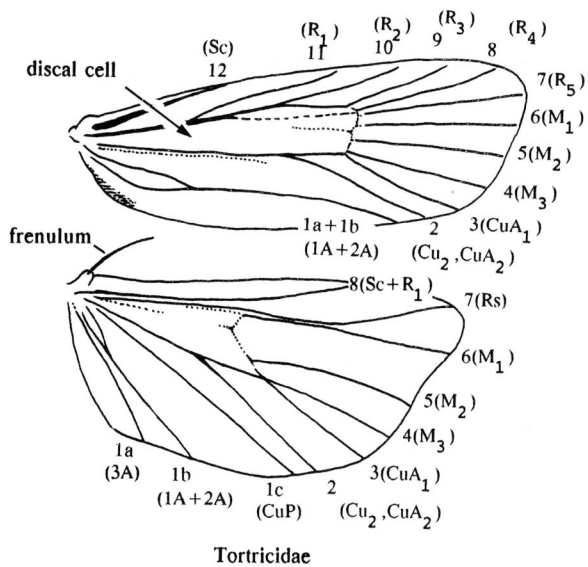

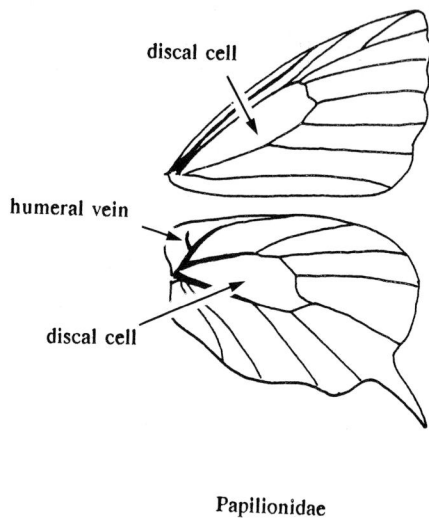

Correlated dual systems of numbering veins

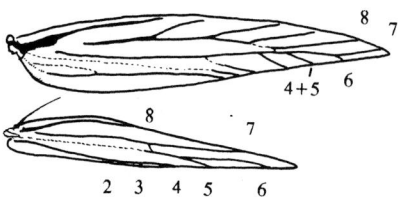

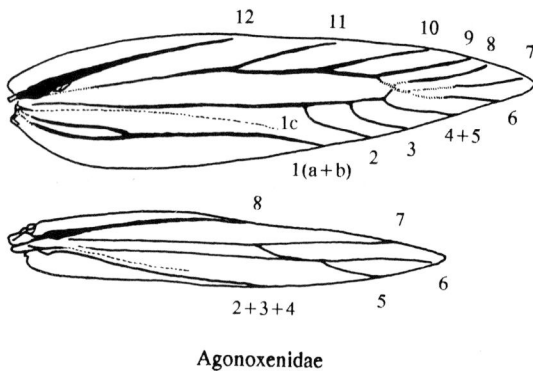

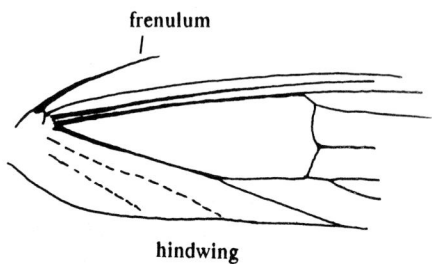

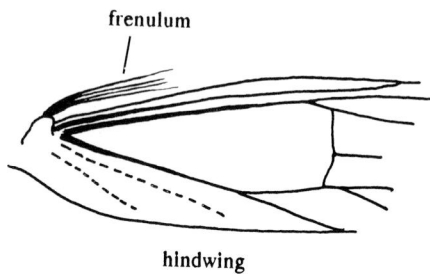

Tortricidae

Papilionidae

Agonoxenidae

Reduced venation as found in some ditrysian micro-moths

36

2.   Minute to small moths (wingspan usually 3-20mm); female with single genital opening
      (cloaca)   on   sterna   9-10   ....**HETERONEURA:**
      MONOTRYSIA (NEPTICULOIDEA, INCURVARIOIDEA; few economic species)

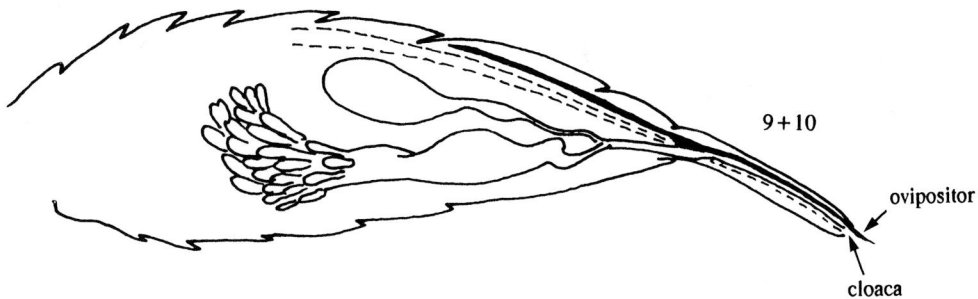

Monotrysian female genitalia

--   Small (micro-moths) to large moths (macro-moths) and butterflies (wingspan ranging
      to 300 mm);   female with separate genital opening (ostium) on sternum 8 (sometimes
      on sternum 7 or intersegmental) .... **HETERONEURA: DITRYSIA**   (99 per
                              cent of Lepidoptera; numerous economic species)

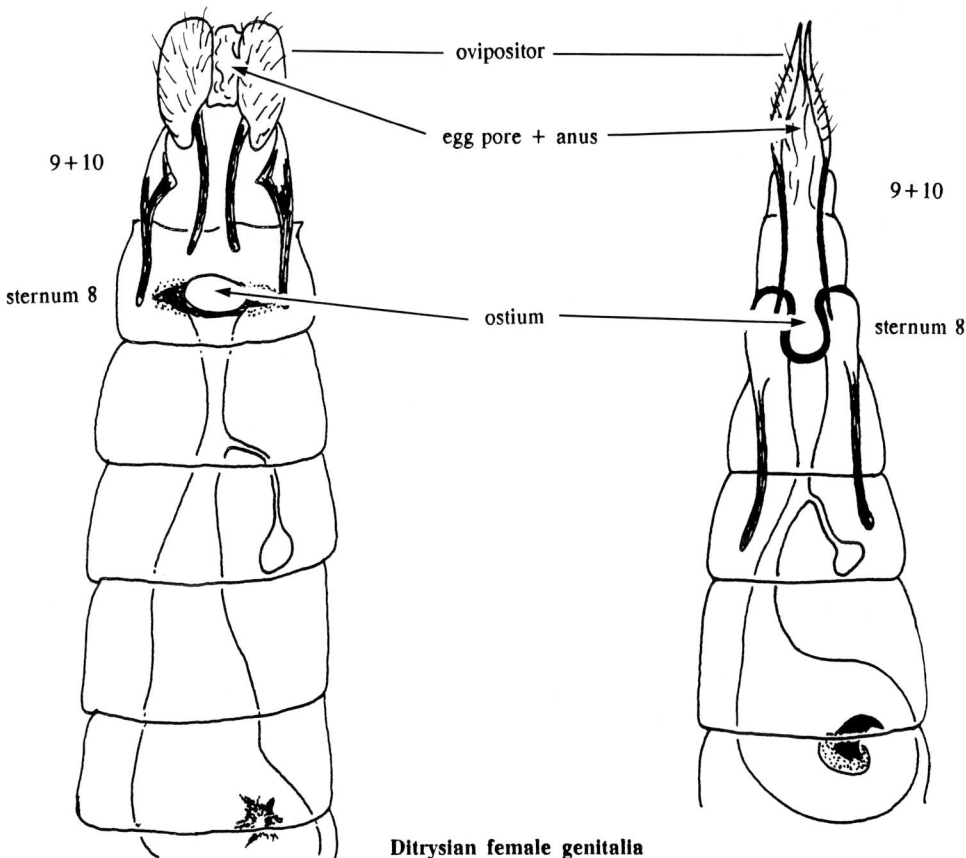

Ditrysian female genitalia

37

3.   Medium-sized to very large and sometimes robust moths (wingspan 25-200 mm): antennae relatively short (usually one-third or less than length of forewing); female with 2 contiguous genital opening on sterna 9-10 ....GLOSSATA: **EXOPORIA**
       (HEPIALOIDEA some economic species; MNESARCHAEOIDEA non-economic)

--   Small moths (wingspan usually under 20 mm); female genital opening on sterna 9-10 ..
      4

4.   With functional mandibles ....**ZEUGLOPTERA**: MICROPTERIGIDAE
                                                                                           (non-economic)

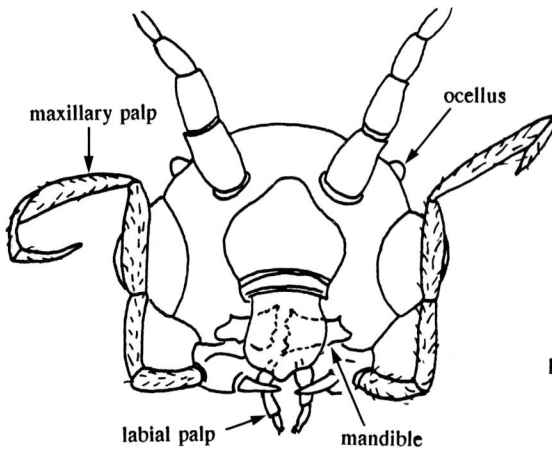

Denuded head of micropterigid imago.

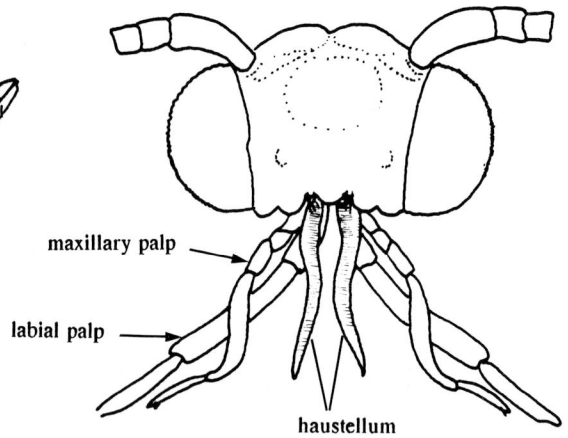

Denuded head of ditrysian imago.

Imago: Head structures

--   Mandibles absent or present as non-functional lobes ....**DACNONYPHA**, **HETEROBATHMIINA**, **AGLOSSATA** (economic), **GLOSSATA: NEOPSEUSTINA**
                                                                                           (non-economic)

KEY TO DITRYSIAN FAMILIES, AND SOME SUBFAMILIES

This key concentrates on readily appreciable characters as far as possible, though a number of these always give difficulty.

Thoracic tympanal organs (couplet 26) are often hard to see in dried, set specimens: the zone where they occur may be distorted, infolded, or obscured by the legs. We recommend that they are first examined in fresh material, preferably of large species. The noctuid subfamilies Stictopterinae and Plusiinae have the tympanum (an opaque whitish membrane) and counter-tympanal hood readily visible.

Wing venation can be studied without destroying a specimen by wetting the wings with absolute alcohol or benzene (the latter is highly toxic), giving them greater transparency, and illuminating them from behind when examining them under the microscope or with a hand lens. Preparations can be made of wings to show up the veins by wetting them in alcohol and placing them in bleach, washing them thoroughly when the wing colours have largely faded away, and then dehydrating them in alcohol and mounting them either dry or under a coverslip with mountant. It is possible to stain the veins by placing the wings in dilute eosin solution overnight.

Characters of male and female genitalia have not been used in the keys, though they are often essential in determination of species; techniques of preparation have been described in an earlier chapter.

1. Antennae gradually or abruptly clubbed, the tip sometimes hooked; hindwing **without** frenulum and with enlarged humeral area, often with humeral vein present ....
.... PAPILIONOIDEA (excluding HEDYLIDAE) 2

Shape of antennae in Papilionoidea

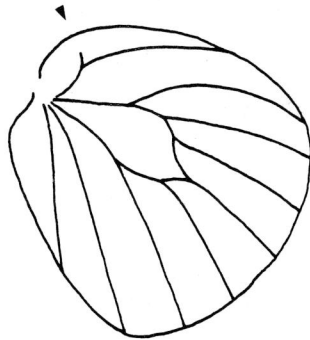

-- Antennae filiform or tapered, often ciliate, pectinate or plumose; if dilated or clubbed, a **frenulum** is present on the hindwing (N.B. *Euschemon*, an Australian hesperiid butterfly, has a frenulum; the CALLIDULIDAE have filiform antennae that broaden very slightly and evenly towards the apex and lack a frenulum in many cases)
....7

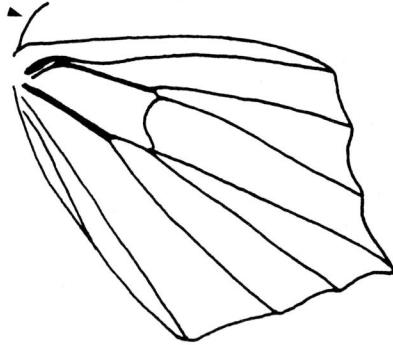

2.   Antennal club usually round-tipped, antennae approximated at base; forewing with at least some veins distal to the cell arising from a common stalk (some LYCAENIDAE are exceptions) 3

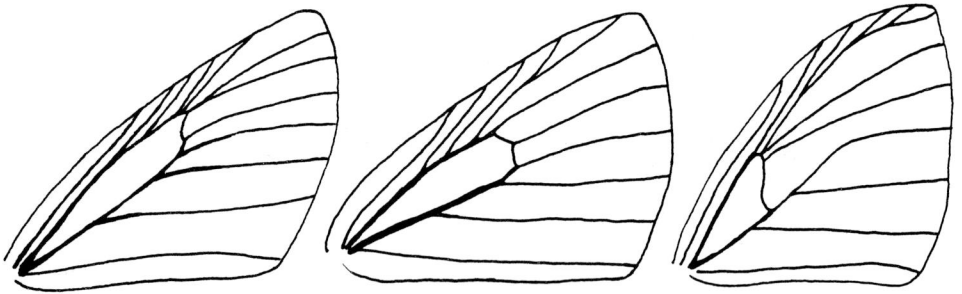

Hesperiidae   Lycaenidae   Nymphalidae

All veins arise independently from cell in Hesperiidae but from a branching system in the other two. After Corbet & Pendlebury (1978) not to scale.

--   Antennal club usually hook-tipped, antennae widely separated at base (width of eye or more); forewing with veins arising from cell separately; robust, fast flying, mainly brown species ....HESPERIIDAE (p.182)

3.   Eyes emarginate (indented at margin) to accommodate base of antenna, or base of antennae in contact with eye margin; small to medium butterflies, often with iridescent (structural) colouring on the upperside, usually blue, green or purple .. ..LYCAENIDAE (incl. RIODINIDAE) (p.183)

-- Eyes not emarginate, antennal base separate from margin of eye; usually large or medium-sized butterflies with browns, yellows, reds or whites predominant, primarily produced by pigments 4

4. Forelegs reduced, at least in male, and not used for walking 5

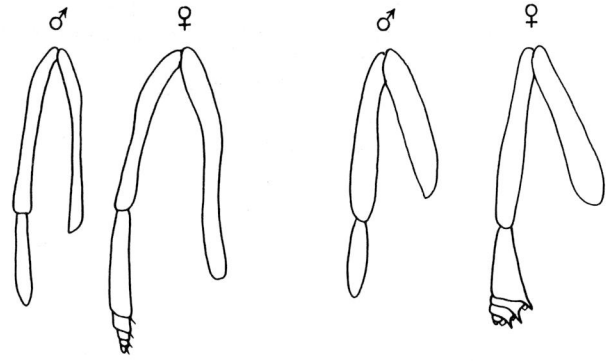

Nymphalidae

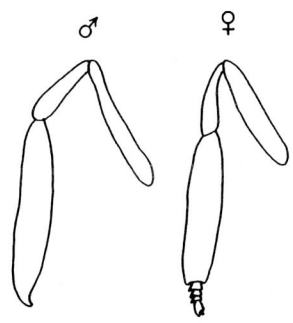

Lycaenidae

Reduced forelegs of Nymphalidae; also seen in Lycaenidae. After Corbet & Pendlebury (1978).

-- Forelegs of both sexes normal 6

5. Forelegs of both sexes reduced; wings in several subfamilies with conspicuous submarginal ocelli; palpi not more than twice as long as head ....
....NYMPHALIDAE sensu lato (p.182) 4

-- Foretarsi imperfect and brushlike in male, normal in female; forewing falcate; no submarginal ocelli on wings; palpi stout, porrect, three times as long as head or more ....LIBYTHEIDAE (p.183) 5

41

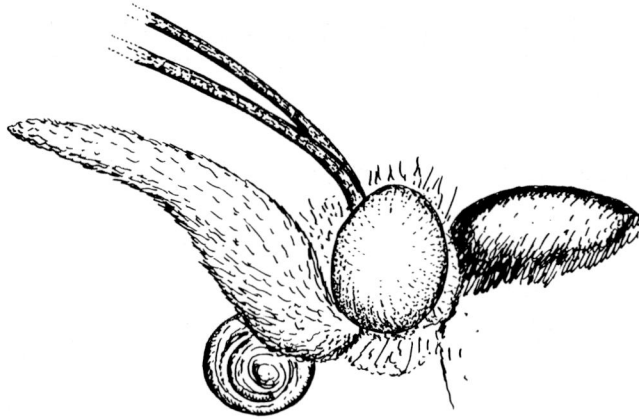

Head of Libytheidae showing large, porrect palps.

6. Hindwing with one anal vein and often with a prominent angle or tail to the centre of the margin; forewing with a short, second anal vein; foretibia with a large medial spine ....PAPILIONIDAE (p.184) 6

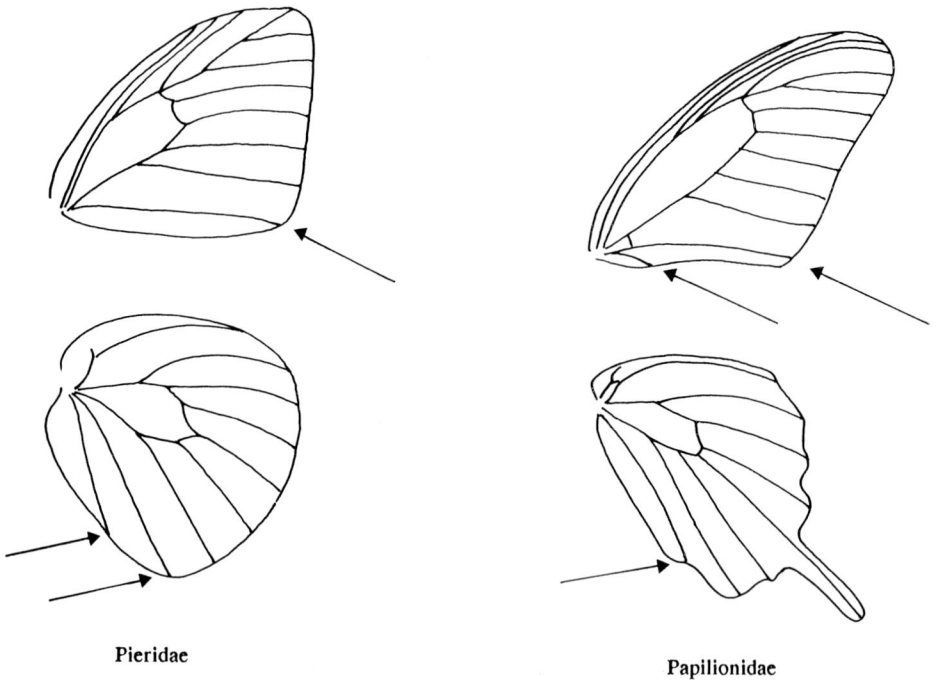

Pieridae

Papilionidae

Differences in anal veins (arrowed) in Papilionidae and Pieridae. Note reduction in number of veins arising from forewing cell in Pieridae.

--   Hindwing with two anal veins, the margin usually rounded; forewing with only one anal vein; usually only 9 (cf.10) veins arising from forewing cell; foretibia without a medial spine ....PIERIDAE (p.185)

7.   Wings usually cleft into two or more plumes 8

--   Wings entire, sometimes vestigial 9

8.   Wings cleft into 2-4 plumes ......................................PTEROPHORIDAE (p.128)
(Agdistinae which have wings entire but very long and narrow are excluded)

Agdistinae

--   Wings cleft into 6 or 7 plumes ....ALUCITIDAE (p.126)

9.   Wings fully developed 10

--   Wings not fully developed (females) 85

10.   Antennae smooth, broadly clubbed; mostly robust moths; chaetosemata absent and hindwing not tailed (presence of these last two characters in absence of venation character below - SEMATURIDAE (part) (p.160); distinctive quadrifid arrangement of veins arising from narrow cell formed from M and Cu on both wings ....
....CASTNIIDAE (p.178) 180

43

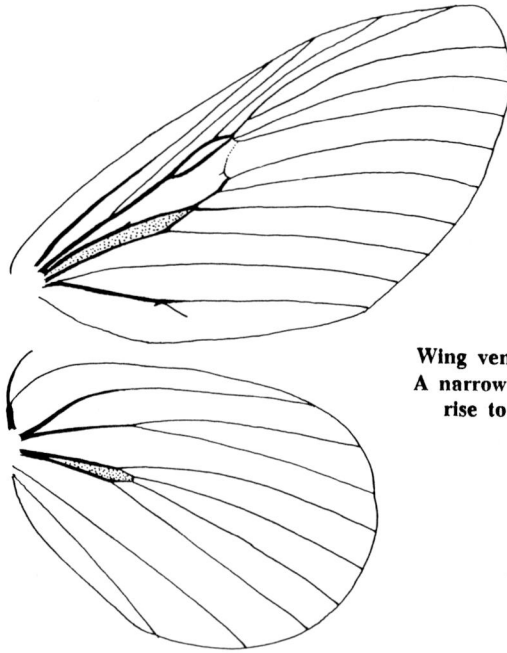

Wing venation of Castniidae (*Synemon*). A narrow posterior cell (stippled) gives rise to quadrifid venation system.

--   Antennae not distinctly clubbed, sometimes dilated, often ciliate or pectinate; venation not as above 11

11.   Vein M moderately to strongly present in cells of both wings (e.g. figure at couplet 18) .....**COSSOIDEA, ZYGAENOIDEA, some TINEOIDEA** 12

--   Vein M weak or absent in cells of one or both wings 26

12.   Chaetosemata present 13

--   Chaetosemata absent 15

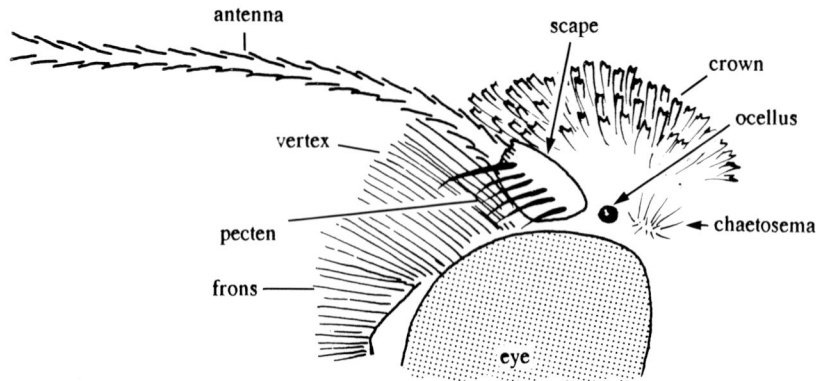

44

13. Ocelli absent; venation as in first half of couplet 25; Neotropics only ....
....MEGALOPYGIDAE (part) (p.139)

-- Ocelli present, or not Neotropics 14

14. Coiled tongue (haustellum) present; antennae often thickened; forewing radial veins often branched; chaetosema extensive, with 'honeycomb' of small scales; usually brightly coloured ....ZYGAENIDAE (p.141)

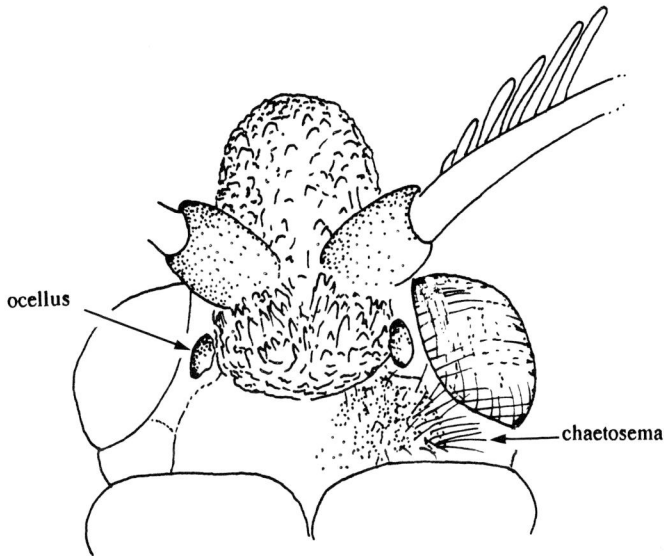

ocellus

chaetosema

Head in Zygaenidae (Chalcosiinae)

-- Tongue absent; antennae bipectinate; forewing radial veins arise directly from cell; small, sparsely scaled brownish species; W.Europe, S.Africa ........HETEROGYNIDAE (p.138)

15. Tympanum present at base of abdomen (sternites 1+2); forewings reddish brown, with clusters of white spots apically and basally .....................DUDGEONEIDAE (p.137) 138

-- Tympanum absent 16

16. Forewing with only one *strong* anal vein, though there may be a weak vein CuP or a fold in its position 17

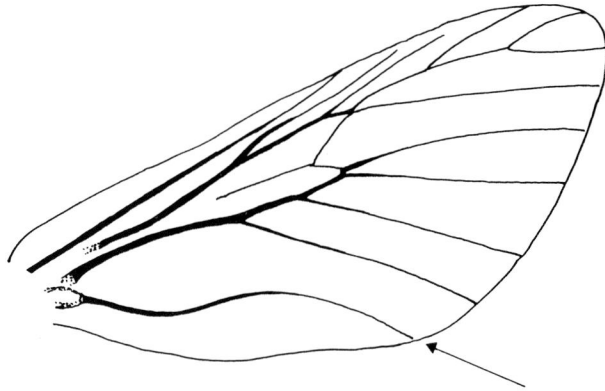

--   Forewing with vein CuP moderate to strong: two anal veins 18

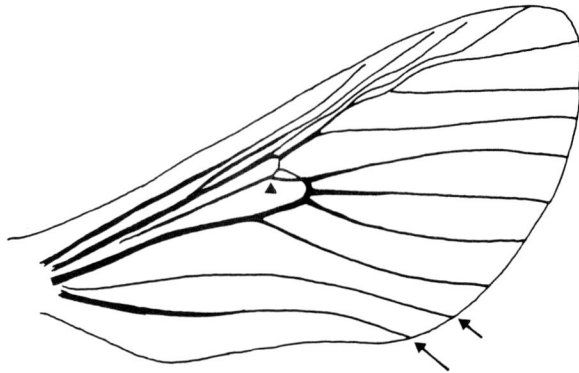

17.   Hindwing with posterior angle of cell with three veins arising from it in closely trifid formation; both wings ovoid with blotchy pattern of white or colour on black; abdomen shorter than hindwing; Oriental ....RATARDIDAE (p.137)

--   Wings narrow with hindwing venation not as above, patterning often reticulate or striate; abdomen longer than hindwing .........................METARBELIDAE (p.137)

18.   M bifurcate within cell on both wings or cell divided into three or more subcells; abdomen extends well beyond hindwings; forewings tend to be elongate, narrow 19

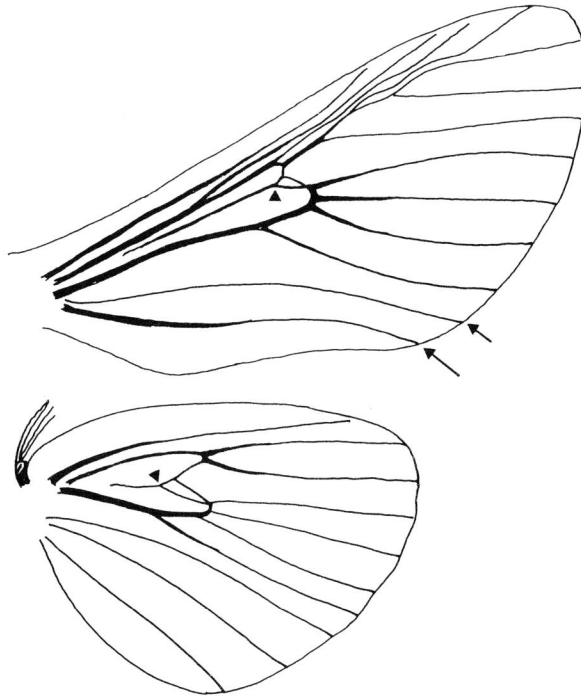

Venation in Cossidae (*Cossus*) with fork of vein M in cell indicated; the two anal veins of the forewing are arrowed.

--   M bifurcate within cell of only one or neither wing; abdomen usually does not extend far beyond hindwings if at all; wings not abnormally elongate, often deep 20

19.   Head with roughened hair-scales; antennae usually bipectinate to apex; wings sparsely scaled, usually black or brown, often transparent; adult rarely larger than 40mm span; larva case-bearing (see couplet 65 for illustrations) ....
    ....PSYCHIDAE (part) (p.79)

--   Head neatly scaled; antennae often bifurcate only over basal half; forewings often very elongate with fine, reticulate or transversely striate patterning; adult often more than 40mm span; larva wood, stem or root boring ....COSSIDAE (p.136)

20.   Wings sparsely scaled, black or brown; larva case-bearing ....
    ....PSYCHIDAE (part) (p.79)

--   Wings otherwise: if black then densely scaled 21

21.   Forewing cell divided more or less longitudinally in three with distal veins all arising directly from it or with, at most, only one distal bifurcation; very small moths with roundedly triangular, often blackish forewings; larvae predatory on Homoptera ....EPIPYROPIDAE (p.136) 138

47

Wing venation in Epipyropidae

-- Forewing cell only divided into two and/or distal veins branching more complexly than a single bifurcation 22

22. Forewing costa markedly concave; Sc and radial veins closely adpressed (but not fused) at cell apex; wings pale, uniform, bone colour, slightly translucent; North Africa ....MEGALOPYGIDAE (*Somabrachys*) (p.139)

-- Forewing costa straight or convex 23

23. Forewing veins R4 and R5 stalked independently from other radial veins, which are usually also stalked, though R4 and R5 sometimes anastomose with them to produce an areole; male forelegs plumose; forewings very deep, generally yellow, cream or white; New World ....DALCERIDAE (p.139)

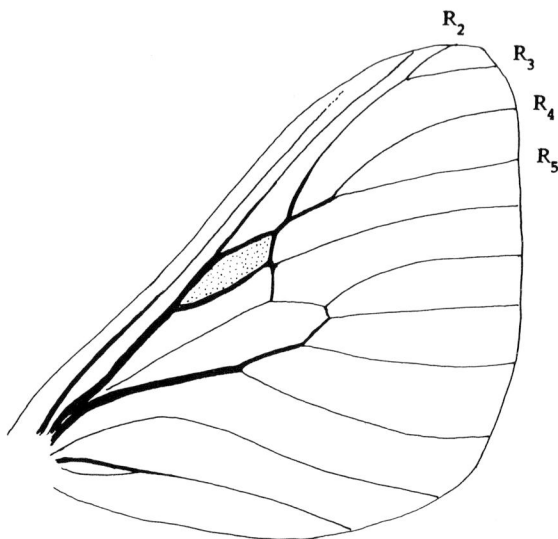

Forewing venation in Dalceridae (*Acraga*). Note separate bifurcate systems of radial veins. The areole or accessory cell is stippled.

48

-- Forewing radials branching in a single system; male forelegs not plumose 24

24. Hindwing area approximately equal to that of the forewing, never less than three-quarters of the area, both wings generally deep, somewhat rounded; forewing densely and coarsely scaled; Africa ................................CHRYSOPOLOMIDAE (p.139)

-- Hindwing area usually less than three-quarters of that of the forewing; forewing usually elongate-quadrate or roundedly triangular, finely scaled 25

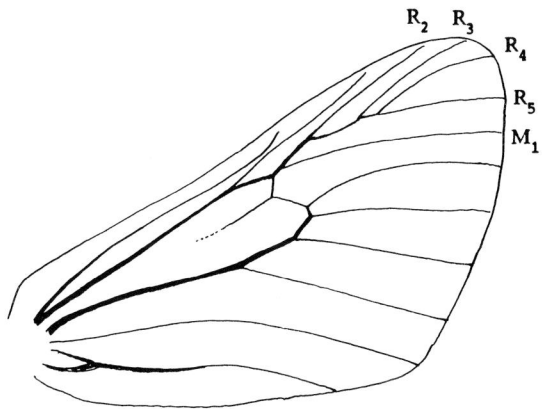

25. Forewing veins R2 to R5 branching from a common stalk, 2-4 arising anteriorly and 5 posteriorly; in a few genera 2 arises from cell or is connate with the stalk of the rest, but 4 is still anterior; antennae usually bipectinate to apex; New World .....
....MEGALOPYGIDAE (part) (p.139)

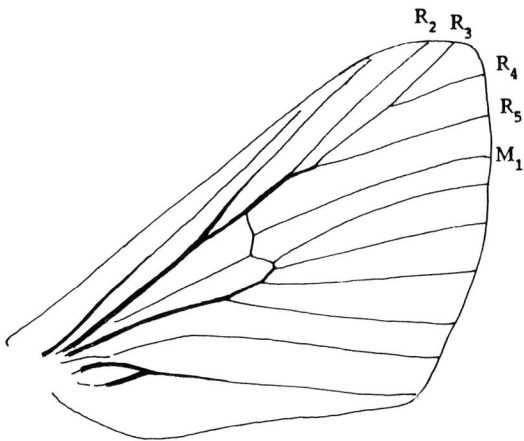

Limacodidae   Megalopygidae

Forewing venation of Limacodidae (*Setora*) and Megalopygidae (*Norape*) showing configuration of radial veins. R$_4$ arises posteriorly in the former and anteriorly in the latter.

-- Forewing veins R3 to R5 arise from a common stalk, 3 anteriorly and 4 and 5 posteriorly; male antennae broadly bipectinate only over basal half to two thirds
....LIMACODIDAE (p.140)
(The genera *Aidos* Hubner and Brachycodella Dyar currently placed in the Megalopygidae, key out as Limacodidae on both venation and antennal characters. *Semyra* Walker and *Prolimacodes* Schaus are New World Limacodidae tending towards the Megalopygidae condition.)

26. Tympanum present on thorax or abdomen 27

-- Tympanum absent 42

49

27.   Tympanum present on metathorax..NOCTUOIDEA...28

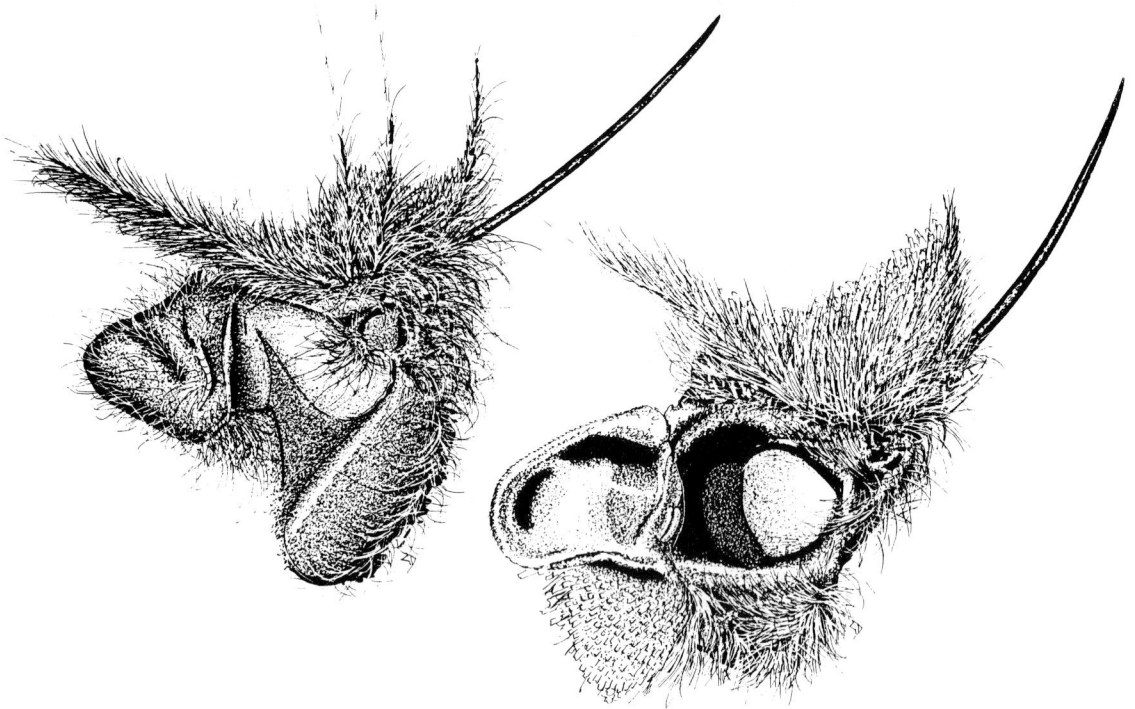

Tympanal organs of Arctiidae (left) and Noctuidae (right). These are to be found laterally at the junction of the thorax and abdomen. There is variability in the size and shape of the various components. See couplet 29 for key to parts.

--   Tympanum present on first, second or seventh abdominal segment 31

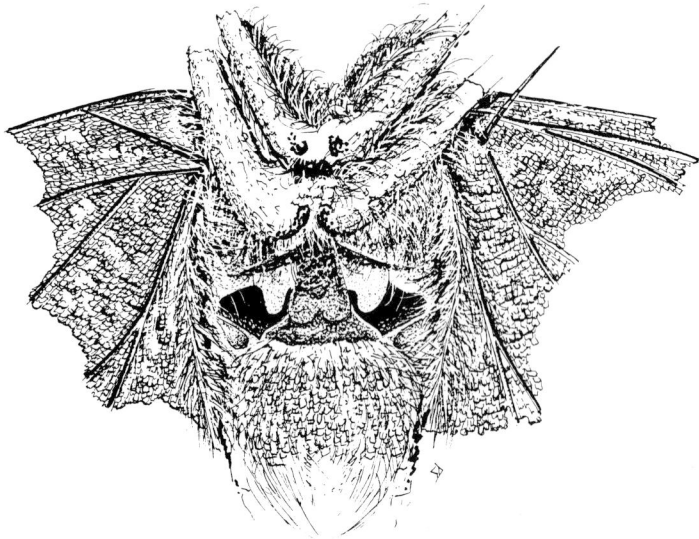

Tympanal organs of Geometridae, a pair at the base of the abdomen best viewed laterally. The pockets are generally less deep in Pyralidae. See couplet 36 for situation in male Uraniidae.

28. Forewing with vein M2 nearer to M3 at its base on the cell than to M1 or veins arising from cell reduced to 9 (some Lithosiinae); tympanum directed posteriorly 29

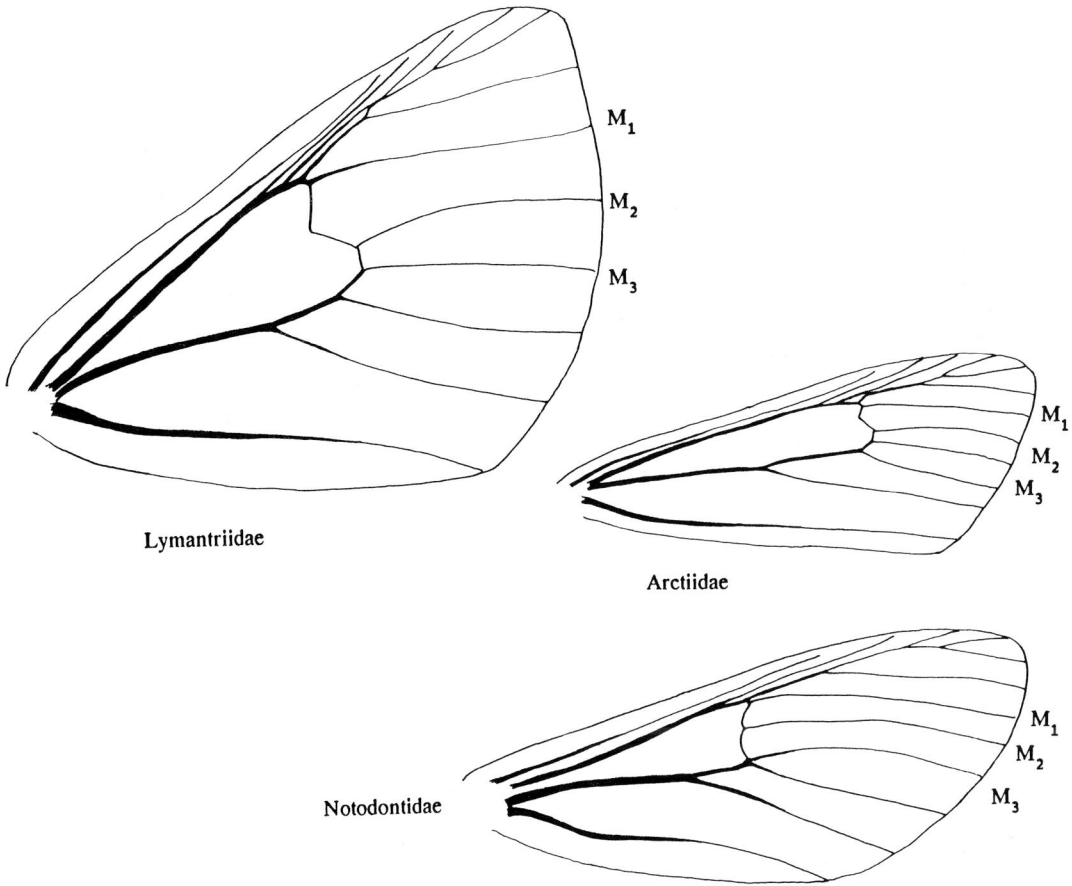

Forewing venation in Noctuoidea showing the relative position of vein M_2 in Notodontidae in comparison with other families.

-- Vein M2 nearer M1 at base or approximately central abdomen long, usually extending well beyond hindwings (not Dioptidae); tympanum directed ventrally ....
....NOTODONTIDAE (including DIOPTIDAE,
THAUMETOPOEIDAE, THYRETIDAE) (p.148-150)

29. Counter-tympanal hood on abdomen post-spiracular or absent .....NOCTUIDAE (p.148) 155

-- Counter-tympanal hood pre-spiracular* 30

*The Noctuidae include a group of species referred to the Herminiinae (included also in Hypeninae) with a pre-spiracular hood; these are mostly delicate, medium-sized, brown moths with an obtusely angled, usually pale submarginal to the hindwing.

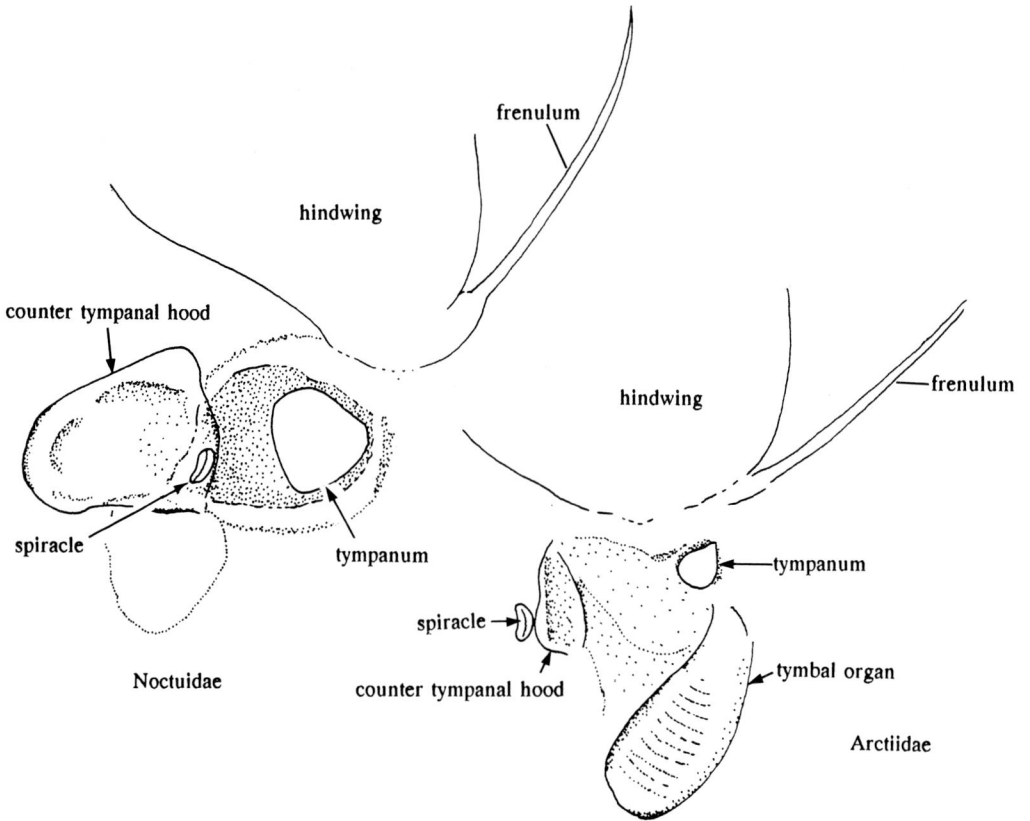

counter tympanal hood

hindwing

frenulum

spiracle

tympanum

Noctuidae

hindwing

frenulum

tympanum

spiracle

counter tympanal hood

tymbal organ

Arctiidae

30. Tongue usually well developed (but not in some Arctiinae); base of hindwing vein Sc
     swollen; tymbal organ usually present (but often reduced in Ctenuchinae); abdomen
     never with pockets between sternites 3 and 4 ....ARCTIIDAE
                                    (incl. CTENUCHINAE, PERICOPINAE) (pp.152)

--   Tongue always vestigial; base of hindwing vein Sc not swollen; tymbal organ absent;
     pair of finely corrugated pockets often present on abdomen between sternites 3 and 4
     ....LYMANTRIIDAE (p.151)

3

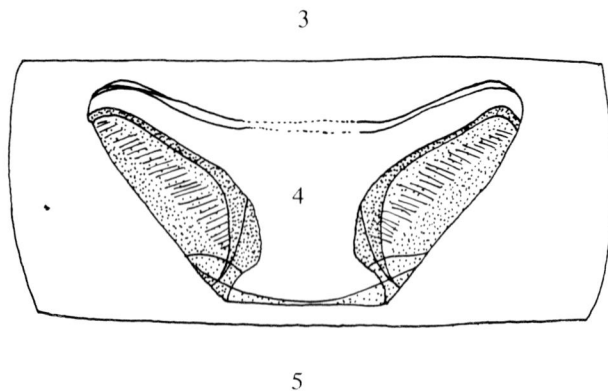

4

5

Lymantriidae (*Redoa*) fourth abdominal segment

31.   Hindwing with veins Rs and Sc fused or at least closely approximated within and for
        some distance beyond the cell; hindwing anal veins usually three; hindwing veins M2
        and M3 often arising from a common stalk; forewing never with areole (if anal veins
        are two or an areole is present on the forewing then choose other half of couplet)
        ....PYRALIDAE (p.113)

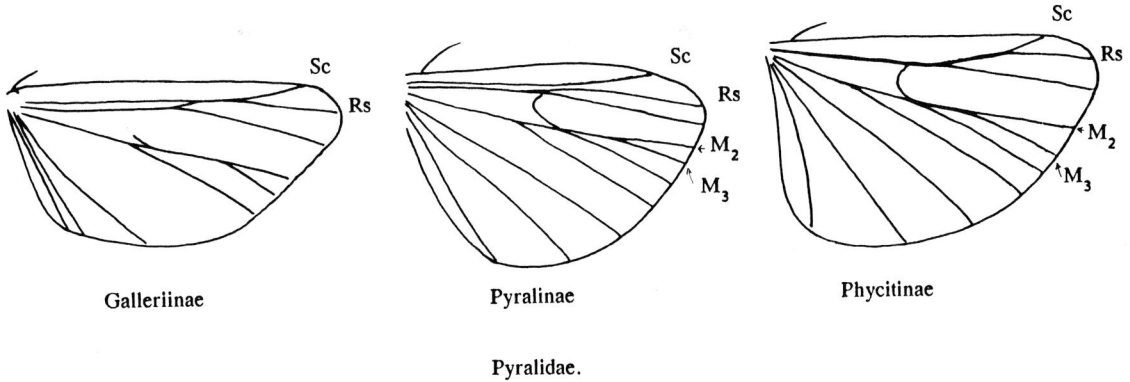

Galleriinae   Pyralinae   Phycitinae

Pyralidae.

--   Hindwing with veins Rs and Sc separate or sometimes approximate but seldom fused
        beyond cell (if so, then anal veins of hindwing two, or areole on forewing present,
        and hindwing veins M2 and M3 arising separately from cell); in GEOMETRIDAE Sc is
        strongly flexed at the base ....GEOMETROIDEA....32

32.   Tympanal organs on seventh abdominal segment; restricted to Mediterranean ....
        ....AXIIDAE (p.169)

--   Tympanal organs on first or second abdominal segment 33

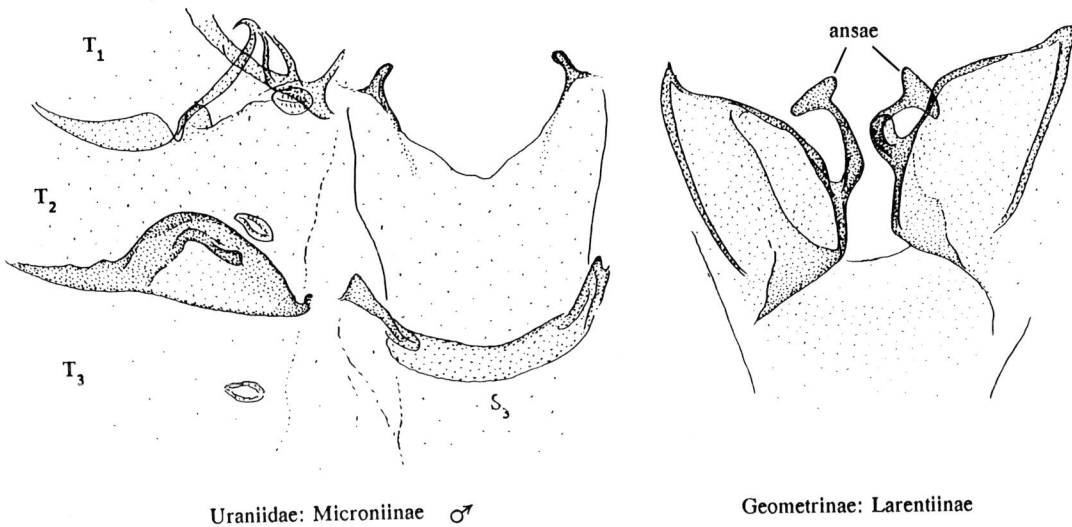

Uraniidae: Microniinae   ♂   Geometrinae: Larentiinae

**Lateral and basal tympana in Geometroidea. Note hammer-headed ansae of
Larentiinae.**

53

33.   Hindwing vein Sc approximating to, or joining Rs beyond the cell 34

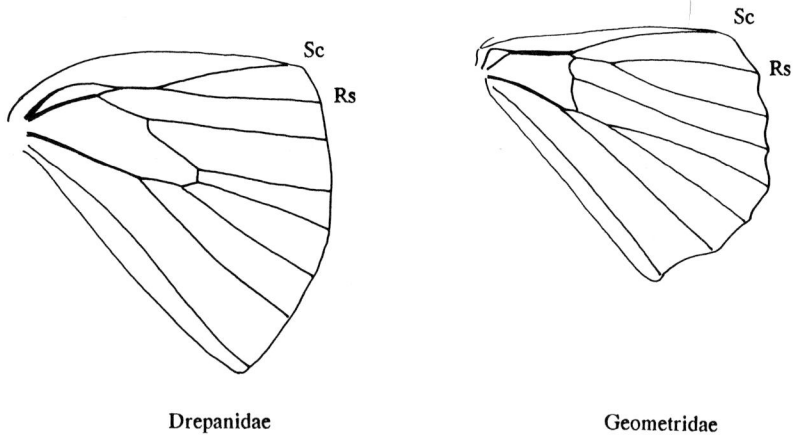

Drepanidae   Geometridae

**Sc anastomoses with Rs distal to the cell in the hindwing of Drepanidae
(*Tridrepana*) and within it, if at all, in Geometridae (*Eois*).**

--   Hindwing vein Sc remote from Rs beyond cell 36

34.   Forewing with vein M2 basally approximate to M3; forewing often falcate (hook-
       tipped) or with central angle at margin; often brightly coloured or intricately
       patterned; uncus of male genitalia entire or bifid, often accompanied by socii .....
       ....DREPANIDAE (p.169)

--   Forewing M2 nearer M1 at base, or at most midway on cross-vein between M1 and M3;
       uncus of male genitalia trifid 35

35.   Forewings triangular, slightly falcate, lacking an areole; build slender; hindwing
       M2 roughly midway between M1 and M3 at cell; Oriental Region ....
       ....CYCLIDIIDAE (p.169)

--   Forewings elongate, roughly twice as long as broad, always with an areole; build
       robust; hindwings with M2 much closer to M3 than to M1 at cell ....
       ....THYATIRIDAE (p.169)

36.   Forewing with R5 remote from R2-4 and usually fused over the basal portion with M1
       leading to two systems of branching veins distal to the cell; areole absent; male
       with tympanal organs on tergum 2, opening posteriorly (at base of abdomen in female)
      37

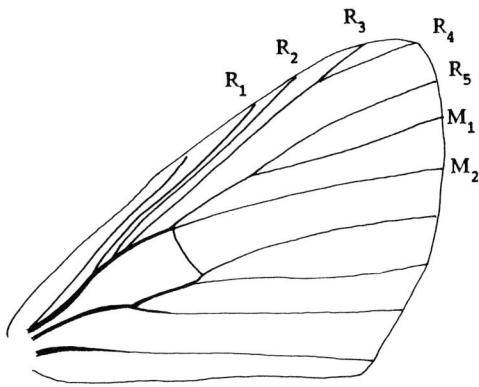

Forewing venation in Epiplemidae and Geometridae, showing the two distinct branching systems in the former ($R_3 + R_4$ and $R_5 + M_1$).

--   Forewing with radial veins 2-5 combined in one branching system of veins distal to the cell; forewing areole often present; male and female with tympanal organs at base of abdomen of first segment ....GEOMETRIDAE...38

37.   Frenulum absent; hindwing usually with single tail or angle; large moths (URANIINAE) or, if moderate to small, hindwing usually white with fine fasciae and striation (MICRONIINAE) ....URANIIDAE (p. 173) 171

--   Frenulum present; hindwing often with two tails or angles; small moths, spanning 20mm or less (except for members of the *Deceiia* Walker group which are moderate in size and mostly lack tails) ....EPIPLEMI ^E (p. 171) 173

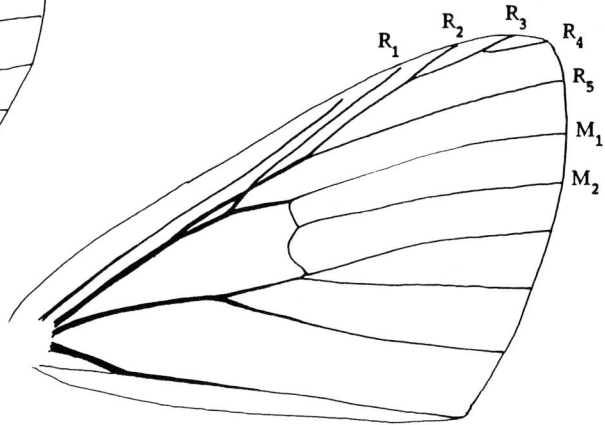

38.   M2 of hindwing lacking or reduced to a fold ....ENNOMINAE (p. 174) 175

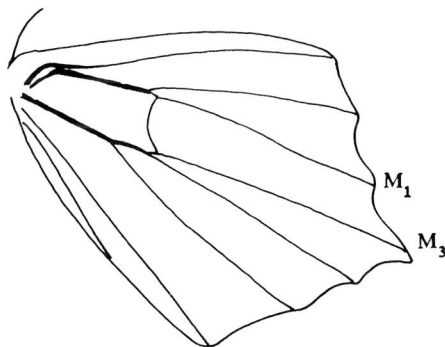

-- M2 of hindwing strong 39

39. Sc of hindwing anastomoses strongly with Rs *within* the cell, or (rarely) connected
with it within the cell by a cross-vein (R1); fasciae of forewing multiple ....
....LARENTIINAE (p.176) 178

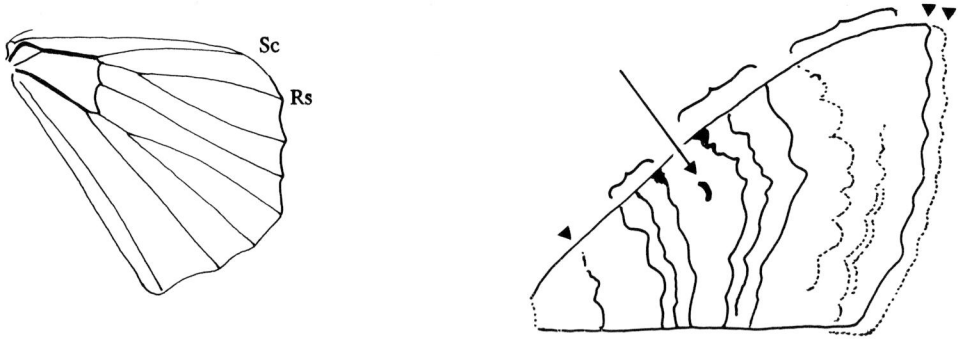

Hindwing venation (*Eois*) and forewing fasciation (*Xanthorhoe*) in Larentiinae.

-- Sc free or approximated to Rs basally or subbasally but rarely fusing with it 40

40. M2 arises very much nearer M1 than M3 on hindwing; the majority of species are
green; uncus of male genitalia bifid or, most commonly, associated with two socii
....GEOMETRINAE (p.174)

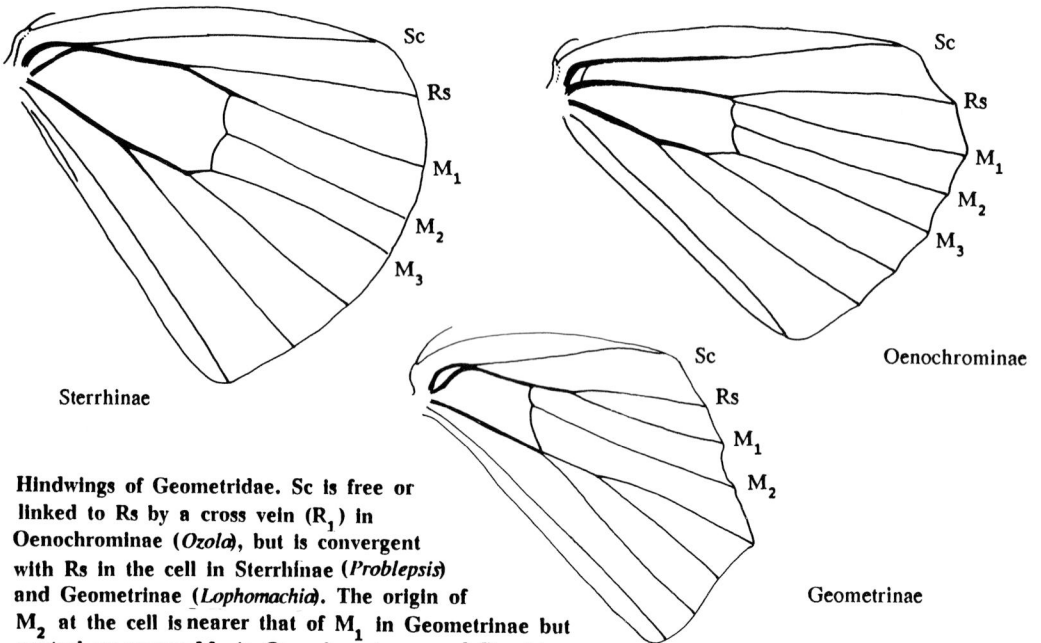

**Hindwings of Geometridae. Sc is free or
linked to Rs by a cross vein (R$_1$) in
Oenochrominae (*Ozola*), but is convergent
with Rs in the cell in Sterrhinae (*Problepsis*)
and Geometrinae (*Lophomachia*). The origin of
M$_2$ at the cell is nearer that of M$_1$ in Geometrinae but
central or nearer M$_3$ in Oenochrominae and Sterrhinae.**

-- M2 on hindwing arises approximately from mid-point between M1 and M3 41

41. Sc on hindwing free though occasionally converging briefly with Rs or connected by a cross-vein (R1) with cell subbasally; grey, or occasionally yellow, brown or white moths with few fasciae; bodies usually slender and legs long; if robust then moths large ....OENOCHROMINAE (p.174)

-- Sc touches Rs over a short distance subbasally but does not fuse with it; moths usually pink or pale fawn, sometimes yellowish or white, the wings often multifasciate but the fasciae not in multiple groups ....STERRHINAE (p.175) 177

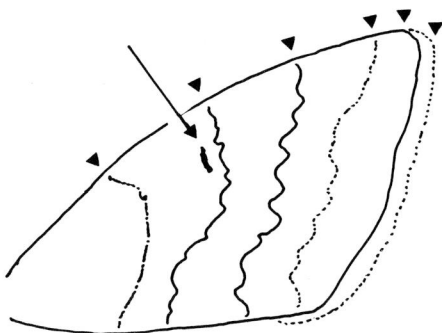

Forewing fasciation **in** *Scopula* **(Sterrhinae)**

42. Hindwing with Sc fused to, or connected with, Rs near base and approximated to or fused with Rs for some distance beyond cell; veins M2 and M3 approximate, connate or stalked at base 43

-- Hindwing venation not as above 44

43. Hindwing with CuP present towards margin; medium sized noctuid-like moths with black and yellow or black and orange hindwings ....HYBLAEIDAE (p.126) 125

-- Hindwing with CuP not present as a tubular vein, but vestigial or absent; small to moderately large moths, often with reticulate pattern ....THYRIDIDAE (p.126) 125

44. Large moths (wingspan over 40mm), or, if small, forewings broad, triangular; tongue often absent or vestigial; frenulum also often absent; ocelli and chaetosemata absent (if chaetosemata present then hindwing conspicuously tailed or antennae filiform, expanding towards apex, combined with orange and brown pattern and butterfly-like build) 45

-- Small to medium-sized moths (wingspan usually under 30mm); wings, with a few exceptions, narrow, the forewing twice as long as broad or more, square-ended or lenticular in many instances; tongue and frenulum usually present; ocelli and chaetosemata sometimes present 62

45. Chaetosemata present 46

-- Chaetosemata absent 47

46. Wingspan greater than 40mm; hindwing tailed, resembling species of Uraniidae; New World ....SEMATURIDAE (part) (p.169)

-- Wingspan less than 40mm; hindwing rounded or, at most, angled; build and field behaviour butterfly-like (wings held erect at rest); pattern of wings orange on dark brown; Indo-Australian tropics ...................................CALLIDULIDAE (p.177)[79]

47. Forewing veins R4 and R5 arising from a common stalk at the cell, with separate stalk giving rise to R2 and R3 at an extremely distal fork; M1 is usually only connate at the base of R4+5; New World (if Old World then Lasiocampidae) ....
....**MIMALLONOIDEA**: MIMALLONIDAE (p.148)

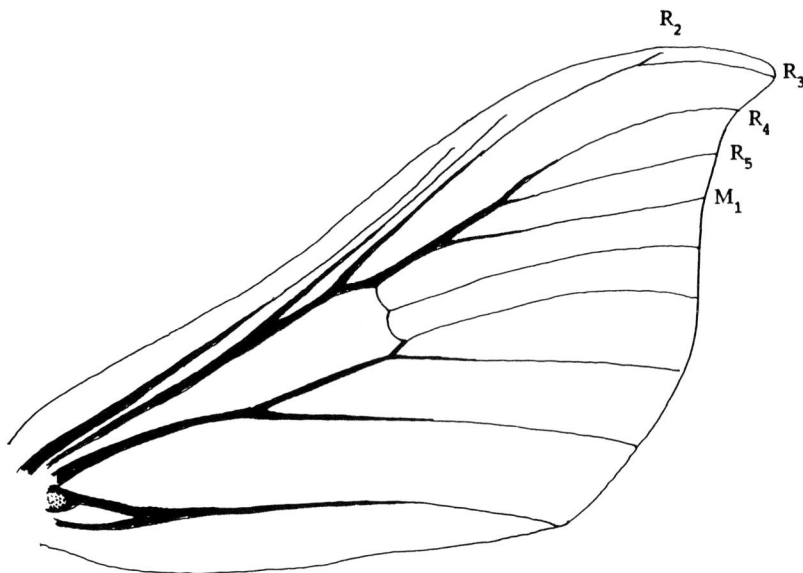

Forewing venation of Mimallonidae (*Cicinnus*); note two branching systems.

-- Forewing veins with different configuration ....**BOMBYCOIDEA**..48

48. M2 arising nearer M3 than M1 in *both* fore- and hindwings, such that the cubitus (posterior vein of cell) appears quadrifid or four-branched 49

-- M2 arising nearer M1 than M3 in at least one pair of wings such that the cubitus appears trifid or three-branched 53

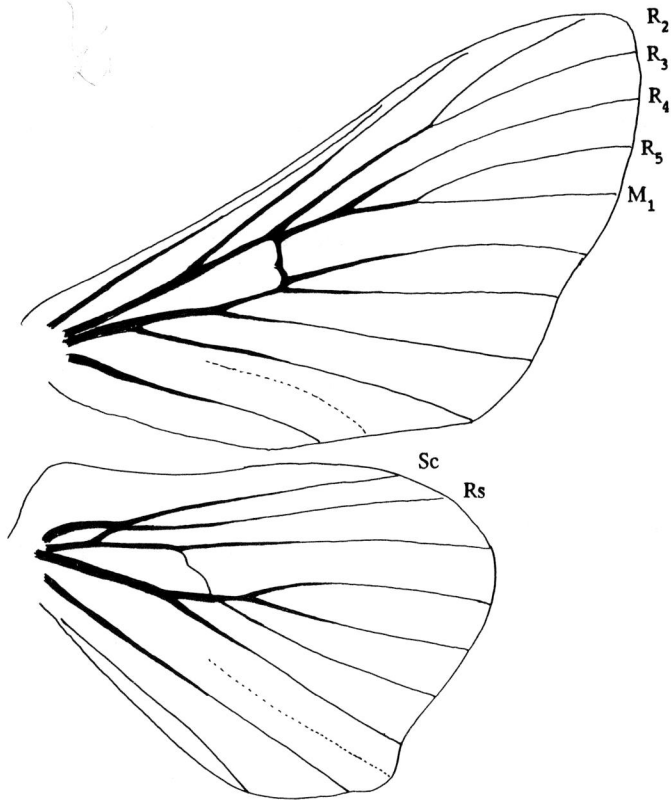

Venation of Lasiocampidae (*Streblote*) showing typical branching systems of
forewing veins. Note also the cubital vein (posterior of cell) is quadrifid
in both wings.

49.   M1 on forewing usually stalked with R5; radial veins usually in two branching
      systems; hindwing veins Sc and Rs share a short common stalk or anastomose,
      separating from the cell near the base; male genitalia often with small valves and
      well developed arms from vinculum ............................LASIOCAMPIDAE (p.143)

--   M1 on forewing not stalked with R5 but usually connate at base; forewing radial
      veins in a single branching system; hindwing vein R5 separates from cell at its
      anterior distal angle and is not associated with Sc when not forming part of cell
     50

50.   Hindwing vein Sc adjacent to, or fused with cell at base only, with no cross-vein;
      fore- and hindwings usually deep 51

--   Hindwing vein Sc linked to cell by cross-vein (R1); fore- and hindwings elongate,
      forewing narrower than long 52

51.   Forewing with a large areole beyond the anterior angle of the cell; Australia ......
      ....ANTHELIDAE (p.142)

--   Forewing without an areole ..................................EUPTEROTIDAE (part) (p.144)

52.   Antennae stout, often hooked at the tip, rarely distinctly bipectinate; tongue often
      very long though sometimes weak; frenulum present ....SPHINGIDAE (p.146)

-- Antennae distinctly bipectinate, not hooked apically; tongue weak; frenulum absent; Palaearctic ....ENDROMIDAE p. 14

53. Frenulum strongly present (though reduced in female EUPTEROTIDAE) 54

-- Frenulum very weak or absent 56

54. Tongue strong 61

-- Tongue reduced or absent 55

55. Medium-sized species; forewings usually falcate or forewings margin angled; markings on hindwing usually distinctly more intense along the dorsum; New World ....
....APATELODIDAE (p.143)

-- Medium to very large species; forewings rarely falcate, margins mostly rounded; hindwing markings not more intense along dorsum; wing scaling coarse; predominantly Old World ....EUPTEROTIDAE (part) (p.142) 144

56. Hindwing finely folded along dorsum which is straight to concave with dark markings and fasciae more intense; Old World ....BOMBYCIDAE (p.145)

Hindwing venation in Bombycidae (*Ocinara*). Note somewhat hooked tornus and closeness of anal veins at dorsum; this zone is usually folded, concave towards the abdomen.

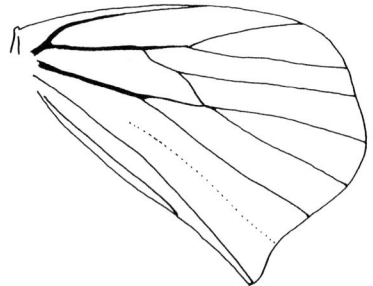

-- Hindwing not folded or with markings conspicuously darker 57

57. Hindwing vein Sc well separate from cell and not linked to it by a cross-vein except almost basally 58

-- Hindwing vein Sc and Rs in cell closely approximate, fused or linked by R1 as a cross-vein 59

58. Tongue fairly strong; male antennae bipectinate; New World .....OXYTENIDAE (p.139) 147

-- Tongue vestigial or absent; often very large moths with hyaline patches or ocelli at the forewing or hindwing discal area; male antennae often quadripectinate ....
....SATURNIIDAE (p.145)

59. Hindwing Sc linked to cell by *strong* cross-vein; New World ........CERCOPHANIDAE
                                                                          (p. 139) 142

--   Hindwing otherwise; Old World 60

60. Sc of hindwing approximate to distal half of cell and to Rs beyond before diverging,
     linked to cell by *weak* cross-vein; hindwing cell short, broad, with medial vein
     weakly present within it; distinctive rippled pattern between postmedial and sub-
     marginal of both wings ....BRAHMAEIDAE (p. 139) 142

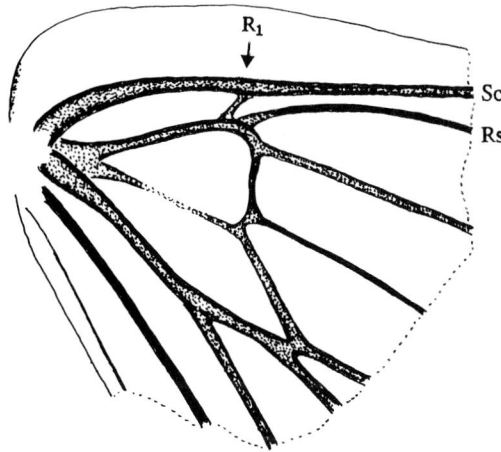

Basal hindwing venation in Brahmaeidae (*Brahmaea*) showing R_1 as a weak cross
vein.

--   Sc and Rs fused over much of length, separating well beyond end of cell; space 1b of
     forewing often with prominent median line of contrasting colour ....
     ....LEMONIIDAE (p. 139) 142

61. Hindwing vein Sc connected to cell by cross-vein (R1); large, robust species with
     discal ocelli; Australia ....CARTHAEIDAE (p. 139) 142

--   Hindwing vein Sc free; forewing veins R2 and R3 sinuous; build fragile; abdomen
     basally constricted; first abdominal tergite strongly pouched; New World ....
     ....HEDYLIDAE (in PAPILIONOIDEA senso lato) (p.182)

62. Tongue scaled, at least near base, usually well developed (rarely reduced), with
     maxillary palpi folded over base; chaetosemata absent; labial palpi usually   upturned
     or ascending, often recurved, reaching vertex or above, sometimes porrect or
     drooping, second segment usually roughened or tufted beneath, third segment usually
     slender, pointed or acute, rarely concealed, palpi sometimes sexually dimorphic;
     ocelli present or absent; antennal scape often with pecten; mostly small to medium-
     sized micro-moths, except some large Australian Xyloryctinae (expanse 60mm or more)
     and some Neotropical Stenomatinae (expanse 40mm or more) ....
     ....GELECHIOIDEA...78

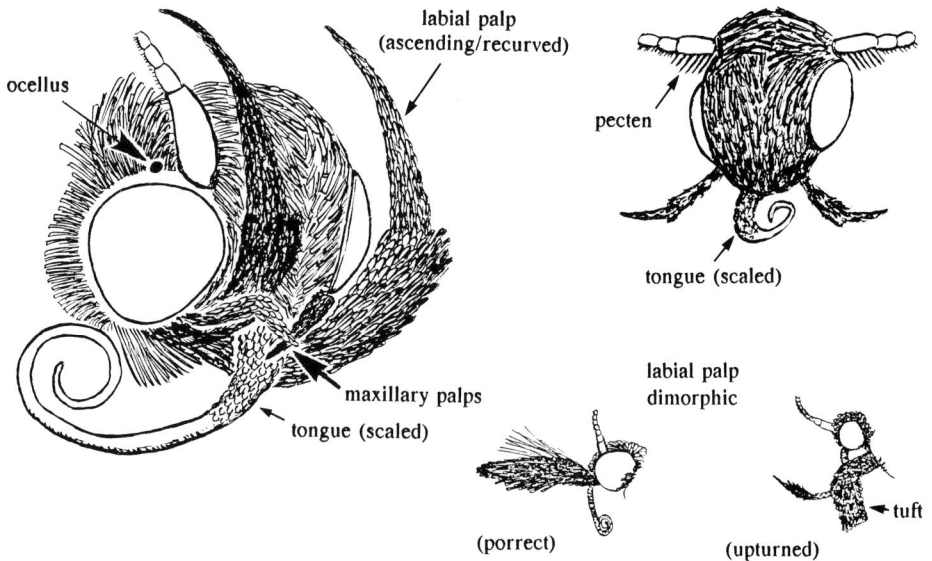

ocellus

labial palp (ascending/recurved)

pecten

tongue (scaled)

maxillary palps

tongue (scaled)

labial palp dimorphic

(porrect)

(upturned)

tuft

-- Tongue not scaled or only exceptionally (Choreutidae (couplet 72), rarely ypono-meutoids), sometimes vestigial or obsolete; maxillary palpi not folded over base of tongue or only exceptionally; labial palpi various; ocelli and chaetosemata present or absent; very small narrow-winged to moderately large broad-winged micro-moths ... .63

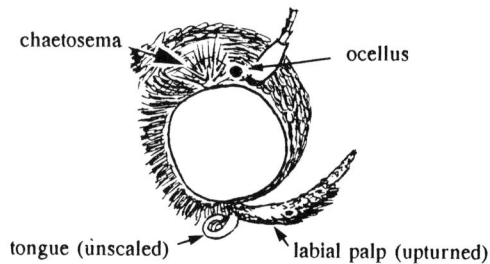

chaetosema

ocellus

tongue (unscaled)

labial palp (upturned)

63. Head entirely rough-haired or rough-scaled, or frons flat and smooth-scaled and vertex (between antennae) with prominent brow-ridge (Hieroxestinae); ocelli and chaetosemata absent; labial palpus rudimentary and partially concealed, or short to moderately long, porrect or drooping, sometimes with laterally projecting pecten bristles on second segment; wings broad or narrow, usually short-fringed 64

-- Head smooth-scaled, or frons smooth and scales loosely appressed or tufted on crown; ocelli and chaetosemata present or absent; labial palpus short to moderately long, curved and ascending, or straight and porrect, or drooping; wings either narrow and long-fringed and venation reduced, or broad/moderately broad with venation developed (TORTRICIDAE) 66

64. Labial palpus usually with prominent bristles laterally on second segment (not Hieroxestinae); forewing with R5 to costa or apex, wing-tip sometimes bent upwards (Erechthiinae); mostly very small to small micro-moths under 30mm wing expanse ....TINEIDAE (p.82)

Tineidae

maxillary
palp

labial palp

bristles

Tinea

Monopis

R_5

Decadarchis

R_5

Tinea

antennal scape

pecten

bristles

Nidutinea

pecten

bristles

Decadarchis

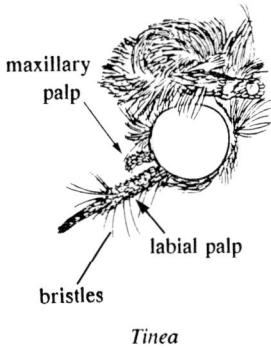

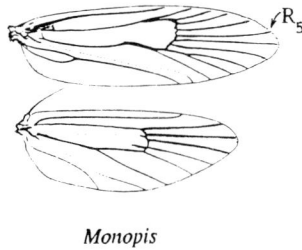

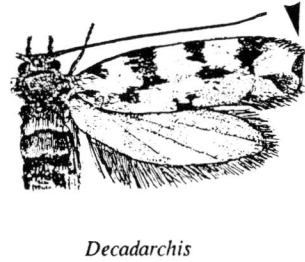

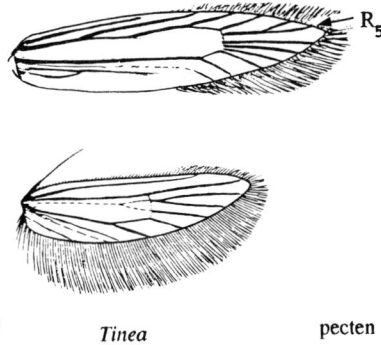

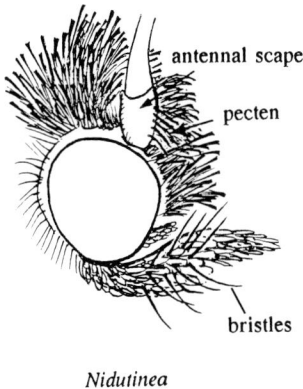

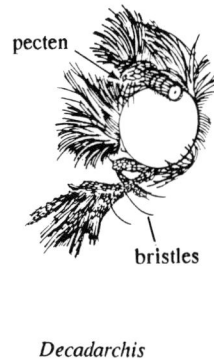

-- Labial palpus without pecten bristles on second segment 65

65. Small to moderately large (expanse 8-40mm), sometimes larger (60mm); wings generally broad or very broad, often rounded, usually more or less unicolorous black or brown, sometimes thinly scaled or hairy, or semi-transparent or transparent; M usually present in both wings, often forked, forewing with R5 to apex or termen, anal veins 1A+2A and CuP curved, often fused medially; antenna simple and ciliate (Psycheoidinae); labial palpus rudimentary or short; female usually apterous and often without legs, remaining in larval/pupal case/bag/cocoon ....PSYCHIDAE (p.79)

63

Psychidae

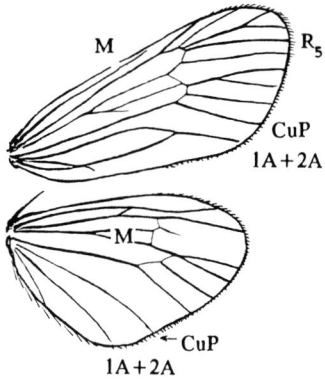

Psychinae

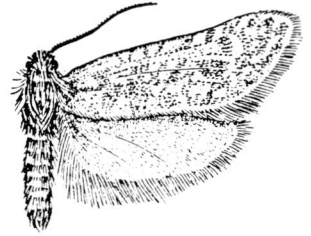

Psychiodinae

Psychinae

-- Small tineid-like moths (expanse 9-12mm); forewing coarsely scaled, hindwing of male usually thinly scaled basally; labial palpus porrect, second segment rough-scaled, third segment partially concealed; antenna usually thickened with shaggy scales towards base; fore- and hindwing with CuA2 usually absent ....
....OCHSENHEIMERIIDAE (p.94)

Ochsenheimeriidae

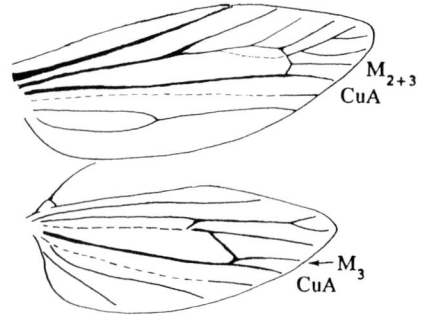

Ochsenheimeria

66.   Antennal scape dilated and forming an eye-cap 67

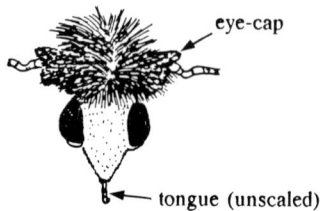

Bucculatrix

--   Scape not forming an eye-cap 68

67.   Eye-cap small; hind tibia with dorsal row of long bristles, usually extending onto tarsi; ocelli and chaetosemata absent; labial palpus straight, porrect; very small (expanse 6-8mm), forewing whitish, finely marked ....
....GRACILLARIIDAE (PHYLLOCNISTINAE) (p.90)

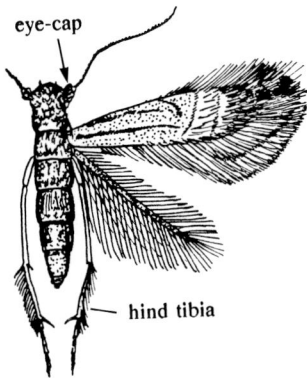

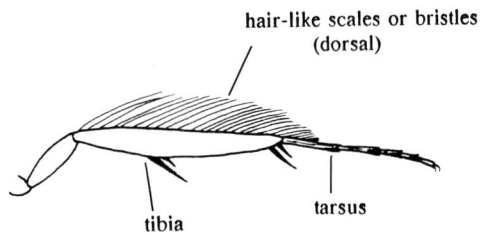

Phyllocnistis

--   Eye-cap usually prominent; ocelli and chaetosemata absent; hind tibia hairy or with long hair-like scales or bristles above and below; labial palpus usually short, often filiform, either porrect or drooping; very small to small species ....
....LYONETIIDAE (p.86)

Lyonetiidae

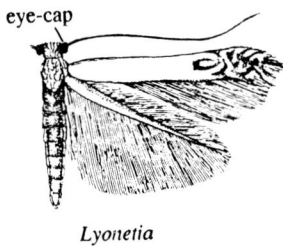

Lyonetia

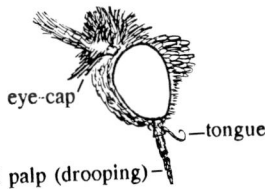

labial palp (drooping)

Lyonetia

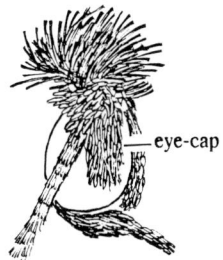

Bedella

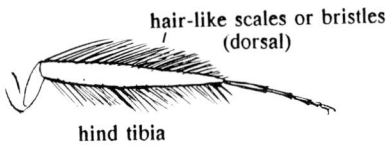

hind tibia

68.   Head smooth-scaled or scales appressed (Gracillariinae), or smooth except for anterior fringe of erect hair-scales (Lithocolletinae); tongue developed, not scaled; ocelli and chaetosemata absent; antenna filiform, either exceeding length of fore-wing up to 1.5x or more (Gracillariinae), or as long or nearly as long as fore-wing (Lithocolletinae); labial palps slender, moderately long, porrect or ascending, moderately pointed, second segment sometimes tufted beneath (Gracillariinae); maxillary palps usually distinct, porrected, sometimes minute; hind legs long, tibia smooth or with dorsal row of bristles or loosely scaled, never hairy, fore and mid tibiae sometimes densely scaled or ornamented (Gracillariinae); slender narrow-winged micro-moths, resting posture either erect with head and body steeply raised on fore and middle legs (Gracillariinae), or nearly horizontal with head lowered, the abdomen slightly raised on hind legs and the fore and middle legs (Gracillariinae), or nearly horizontal with head lowered, the abdomen slightly raised on hind legs and the fore and middle legs stretched forwards (Lithocolletinae, Phyllocnistinae) ....GRACILLARIIDAE (p.88)

Gracillariidae

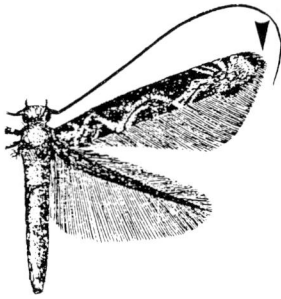

maxillary palp (porrect)

labial palp (upturned)

Acrocercops

Phyllonorycter

Conopomorpha

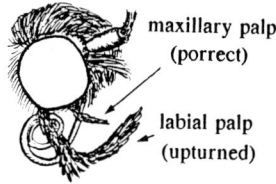

Lithocolletinae

Gracillariinae

Acrocercops

--   Not so; antenna usually shorter than forewing, but if longer then comparatively robust, broad-winged micro-moths (some Gelechiidae) 69

69.   Head usually smooth-scaled, sometimes with raised hair-like scales on crown; forewing elongate, distinctly narrowed basally and hymenopterous-like (Sesiidae), or elongate-ovate to moderately broad, often with raised scales over wing or spread along dorsal margin, or wings broad, forewing subtruncate, M rarely present, pterostigma sometimes present; chaetosemata usually absent; ocelli present, sometimes prominent, or absent; legs sometimes bristly 70

--   Head with frons smooth or with scales loosely appressed, crown with appressed, loosely appressed or semi-erect scales; forewing usually broad, with M and chorda often present (Tortricidae); ocelli and chaetosemata present or absent; legs never bristly 73

70.   Small to medium-sized or large, often with wings only partially scaled or hyaline (Sesiidae), or if wings fully scaled, except sometimes hindwing (Brachodidae), then somewhat tortricoid in appearance (Choreutidae) but lacking M in forewing and without chaetosemata 71

--   Small to medium-sized, wings fully scaled 74

71.   Wings partially hyaline (mimics of Hymenoptera), forewing elongated, usually strongly narrowed in basal half; wings secondarily coupled by bristles along dorsal margin of forewing and costal margin of hindwing; frenulum and reticulum present; ocelli prominent; antenna clavate, usually with tuft of bristles at apex; small to medium-sized, sometimes larger (a few species up to 60 mm expanse) ....
....SESIIDAE (p.96)

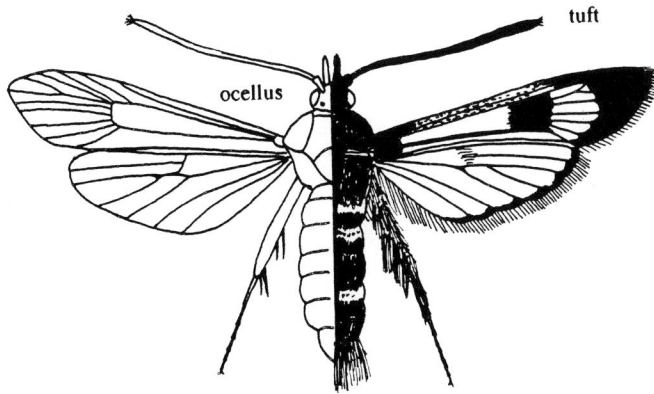

Sesiinae
Seslidae

-- Wings fully scaled, or only with hindwing hyaline basally, without secondary marginal coupling, forewing not strongly narrowed basally 72

72. Tortricoid-like moths, with small ocelli but lacking chaetosemata (present in Tortricoidea) and with tongue scaled at base (not scaled in Tortricoidea); wings broad (costal margin wide), forewing subtruncate, often with contrasting pattern or areas of metallic scales ....CHOREUTIDAE (p.97)

Choreutidae

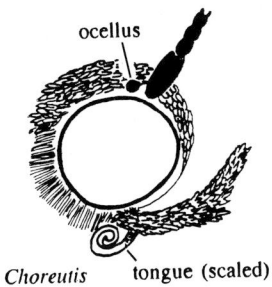

Choreutis tongue (scaled)

Tebenna

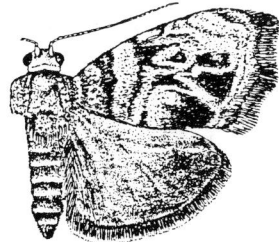

Simaethis

-- Small to medium-sized robust micro-moths (expanse 10-30mm), relatively narrow-winged, hindwing trapezoidal, sometimes thinly scaled or hyaline basally; ocelli present (large); chaetosemata absent; tongue not scaled ....BRACHODIDAE (p.98)

Brachodidae

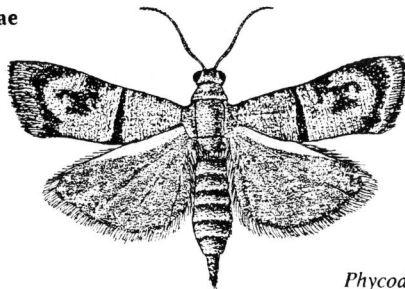

Phycodes

73. Chaetosemata usually well developed and distinct; ocelli present (rarely absent); labial palpus short to long, usually porrect, beak-like, third segment usually short, obtuse, downturned; forewing with M usually present in cell, CuP usually present, male often with costal fold; hindwing usually as broad or broader than forewing, CuA sometimes with hair-pecten, CuP usually present or weakly developed ....TORTRICIDAE (p.129)

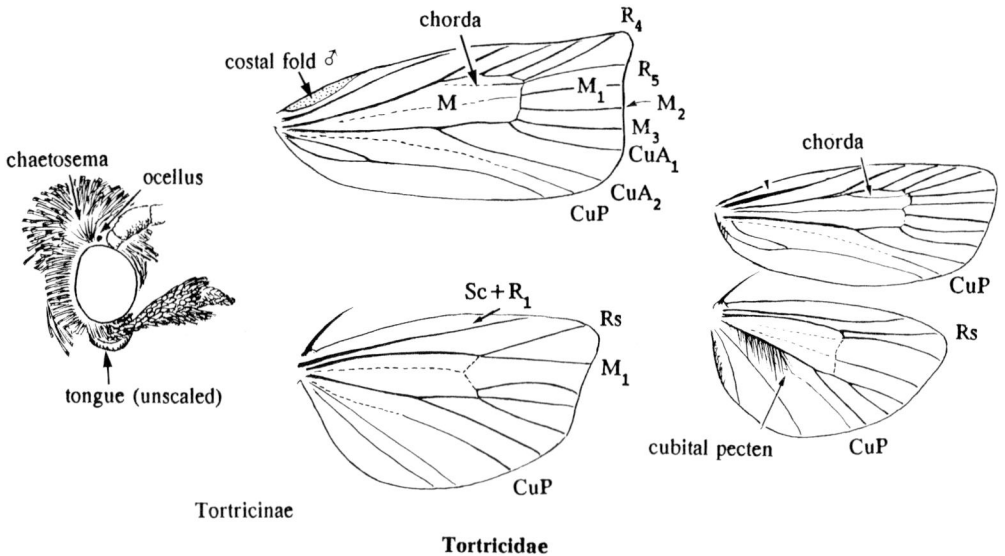

Tortricinae

Tortricidae

-- Chaetosemata usually present but simple; ocelli usually absent; frons smooth, crown with appressed scales; labial palpus upturned, second segment usually densely scaled and long, male sometimes with meso-apical spined protrusion and with bristle-like third segment; forewing with M absent; broad-winged micro-moths ....IMMIDAE (p.99)

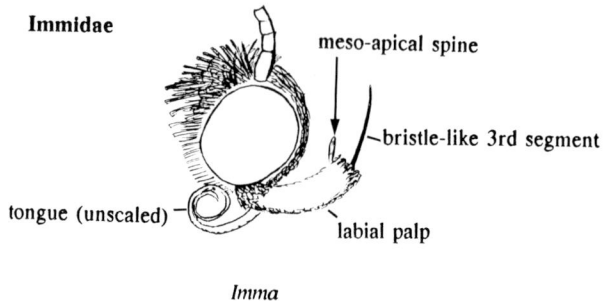

Imma

74. Forewing smooth-scaled, except scale-tufts usually present on dorsal margin only 75

-- Forewing coarsely scaled and usually with scattered scale-tufts 77

75. Forewing narrowly elongate-ovate, usually with small, sometimes prominent tooth-like scale-tufts on dorsal margin, without pterostigma; hindwing elongate-lanceolate; head with scales appressed; ocelli and chaetosemata absent; hind tibia and tarsi with long stiff hairs or bristles above and below; antennal scape with pecten; tongue unscaled, with maxillary palpi folded over base ..........EPERMENIIDAE (p.94)

Epermeniidae

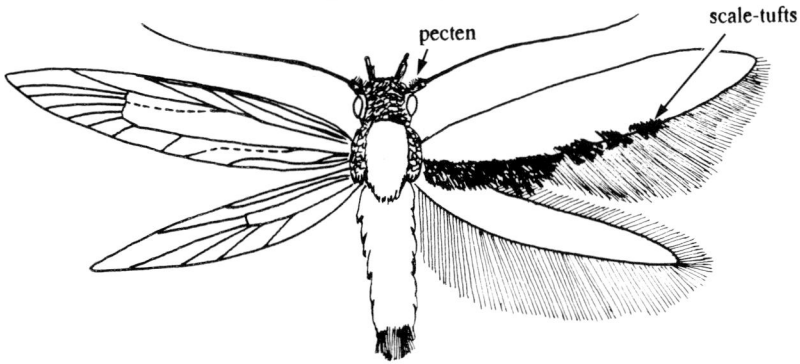

pecten

scale-tufts

Epermenia

--   not so 76

76.   Forewing elongate-ovate, often moderately broad and truncated distally, often with spotted pattern (Yponomeutinae), or comparatively narrow and tapered, usually with obscure or partial fasciae (Plutellinae) or with distinct oblique median fascia (Argyresthiinae); pterostigma more or or less perceptibly developed; labial palpus variable, usually moderate or long ascending or upturned, sometimes short or drooping (Yponomeutinae, Plutellinae) or moderate, somewhat curved and porrect (Argyresthiinae); ocelli rarely present (small); chaetosemata absent or weakly developed ....YPONOMEUTIDAE (p.91)

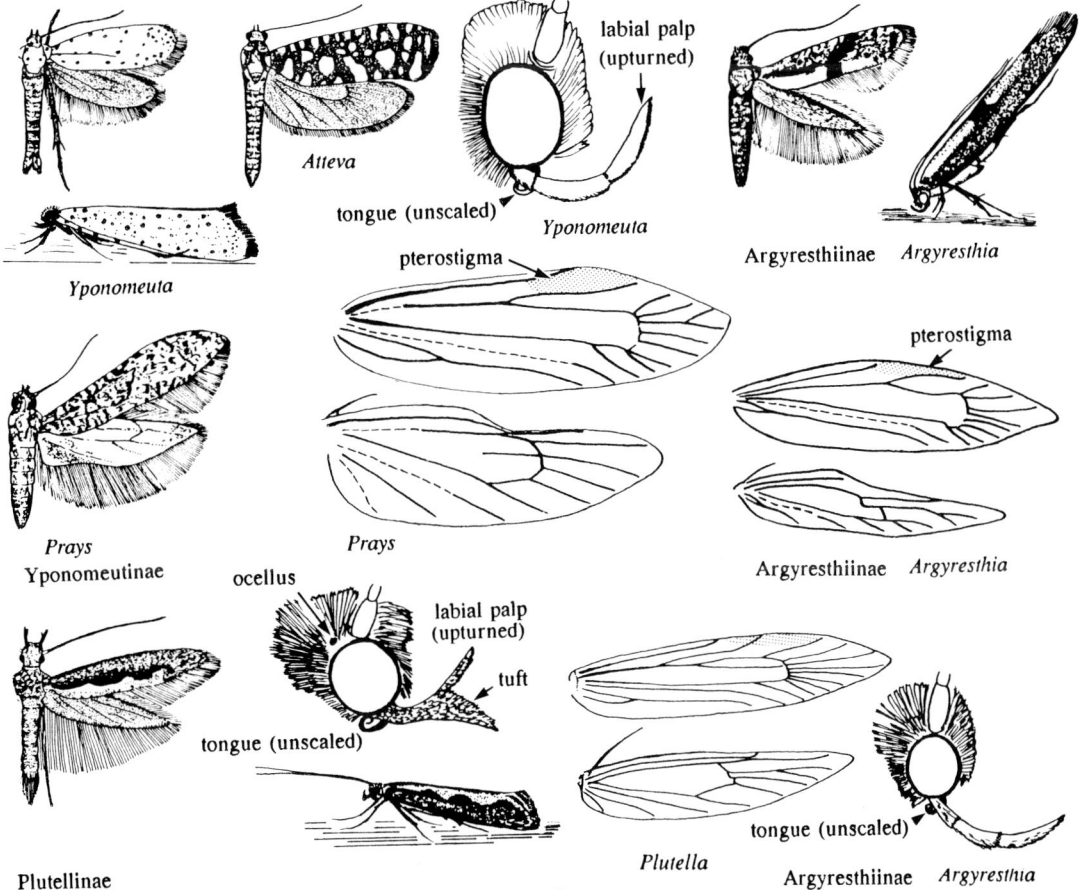

Yponomeutidae

Atteva

labial palp (upturned)

tongue (unscaled)

Yponomeuta

Yponomeuta

Argyresthiinae   *Argyresthia*

pterostigma

pterostigma

Prays

Yponomeutinae

Prays

Argyresthiinae   *Argyresthia*

ocellus

labial palp (upturned)

tuft

tongue (unscaled)

Plutellinae

Plutella

tongue (unscaled)

Argyresthiinae   *Argyresthia*

69

-- Forewing elongate-sublanceolate, termen often notched below apex, markings usually sharply defined; pterostigma often present; hindwing elongate or narrow-oblong; head smooth-scaled; ocelli present or absent; chaetosemata rarely present; scape without pecten; labial palpus moderately long, usually ascending, smooth-scaled or with scales loosely appressed; mostly sun-basking micro-moths with metallic coloration ....GLYPHIPTERIGIDAE (p.95)

Glyphipterigidae

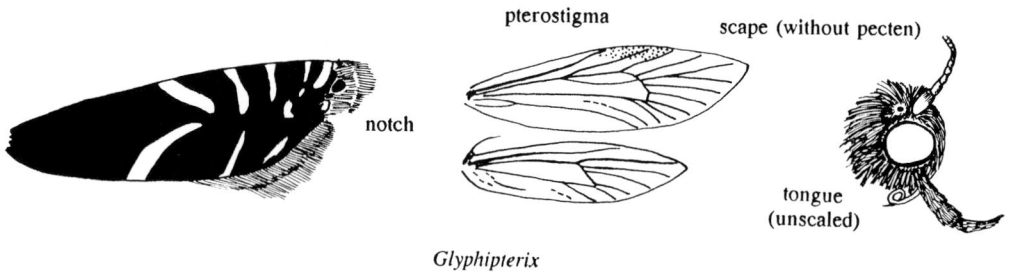

Glyphipterix

77. Small to near medium-sized moderately broad-winged micro-moths; forewing coloration predominantly black, white and grey, often with scattered raised scales or scale-tufts, CuP absent; hindwing with one or usually two median veins absent; ocelli and chaetosemata absent; scape without pecten, flagellum densely ciliated in male; labial palpus in male recurved or apical segment porrect, in female long, porrect ....CARPOSINIDAE (p.126)

Carposinidae

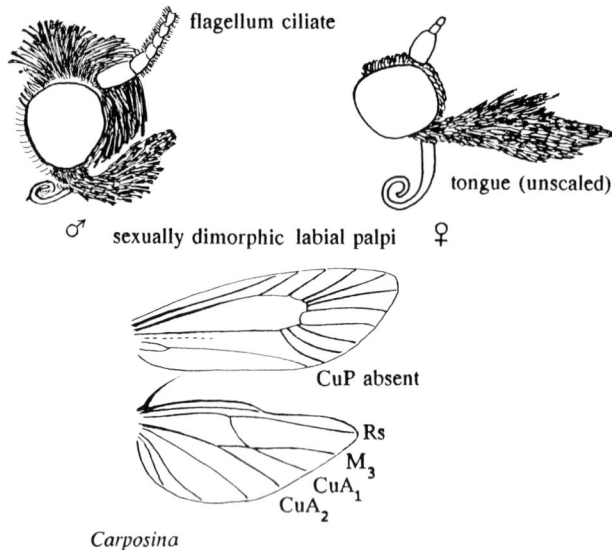

Carposina

-- Small to medium-sized micro-moths, wings broad to moderately broad; forewing coloration predominantly brown or yellowish brown, sometimes green, with scattered scale-tufts and/or roughened scales, distal area sometimes thinly scaled between veins, CuP vestigial; hindwing with three median veins present, Rs and M1 usually parallel; scape without pecten, flagellum simple or unipectinate, rarely bipectinate; ocelli present or absent; chaetosemata absent; labial palpus usually upturned or ascending, sometimes recurved or porrect ..............................COPROMORPHIDAE (p.127)

70

Copromorphidae

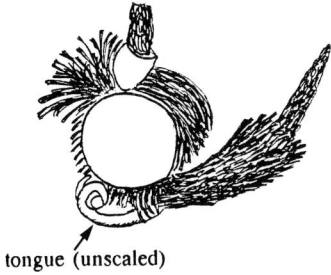

tongue (unscaled)   *Copromorpha*

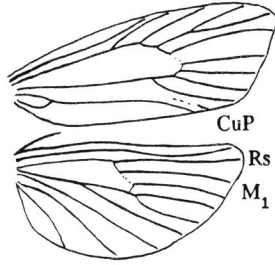

78.   Hindwing narrow, lanceolate to linear, narrower than its fringe of cilia 79

--   Hindwing ovate-lanceolate, ovate or trapezoidal with sinuate or emarginate termen
   84

79.   Forewing with pterostigma present, usually well developed, R5 to costa; antennal
    scape usually enlarged and with pecten; ocelli absent; labial palpus often sexually
    dimorphic (diagnostically specific for male); mostly small dull-coloured micro-moths
    ....BLASTOBASIDAE (p.105)

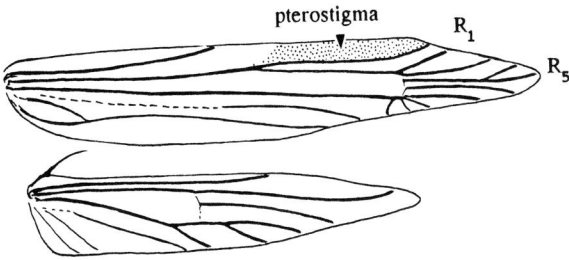

Blastobasidae

Blastobasis

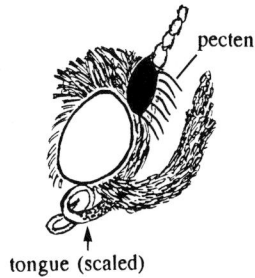

--   Forewing without pterostigma or with weak indication only; hindwing venation reduced
   80

80.   Labial palpus divergent (explanate); maxillary palpi vestigial; head and abdomen
    flattened; ocelli absent; forewing ovate-lanceolate, coloration pale yellowish brown
    with brownish or reddish longitudinal patches or streaks; hindwing venation reduced
    ....AGONOXENIDAE (p.106)

Agonoxena

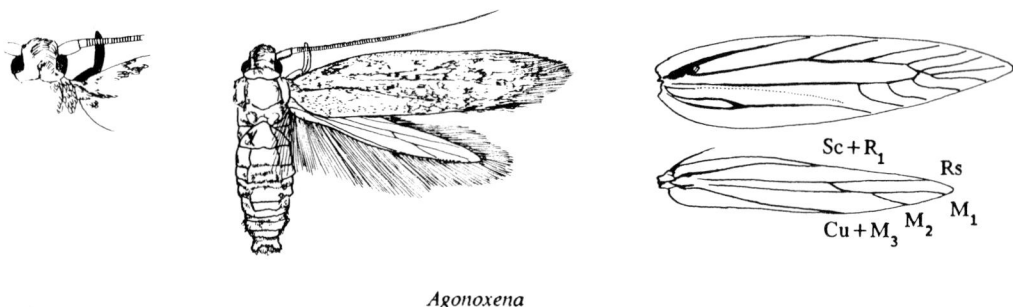

-- Labial palpi not divergent; head and abdomen not flattened 81

81. Forewing narrow to elongate-lanceolate, often unicolorous or almost so, markings usually longitudinal streaks or radiating lines (transverse fasciate markings lacking); hindwing linear-lanceolate; labial palpus moderately long, recurved, second segment usually roughened or tufted beneath at apex; abdominal terga each with a pair of spinose patches, often specific; antenna sometimes thickened with scales towards base, held porrect in repose; larva usually living in portable case constructed from seed, leaf, etc. of foodplant .................COLEOPHORIDAE (p.105)

Coleophora

tongue (scaled)

Coleophoridae

spinose patches on abdominal terga

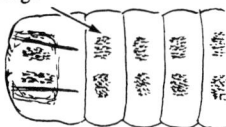

-- Forewing narrow to very narrow and acuminate, usually with at least indication of fasciate markings, sometimes with metallic scales 82

82. Wings narrow with apices acutely rounded, or very narrow with apices acuminate, forewing without pterostigma, with R2 arising well before end of cell, R4 and R5 stalked, R5 to costa or apex, CuP usually absent, or present at margin; labial palpus long, slender, recurved, second segment usually smooth; ocelli present or absent; scape usually with pecten; genitalia usually asymmetrical, male with aedeagus anchylosed with anellus, uncus and sometimes gnathos absent, gnathos when

present usually in form of two asymmetrical lobes (brachia), valvae with often strongly asymmetrical lobes or processes from base or from base and diaphragm, anellus sometimes with pair of lobes, segments 7-8 often modified to form valve-like structures (epiptygmata), female with ostium on 7th sternite, often protruding ....COSMOPTERIGIDAE (p.109)

Cosmopterigidae

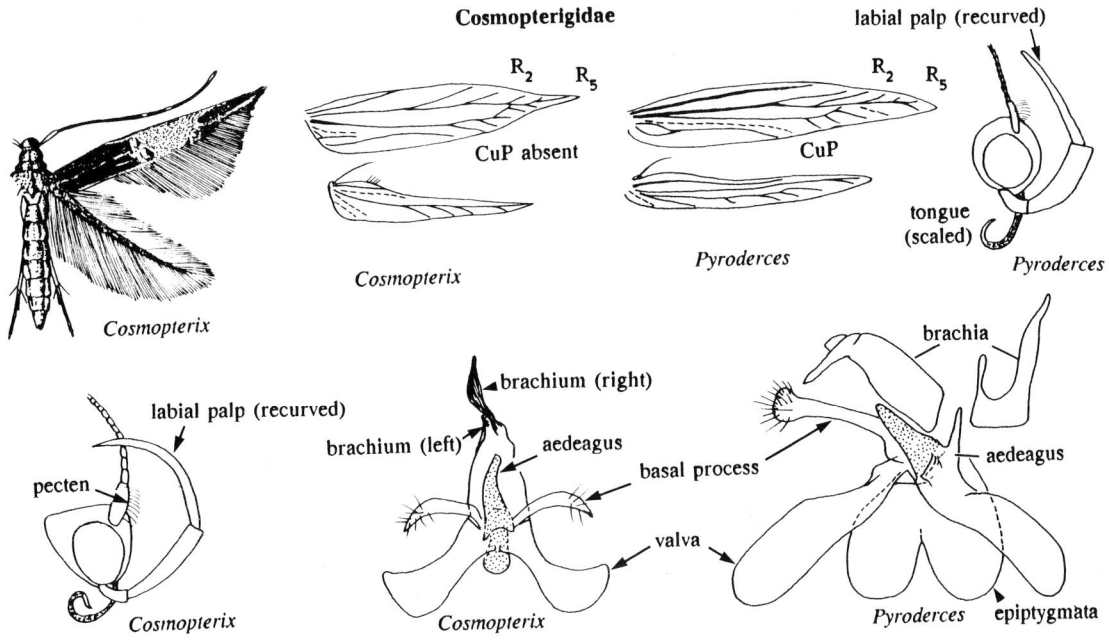

Cosmopterix

R₂ R₅

CuP absent

Cosmopterix

R₂ R₅

CuP

Pyroderces

labial palp (recurved)

tongue (scaled)

Pyroderces

labial palp (recurved)

pecten

Cosmopterix

brachium (right)

brachium (left)

aedeagus

basal process

valva

Cosmopterix

brachia

aedeagus

Pyroderces epiptygmata

--   Wings narrow or lanceolate, not markedly attenuate or acuminate; forewing with apex sometimes upturned, CuP present or absent; labial palpus moderate, upturned, porrect or drooping, second segment usually loosely scaled beneath; genitalia symmetrical, male with aedeagus articulated (not anchylosed), female with ostium on 8th sternite 83

83.   Micro-moths (expanse 8-10mm), forewing lanceolate but appearing broad because of dense fringe, sometimes slightly attenuated, apex occasionally upturned; antennal scape large, sometimes with pecten; labial palpus moderate, upturned, usually only slightly curved, sometimes short, porrect or slightly drooping, second segment usually roughened beneath; male genitalia with uncus usually bifid, juxta and digitate process present, gnathos spinose, with terminal knob/pair of knobs, or reduced; female genitalia with ostium simple, signum dentate ..............ELACHISTIDAE (p.104)

Elachistidae

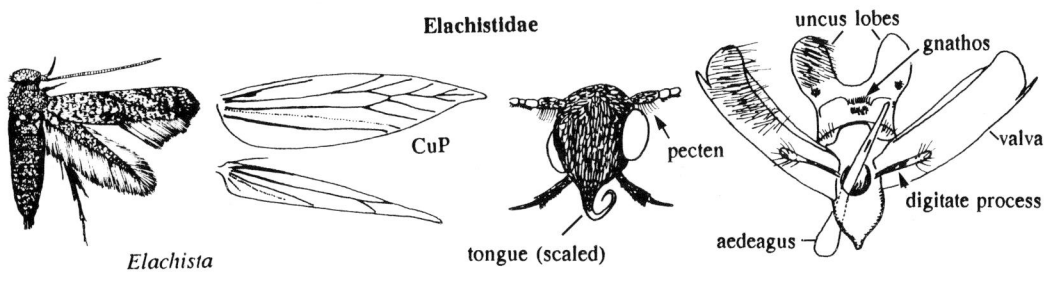

CuP

Elachista

tongue (scaled)

pecten

uncus lobes

gnathos

valva

digitate process

aedeagus

73

-- Micro-moths (expanse 10-15mm); forewing narrow (Momphinae) to very narrow (Batra-chedrinae, Blastodacninae), sometimes with scale-tufts; CuP usually absent; labial palpus moderate to long, ascending, often recurved; genitalia symmetrical, male with uncus usually long, slender, articulated with tegumen (Momphinae, Batrachedrinae), or uncus absent, gnathos present and recurved (Blastodacninae)....................
....MOMPHIDAE (p.107)

Momphidae

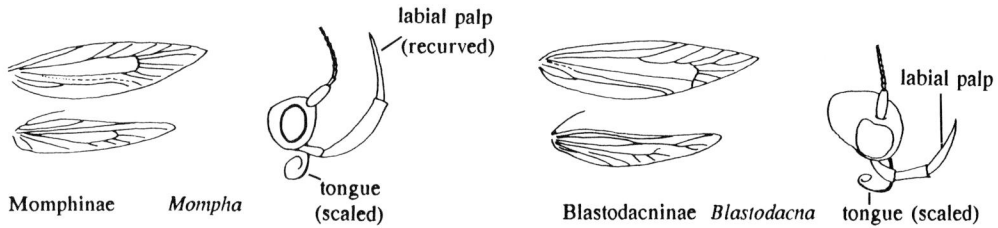

Momphinae   *Mompha*   labial palp (recurved)   tongue (scaled)   Blastodacninae *Blastodacna*   labial palp   tongue (scaled)

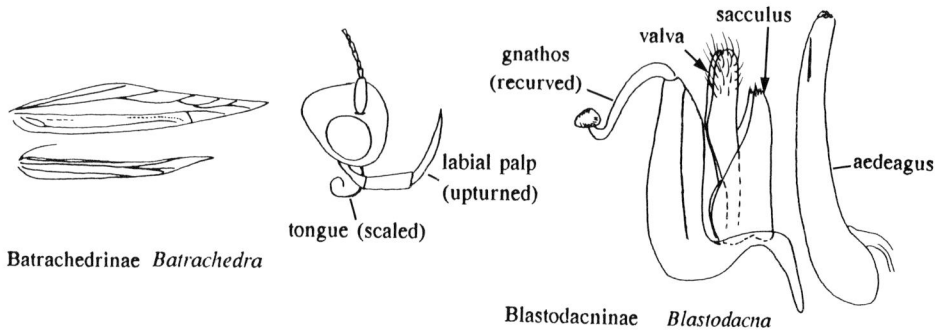

Batrachedrinae *Batrachedra*   labial palp (upturned)   tongue (scaled)   gnathos (recurved)   valva   sacculus   aedeagus   Blastodacninae   *Blastodacna*

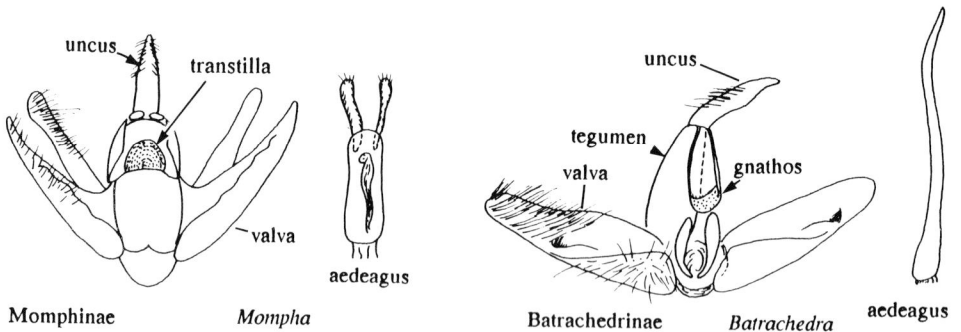

Momphinae   *Mompha*   uncus   transtilla   valva   aedeagus   Batrachedrinae   *Batrachedra*   uncus   tegumen   valva   gnathos   aedeagus

74

84. Forewings with CuP absent; hindwing trapezoidal, with apex often produced, seldom elongate-ovate, R5 and M1 stalked or closely approximate; small to medium-sized micro-moths ....GELECHIIDAE (p.110)

Gelechiidae

tongue (scaled)

Sitotroga

CuP absent

M₁

apex produced

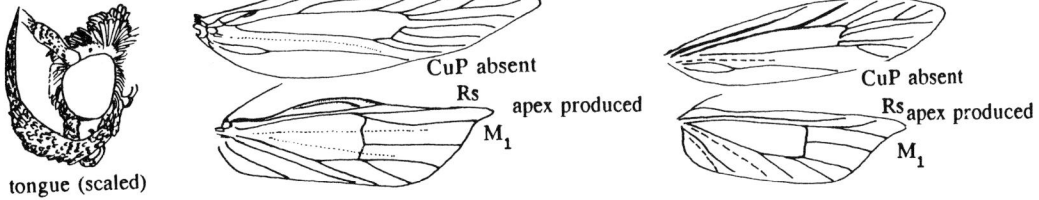

tongue (scaled)

Phthorimaea

CuP absent

Rs

apex produced

M₁

CuP absent

Rs apex produced

M₁

-- Forewing with CuP present at least towards margin; hindwing broadly ovate to lanceolate, with Rs and M1 separate and more or less parallel (Oecophorinae), or approximate, connate or stalked (Stenomatinae, Xyloryctinae); mostly small to moderately large, sometimes very large micro-moths .........................OECOPHORIDAE (p.100)

Oecophoridae

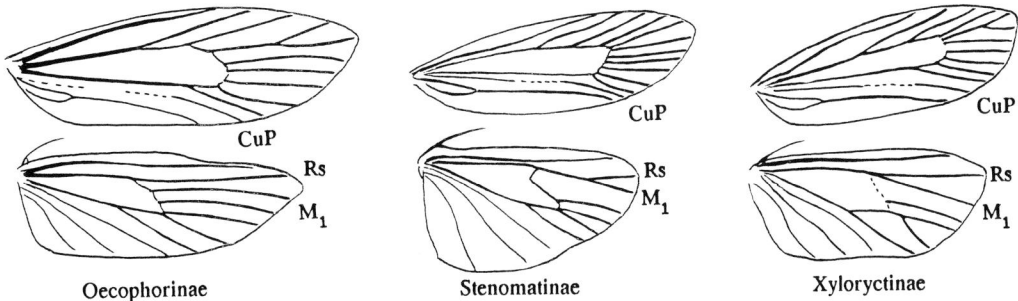

CuP

Rs

M₁

Oecophorinae

CuP

Rs

M₁

Stenomatinae

CuP

Rs

M₁

Xyloryctinae

85. Wings not developed (rudimentary, vestigial or obsolete); brachyptery (loss of use of wings) occurs sporadically in several lepidopterous families, sometimes induced as a result of harsh environment. The economically most important species with apterous or brachypterous females are found in ....
PSYCHIDAE, LYMANTRIIDAE, GEOMETRIDAE

-- Legs vestigial, female confined to larval case/bag/cocoon; without tympanum on thorax or abdomen..PSYCHIDAE (p.79)

Orgyia
Lymantriidae

Luffia
Psychidae

--   Legs developed, female not confined to larval case/cocoon; tympanum present on meta-
        thorax ....LYMANTRIIDAE (p.151)

--   Tympanum present on first abdominal segment ....GEOMETRIDAE (p.173)

Geometridae

Operophthera

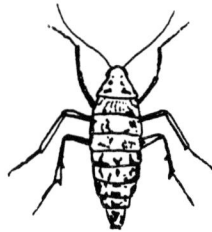

Erannis

THE DITRYSIAN 'MICROLEPIDOPTERA' SUPERFAMILIES
(JDB)

The 'microlepidoptera' superfamilies are no exception within the Order as a whole in being in a process of revision and reclassification. Thus the outline account of super-families and families presented here is likely to become progressively modified. For this reason the discussion is restricted largely to those families indicated in the checklist (p.16) as containing genera and species of economic importance. Subfamilies and genera that are of economic importance are indicated by asterisks in this chapter. Reference to key works on important families and genera are given with each familial account in this section of the manual.

TINEOIDEA

PSYCHIDAE

Psychids are important chiefly as defoliators in the tropics, and are remarkable for their often elaborately adorned larval cases or bags. Larvae are polyphagous or oligo-phagous, and are generally browsers on leaves, buds, flowers and fruits (Psychinae), or on lichens (Psycheoidinae). The bag or case constructed by the larva has a silk inner lining to which are attached bits of leaves, twigs or other material, often cut into regular shapes. Additional pieces are added and the case enlarged as the larva grows, and when the larva is ready to pupate the case serves as a cocoon.

Although there do not seem to be any definitive derived characters, the family may be conveniently divided into two main groups which are sometimes recognised as subfamilies: Psycheoidinae (primitive) and Psychinae (specialised). Psycheoidinae are mostly small and restricted mainly to temperate regions. They often occur in large concentrations on the bark of trees or on walls of buildings, sometimes causing a minor nuisance. Psychinae are generally much larger and are well represented in the tropics where some species have periodic outbreaks and cause indiscriminate defoliation of ornamental trees and plantation crops, such as tea, coffee, oil palm and cocoa.

The male genitalia are relatively simple: tegumen broad, elongated, hood-shaped; uncus vestigial or absent, sometimes represented by membranous pads; gnathos absent; valva simple and elongate; sacculus developed and distally often separated; aedeagus simple and usually without cornuti.

The female genitalia are generally simple and without internal diagnostic characters; hair-tuft (corethrogyne) sometimes present (also found in some Tineidae).

PSYCHEOIDINAE

Non-economic. Adults mostly small, primitive tineid-like moths (wingspan 8-30 mm). Male fully winged; female usually brachypterous or apterous, but wings fully developed in some genera (e.g. *Melasina, Narycia*), and nearly always with legs and long ovipositor. Parthenogenesis occurs in some genera (e.g. *Dahlica*), eggs being laid by unimpregnated females for several generations. Larvae are mostly lichen-feeders and construct relatively simple cases which are usually covered with minute particles of detritus, sand grains, etc. They are most frequently found in open habitats, living on lichen-covered rocks, walls, trunks of trees, etc.

Main genera: *Dahlica, Fumea, Melasina, Narycia, Taleporia.*

PSYCHINAE

Many economic species, mostly occurring in thinly wooded or bushy cultivated areas and savannah in the tropics. Adults are small to medium-sized, with a few larger species reaching about 50 mm wingspan. Usually only the male is fully winged and the female apterous or brachypterous. Coloration is typically blackish or brownish, sometimes enhanced with whitish scaling, the wings rounded, thinly scaled, sometimes hyaline or transparent, the vestiture of head and body shaggy, antennae strongly pectinated, and vein M is often forked in both fore- and hindwing. Wingless females do not leave the bag or case, and because of their confined life-style the eyes, antennae and legs have degenerated, and the abdomen is often enlarged into little more than an enormous egg-sac, in some species containing 3000+ eggs.

Males of the larger species fly strongly, often in sunshine. Their wings soon become worn and this can lead to difficulty in identification. The case or bag is often diagnostic and should be kept with the specimen. The males assemble outside the case to copulate with the female within; the abdomen of the male is often extensile and can reach deep into the case and into the pupal exuviae in which the female may still be encased. Oviposition occurs inside the case. Pupae of males and winged or brachypterous females are protruded at eclosion, but not those of apterous females.

Main genera: *Acanthopsyche**, *Clania**, *Eumeta**, *Kotochalia**, *Mahasena**, *Metisa**, *Monda*, *Oiketicus**, *Pagodiella**, *Pteroma* (= *Cremastopsyche*)*.

The species illustrated here are *Eumeta variegata* Snellen, *Mahasena corbetti* Tams and *Pteroma plagiophleps* Hampson, all defoliating a range of crops in the Oriental tropics; all except the first extend to the Papuan Subregion. Taxonomy of Psychidae is often very difficult, as complexes of very similar species may be involved.

References: Bourgogne, 1955, et seq.; Common, 1970; Conway, 1971; Davis, 1964; Dierl, 1965, etc.; Entwistle, 1963; Kozanchikov, 1969.

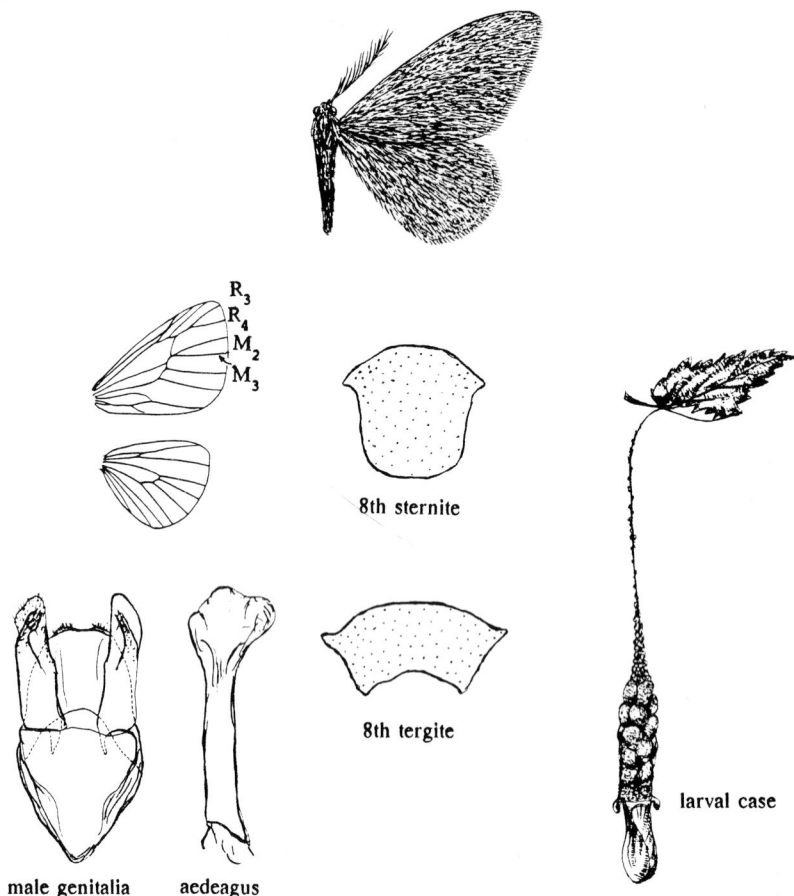

8th sternite

8th tergite

male genitalia   aedeagus

larval case

Pteroma plagiophleps

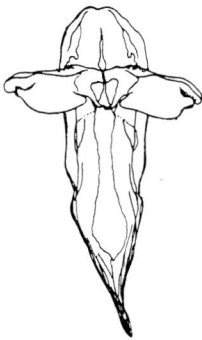

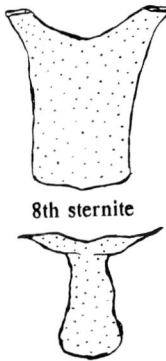

8th sternite

male genitalia   aedeagus

8th tergite

larval case

Eumeta variegata

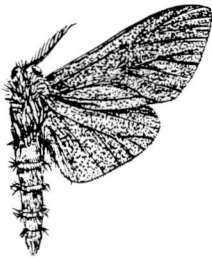

male genitalia   aedeagus   larval case

Mahasena corbetti

TINEIDAE

Economic species in this family include keratin-feeders (Tineinae), fungus-feeders (Nemapogoninae and Scardiinae) and scavengers on dried vegetable and animal matter. Adults are generally small (wing span mostly under 30mm), dull-coloured, often with a silky sheen, but sometimes the forewing pattern is intricate and sometimes conspicuous. Characteristic features are: roughened head (except Hieroxestinae), presence of stiff pecten bristles on the second segment of labial palpus, absence of ocelli and chaetosemata, short or absent tongue, simple antennae (usually finely pubescent, very rarely ciliate or pectinate), and rough-scaled hind tibiae. Tineids are generally feeble fliers but fast, furtive runners.

The adult wing venation is often unstable: variation occurs especially in the position and branching of veins arising from the cell in both fore- and hindwing. Fore- and hindwing have Cu2 present; forewing often with M and chorda present or indicated in cell, and R5 extending either to costa or apex.

Larvae feed mostly on dried or decaying plant or animal matter, especially fungi and keratinous material; some are inquilines in nests of social insects, e.g. termites and ants, and some live in vacated nests of birds, mammal burrows, etc. They usually construct a portable case, or live beneath silken webbing, or tunnel in the substrate.

The male genitalia vary from simple (Tineinae) to complex (Nemapogoninae). Wide differences in structure distinguish the subfamilies, but differences between related species are often subtle, e.g. comparative shape of valve, denticulation of aedeagus.

Female genitalia: ovipositor usually long and extensile; ostium with lamella antevaginalis (lower lip) usually well sclerotised and often diagnostic; ductus bursae and bursa copulatrix membranous, ductus usually thin and narrow, bursa often lacking signum or with multiple signa. Corethrogyne sometimes present (Hapsiferinae, Myrmecozelinae, Scardiinae, Perissomasticinae).

Subfamilies include: Acrolophinae*, Dryadaulinae*, Erechthiinae*, Hapsiferinae*, Hieroxestinae*, Meessiinae, Myrmecozelinae, Nemapogoninae*, Perissomasticinae*, Phthoropoeinae*, Scardiinae*, Setomorphinae*, Teichobiinae, Tineinae*.

References: BMNH, 1980; Carter, 1984; Corbet & Tams, 1943; Davis, 1975; Gozmany & Vari, 1973; Hannemann, 1977; Hinton, 1956; Robinson, 1979, 1986; Zagulyaev, 1960, 1964, 1975, 1979; Zimmerman, 1978.

TINEINAE

Includes synanthropic clothes moths and other tineines whose larvae feed on keratinous and cartilaginous animal material, e.g. wool, hair, horn, hide, feathers, guano (cave-dwelling species) or on fungi and decaying plant material.

Main genera: *Tinea*, *Tineola*, *Monopis*, *Niditinea*, *Praeacedes*, *Phereoeca*, *Trichophaga*, *Ceratophaga*.

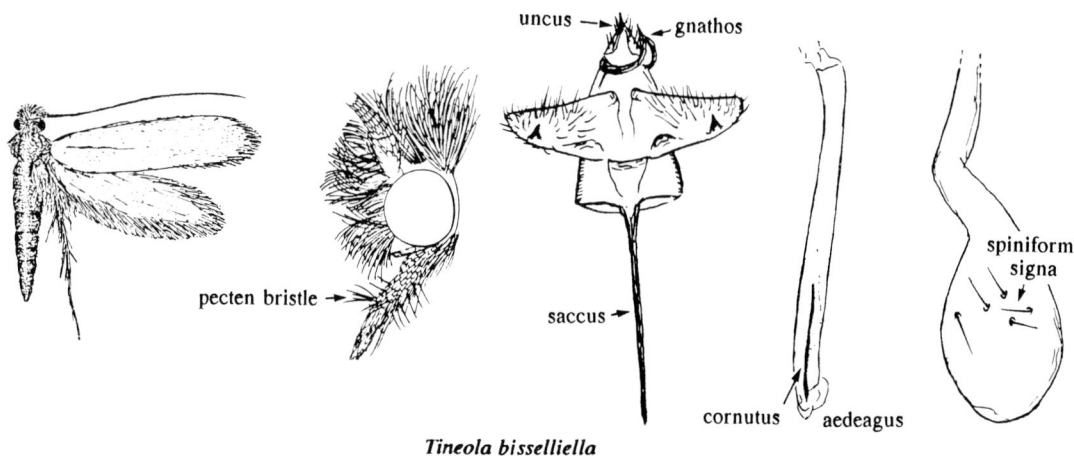

Tineola bisselliella

82

The species illustrated above is *Tineola bisselliella* Hummel, the common clothes moth of temperate regions, and below is *Monopis crocicapitella* Clemens, a cosmopolitan species that feeds on organic detritus, stored products, and is often found in birds' nests.

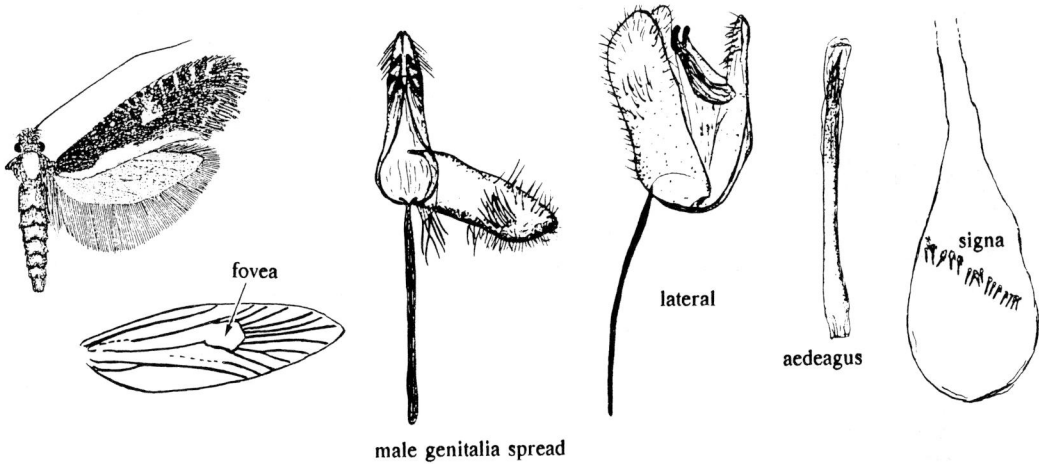

fovea

male genitalia spread

lateral

aedeagus

signa

Monopis crocicapitella

NEMAPOGONINAE

Primary fungivores, secondarily on stored food products. Adults have characteristically strong, fractured fasciate forewing markings. Male genitalia are distinctive structurally, the valve bearing various appendages and usually fused with juxta/anellus.

Main genus: *Nemapogon**.

Illustrated below is the corn moth, *Nemapogon granella* Linnaeus, a cosmopolitan species associated with households, granaries etc.

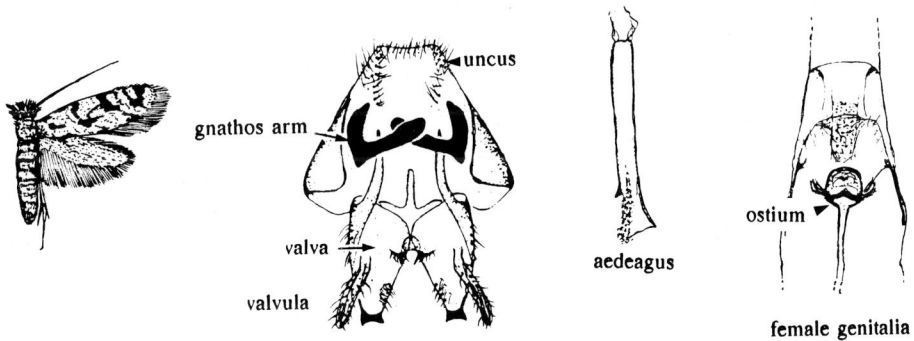

uncus

gnathos arm

valva

valvula

aedeagus

ostium

female genitalia

Nemapogon granella

SETOMORPHINAE

Main genera: *Lindera**, *Setomorpha**.
Feed on dried vegetable matter. *Setomorpha rutella* Zeller, illustrated, is known as the tropical tobacco moth, though is not restricted to stored tobacco.

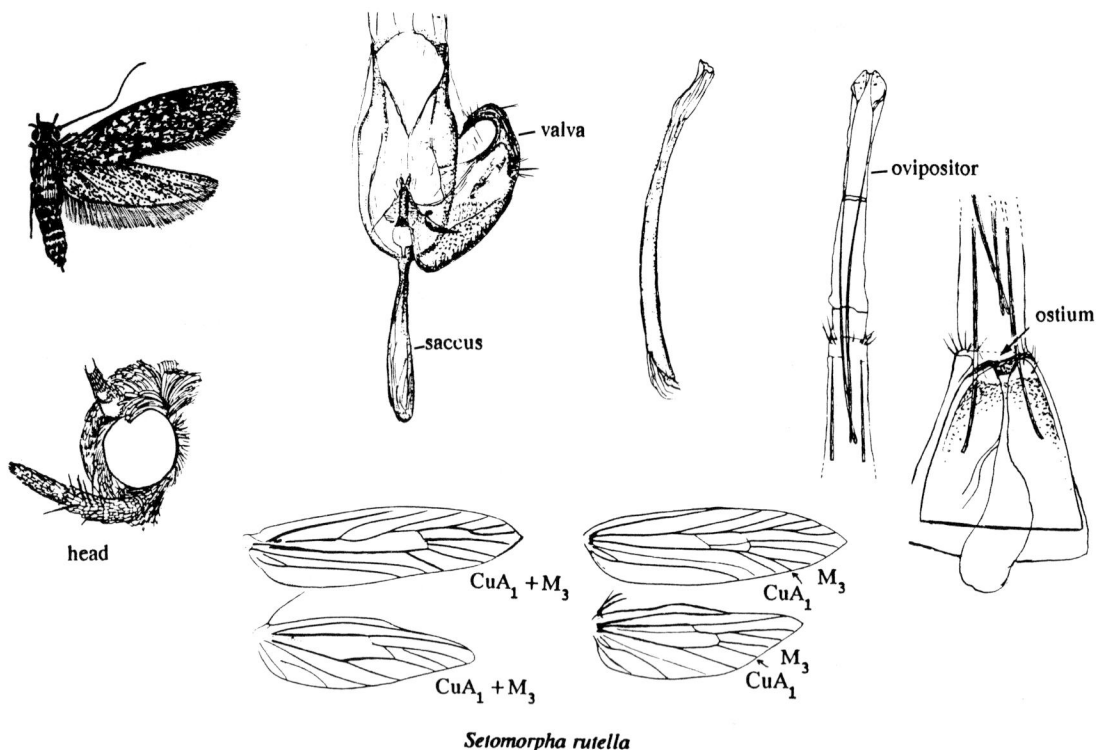

Setomorpha rutella

HAPSIFERINAE

Main genera: *Dasyses**, *Hapsifera*, *Tiquadra*.
This subfamily also includes species that feed on dried vegetable matter, the larvae living in silken galleries.

HIEROXESTINAE

Mostly tropical and subtropical; scavengers on dried/harvested vegetable material. Hieroxestine adults are distinguished from other tineids by the flat smooth-scaled frons, 'brow-ridged' vertex, narrow wings and reduced venation. Pecten bristles on the second segment of the labial palpus are sometimes short and concealed amongst scales. Forewing coloration varies from uniform brown to a contrasted bipartite pattern of bright yellow and brown or black.
Main genera: *Opogona**, *Oinophila**
The species illustrated below is *Opogona dimidiatella* Zeller, the larva of which lives in coconut fronds in the African and Oriental tropics.

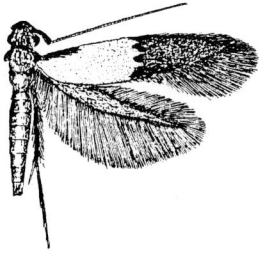
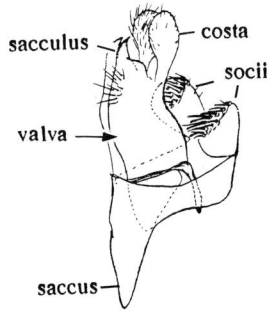

sacculus

costa

socii

valva

saccus

aedeagus

Opogona dimidiatella

ERECHTHIINAE

Mainly tropical plant detritus-feeders, scavengers on harvested crops, tunnellers in bark, stems, bolls, etc., and occasionally on growing plants. Adults are small and usually have the tip of the forewing characteristically turned sharply upwards and the head roughened. *Erechthias zebrina* Butler is a tropicopolitan species, the larvae feeding on plant detritus.

PHTHOROPOEINAE

A small as yet poorly researched group known from the Mediterranean region, Africa and India. The Afrotropical *Phthoropoea oenochares* Meyrick, illustrated, is a seed, pod and fruit feeder attacking cocoa, coffee, cotton, legumes and other crops.

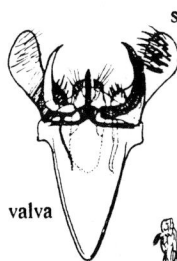

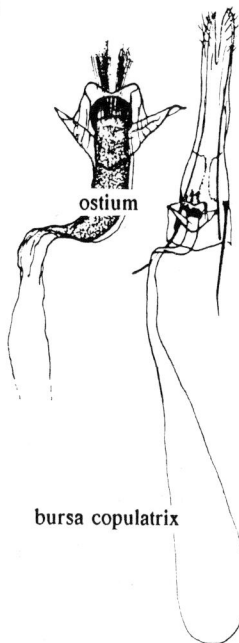

sacculus

valva

aedeagus

ostium

bursa copulatrix

Phthoropoea oenochares

85

LYONETIIDAE

Worldwide but most numerous in the Afrotropical and Indo-Australian regions, and containing a number of economically important leaf-mining species. Adults are mostly delicate micro-moths (wingspan 5-10 mm), with long-fringed lanceolate wings, the venation greatly reduced. The forewings are typically shining white ornamented with fine linear markings, the apices often slightly upturned or downturned (Lyonetiinae, Cemiostominae), or grey-brown (Bedelliinae), or whitish or greyish ochreous, often much irrorated and with stigmata and subdued oblique markings (Bucculatricinae).

Larvae are generally leaf miners, making blotch or serpentine mines, or mine bark, or sometimes skeletonise leaves (Bedelliinae). Pupation occurs outside the mine, the pupa being either in a slight cocoon cradled amid strands of silk attached to the underside of a leaf (Lyonetiinae), or in a silk cocoon attached to a leaf or twig (Cemiostominae), or suspended naked from a few strands of silk spun on a leaf of the foodplant (Bedelliinae) or in a dense fusiform cocoon firmly attached to the substrate (Bucculatricinae).

The male genitalia are often extremely small (Cemiostominae, Lyonetiinae) requiring careful dissection because of the modified 8th somite. Tegumen usually moderately broad but weak; vinculum rounded, saccus undeveloped; uncus absent (Cemiostominae) or represented by widely spaced setose lobes or socii (Bucculatricinae) or reduced (Bedelliinae,Lyonetiinae); valva simple, either fused basally (Cemiostominae, Lyonetiinae) or comparatively broad and freely articulated (Bedelliinae, Bucculatricinae); aedeagus cylindrical, usually loosely attached, sometimes with bulbous base (Cemiostominae).

Female genitalia: ovipositor short, membranous, sometimes with spinose pads (Bucculatricinae), or beset with file-like teeth adapted for piercing (Lyonetiinae); ostium weak, usually cup-shaped; ostial plate (lamella antevaginalis) sometimes projecting (Cemiostominae); signum absent or minute, or (Bucculatricinae) very large, covering the bursa as mass of coarse spines.

There are four subfamilies: Bedelliinae*, Bucculatricinae, Cemiostominae*, Lyonetiinae*. The relationship of these subfamilies is tenuous. Recent work (Kyrki, 1984) indicates that the affinities of the Bucculatricinae are with Tineoidea but as a distinct family group, while those of the Lyonetiinae, Cemiostominae and Bedelliinae lie with the Yponomeutoidea as the family Lyonetiidae. This arrangement is followed by Minet (1986).

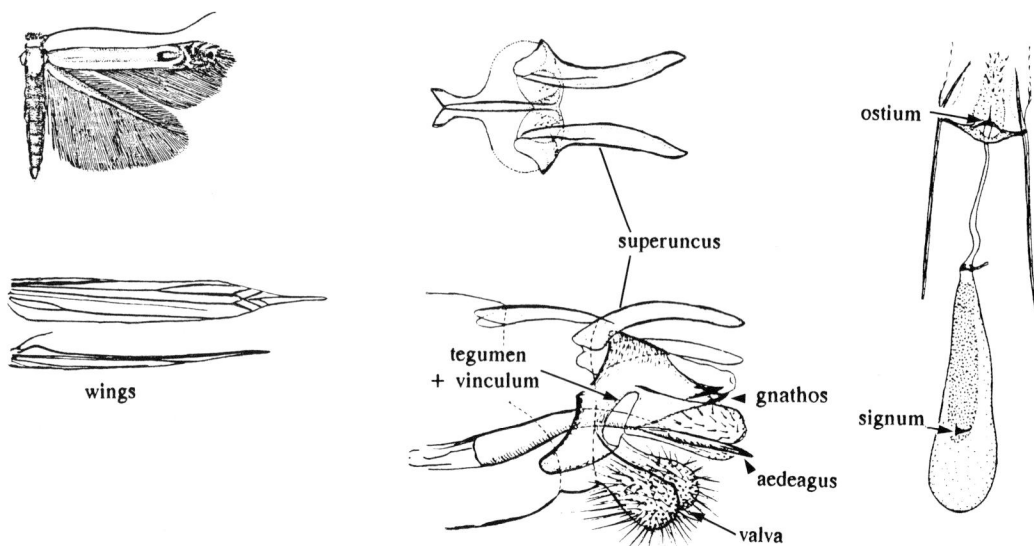

Lyonetia clerkella

LYONETIINAE

An apparently monogeneric group (*Lyonetia*) of few species. Adult with vertex of head rough-haired, forewing narrow-lanceolate, with apex strongly produced, antenna about as long as forewing.

The apple leaf miner, *Lyonetia clerkella* Linnaeus, is illustrated above. It ranges throughout the Palaearctic and also occurs in Madagascar, mining the leaves of various Rosaceae, but is also found on other plant families during outbreaks.

CEMIOSTOMINAE

A subfamily of worldwide distribution. Head with vertex usually rough-haired, forewing broad-lanceolate with apex pointed and often slightly caudate and upward-flexed. Cemiostomine adults tend to be very alike superficially, with the forewing predominantly white or grey, with costal and apical areas variably patterned with yellow and fuscous bars, the apical fringe often with one or more stark black pencils, and the tornal area heavily embossed with a leaden-metallic spot. Dissection of the genitalia is usually necessary for accurate identification of species. The male genitalia are remarkably small and complicated in relation to the size of the moth; the generalised structure is as illustrated below for *Leucoptera psophocarpella*.

Main genera: *Leucoptera**, *Paraleucoptera*, *Perileucoptera*, *Proleucoptera*.

The species illustrated is the recently described *Leucoptera psophocarpella* Bradley & Carter, a miner of winged bean (*Phosphocarpus tetragonolobus*) in Papua New Guinea.

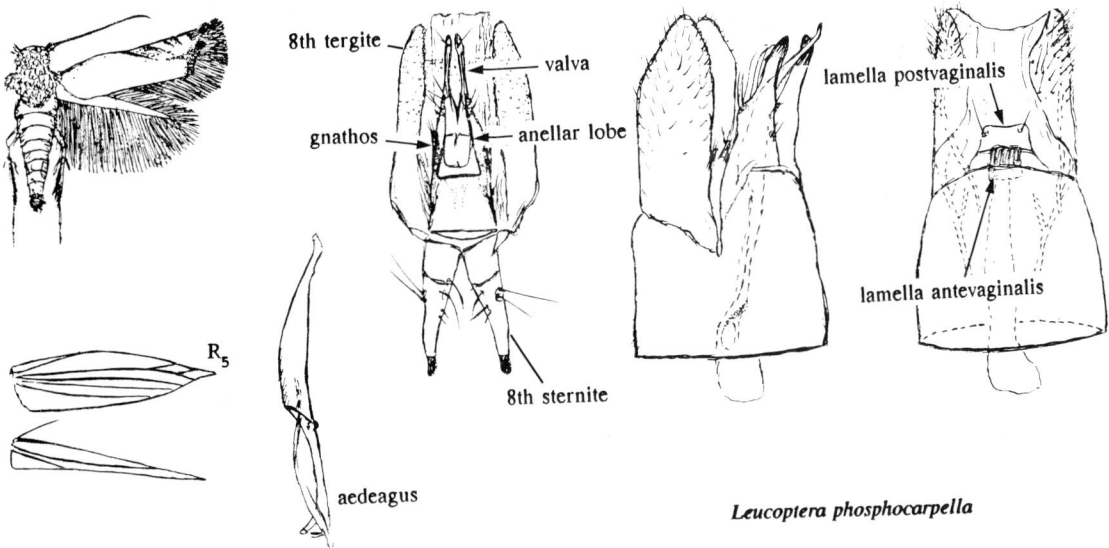

8th tergite — valva

gnathos — anellar lobe

R_5

8th sternite

aedeagus

Leucoptera phosphocarpella

lamella postvaginalis

lamella antevaginalis

BUCCULATRICINAE

A homogeneous and apparently mongeneric (*Bucculatrix*) group of micro-moths (wingspan 7-10 mm), now considered to represent a separate family within the Tineoidea. Over 150 described species are known, occurring in most regions, but most numerous in North America. A few are minor pests on cotton (*Gossypium*), attacking the leaves. Adults have the forewing lanceolate, crown of the head tufted but frons smooth, antenna shorter than forewing, scape developed as a large eyecap and, in the male, the first segment of flagellum deeply notched.

The larvae at first mine in the leaves of their foodplant, but later they leave the mine and spin "moulting cocoons" on the surface of the leaves. They then feed externally, generally from the underside, eating small patches of parenchyma but leaving the cuticle intact. The pupa is spined on the abdomen, and is spun up in a ribbed, fusiform cocoon which is firmly attached to the substrate and is sometimes protected by a "palisade" of upright "pegs".

Pest species on cotton: *Bucculatrix gossypii* Turner (Australia); *B. loxoptila* Meyrick (India, Africa); *B. thurberiella* Busck, the cotton leaf perforator (Neotropics, Hawaii - introduced).

References: Braun, 1963; Zimmerman. 1978.

BEDELLIINAE

As far as known, a very small but widely distributed monogeneric group (*Bedellia*) associated with sweet potato and other species of *Ipomoea* and also with bindweed (*Convolvulus*). Apparently much speciated in the Hawaiian islands. Adult with head vertex rough-haired, labial palpus short, porrect antenna about as long as forewing with scape elongate and somewhat thickened, without eyecap (present in other lyonetiid subfamilies) but with dense pecten.

Bedellia somnulentella Zeller, illustrated, is an almost cosmopolitan leaf miner of *Convolvulus* and *Ipomoea*, attacking sweet potato (*I. batatas*) in the latter genus.

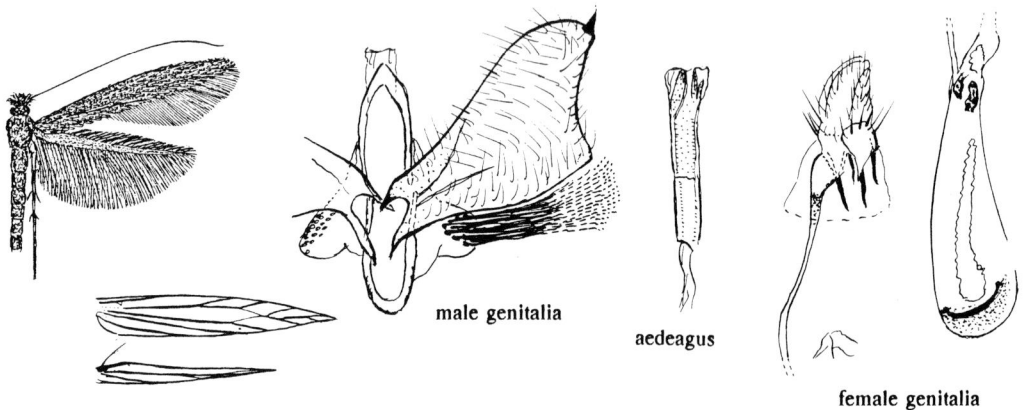

male genitalia

aedeagus

female genitalia

Bedellia somnulentella

GRACILLARIIDAE

The family includes a number of species of considerable economic importance as mining pests, mainly on the leaves and fruits of trees and shrubs. Adults are typically brightly coloured or shining white micro-moths with long-fringed lanceolate wings, the forewings often intricately patterned. Three rather contrasting subfamilies are recognised: Gracillariinae, Lithocolletinae and Phyllocnistinae. Gracillariine adults mostly have the antennae distinctly longer than the forewings and have a peculiar resting posture, sitting steeply on the tail with the head raised high above the prominently displayed fore and middle legs which are often conspicuously ornamented. Lithocolletinae have antennae only about as long as the forewing and adopt a horizontal resting posture, the head being kept low and the abdomen being slightly lifted and supported by the hind legs, the front and middle legs stretched forward. Phyllocnistinae have the antennae shorter than the forewing and also adopt a horizontal resting posture, but are smaller and more delicate than most lithocolletines and have characteristically shining white background coloration on the forewings.

Larvae are miners (at least up to the second or third instar), usually in leaves, fruits or occasionally green bark, and are mostly host-specific (monophagous), but some-times oligophagous. The gracillariid larva characteristically has two distinct stages or morphs (hypermetamorphosis): a sap-drinking stage and a tissue-feeding stage. In general, the Gracillariinae leave the mine after completing the sap-feeding stage and live extern-ally in a rolled or folded leaf, but the Lithocolletinae generally remain within the mine and continue mining and feeding in both these stages. In the Phyllocnistinae the second morph does not occur until after feeding has finished, the mouth parts are atrophied and it exists solely for cocoon-spinning. Pupation occurs either in the mine or outside, usually beneath a glossy membrane spun on the substrate; the pupa is protruded at eclosion.

Male genitalia: tegumen elongated, broad or narrowed apically, often weak and membranous; uncus rudimentary; gnathos absent; valva relatively simple, very slender to broad, sometimes asymmetrical (Lithocolletinae); sometimes with small comb(s) on inner surface; vinculum V- or U-shaped; saccus short when developed, or indeterminate; aedeagus long, with linear rows of small cornuti, or without cornuti. Eighth sternite produced into a wide flap covering genitalia.

Female genitalia: ovipositor short, weak; ostial plate usually sclerotised, ostium well defined, sometimes projecting, or membranous (Phyllocnistinae); signum single or double minute points, or small sigmoid or cornute structures, or diversely-shaped plates, often unequal.

References: Common, 1970; Emmet, 1985; Vari, 1961.

GRACILLARIINAE

Main genera: *Acrocercops**, *Callisto*, *Caloptilia** *Conopomorpha**, *Crematobombycia**, *Cryphiomystis**, *Epicephala**, *Marmara**, *Parectopa**, *Parornix*, *Spulerina**, *Stomphastis*.

The species illustrated is *Conopomorpha cramerella* Snellen, the cocoa pod borer of south-east Asia and the western Pacific. It is the most important member of a complex of species ranging through the Indo-Australian tropics (Bradley, 1986) attacking cocoa (*Theobroma cacao*) and also rambutan (*Nephelium lappaceum*), nam-nam (*Cynometra*), and *Cola*.

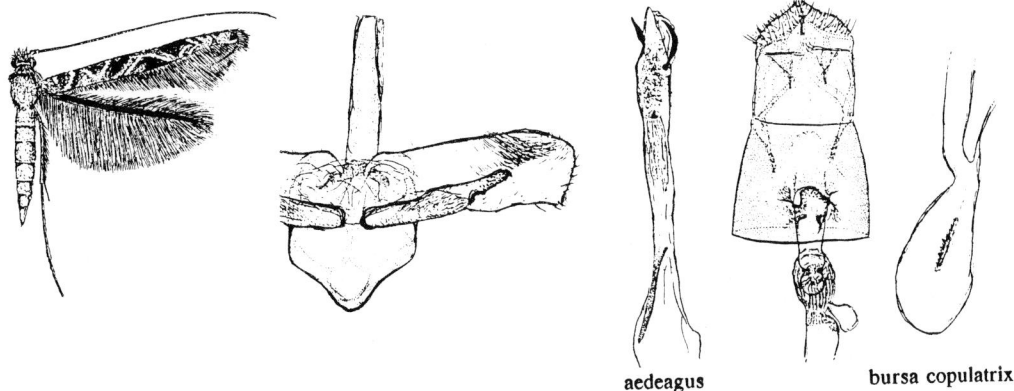

aedeagus   bursa copulatrix

Conopomorpha cramerella

LITHOCOLLETINAE

The three hundred or so described species in this subfamily are assigned with few exceptions to the genus *Phyllonorycter* (=*Lithocolletis*). Most occur in the Holarctic

Region, and comparatively few south of the tropics. Adults are typically colourful and strikingly patterned micro-moths (wingspan averaging 6-10mm), the rough-scaled head varying in colour from front to back, the smooth frons nearly always silvery white, the generally rather oval forewing with strong fasciate or wedge-like markings, often clear white and contrasting sharply with metallic brassy or orange-brown ground coloration. The species tend to form hostplant groups and in some groups the adults are difficult to distinguish without dissection of the genitalia. Most species are host-specific (monophagous) or live on closely related plant species (oligophagous), and hostplant data can aid in identification.

The mine formed by a lithocolletine larva after the sap-feeding stage has passed is often highly diagnostic specifically. It is usually a small blotch on the upper or lower side of the leaf, made by separating the cuticle from the parenchyma; the separated cuticle is lined with silk which contracts and forms a tentiform pucker in the leaf. The side of the leaf mined (upper or lower side) is constant for each species. Pupation occurs within the mine; pupa extruded at eclosion.

Many pest species are found in the genus *Phyllonorycter* (Balachowsky, 1966; Emmet et al, 1985; Pottinger & LeRoux, 1971). *P. blancardella* (Fabricius), illustrated, is a Palae-arctic leaf miner of apple and other rosaceous fruit-trees and has been introduced into N.America.

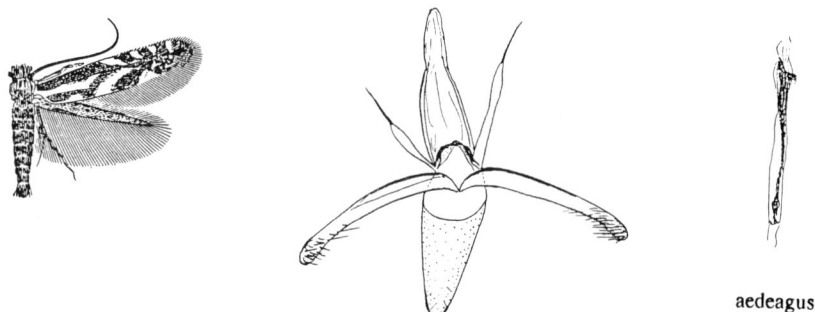

aedeagus

Phyllonorycter blancardella

PHYLLOCNISTINAE

Monogeneric (*Phyllocnistis*), comprising probably less than 100 described species and mainly tropical in distribution. Adults are among the smallest (wingspan 6-8 mm) and most delicate of the Lepidoptera, the forewing generally shining white finely marked with yellow or orange and with a black spot at the tip. Excepting *Phyllocnistis citrella* (on *Citrus*) the group is of little economic importance.

Phyllocnistis citrella Stainton, illustrated below, is a citrus leaf miner found in most places where citrus crops are grown in the Old World (CIE map 274).

References: Common, 1970; Emmet, 1985.

aedeagus

signa

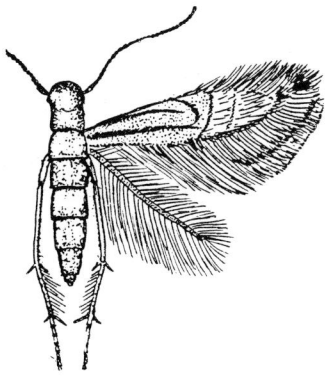

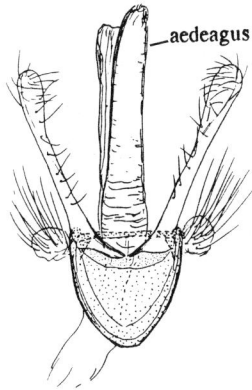

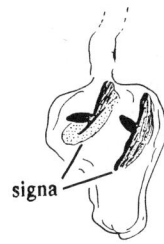

bursa copulatrix

Phyllocnistis citrella

YPONOMEUTOIDEA

YPONOMEUTIDAE

As treated here - in a broad concept - the taxon Yponomeutidae is a heterogeneous family group comprising the subfamilies Acrolepiinae, Argyresthiinae, Plutellinae and Yponomeutinae. These subfamilies differ markedly, especially the Argyresthiinae and Acrolepiinae, and are sometimes separated as families. In general, yponomeutid adults are small to medium-sized narrow-winged micro-moths, often brightly coloured, with a few larger moderately broad-winged species, notably in the Yponomeutinae. Larval feeding habits differ according to the subfamily, some feed exposed (*Plutella*, Plutellinae), some live gregariously in a communal webbing or nest (*Yponomeuta*, Yponomeutinae), some are concealed-feeders, tunnelling in fruit and shoots (*Argyresthia*, Argyresthiinae), or living between the leaves (*Acrolepiopsis*, Acrolepiinae).

The structure of the male genitalia varies greatly with the different subfamilies: tegumen usually narrow, weak; uncus usually reduced, often indefinite; socii strongly developed, prominent and setose, often erect (Yponomeutinae), sometimes with specialised scales on internal surface (Argyresthiinae); gnathos usually present, tongue-like (Yponomeutinae), or recurved and with spines at tip (Argyresthiinae), or absent (Acrolepiinae); anellus present as sclerotised ring (Plutellinae), or membranous; vinculum narrowly rounded, or rectangular, usually with slender saccus, or W-shaped or nearly so in ventral aspect (Argyresthiinae); sternite 8 sometimes in form of V- or Y-shaped plate (Argyresthiinae), or entire segment 8 sometimes divided, forming two lobes enclosing genitalia (Yponomeutinae, *Zelleria*).

Structural differences between the subfamilies are less marked in the female genitalia than in the male: ovipositor sometimes partially extensile, lobes short or reduced; sternite 8 often with spinulose pads on either side of ostium; signum single or absent; coremata sometimes present.

References: Balachowsky, 1966; Carter, 1984; Freise, 1969; Gaedike, 1970; Hannemann, 1977; Kyrki, 1984; Moriuti, 1969; Zagulyaev, 1981; Zimmerman, 1978.

YPONOMEUTINAE

Main genera: *Atteva, Prays*, Swammerdammia*, Yponomeuta*, Zelleria.*

Yponomeuta malinellus Zeller is illustrated. It is widespread in the Palaearctic and has been introduced into N. America. The larvae feed on apple leaves (*Malus*), living and pupating in a communal web. The species belongs to a complex of host-specific and sympatric species, or perhaps biological races of a single species, as morphological distinction is unclear (see Beirne, 1943; Emmet, 1979; Friese, 1960).

Yponomeuta malinellus

PLUTELLINAE

Main genera: *Plutella*, Ypsolopha*

The species illustrated is the almost cosmopolitan diamond-back moth, *Plutella xylostella* Linnaeus, a migratory pest on crucifers. The Larva skeletonises the leaves; pupa in an open network cocoon.

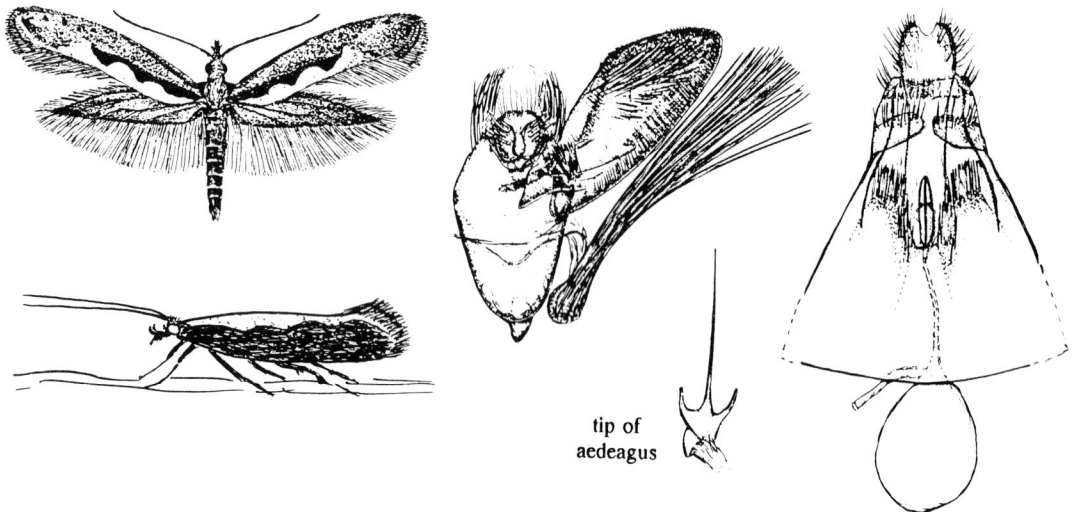

tip of
aedeagus

Plutella xylostella

ARGYRESTHIINAE

About 100 species, generally classified in the genus *Argyresthia*, are known, the majority Holarctic; only a few occur in other regions, including *A. notoleuca* (Turner) from Australia and *A. conspersa* Butler from Chile, but none from New Zealand. Adults have a characteristic resting posture, sitting with the head depressed and the body raised at a steep angle.

Main genus: *Argyresthia**.

There are a number of pest species in the genus, that illustrated below being the Palaearctic apple fruit moth *Argyresthia conjugella* Zeller, introduced into N. America.

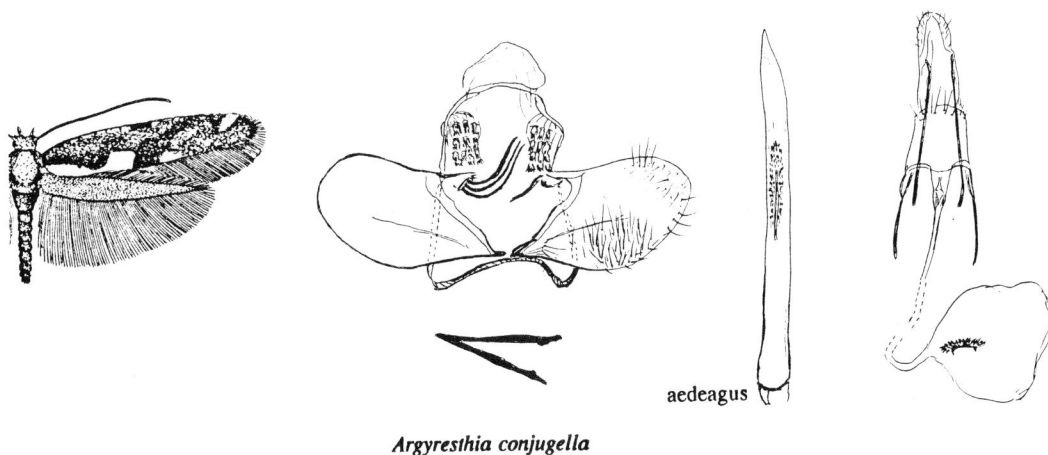

aedeagus

Argyresthia conjugella

ACROLEPIINAE

Main genera: *Acrolepiopsis**, *Digitivalva*.

The illustrated species, *Acrolepiopsis assectella* Zeller, is a Palaearctic pest of onion, leek, garlic and chives, mining the leaves and sometimes boring into the bulb.

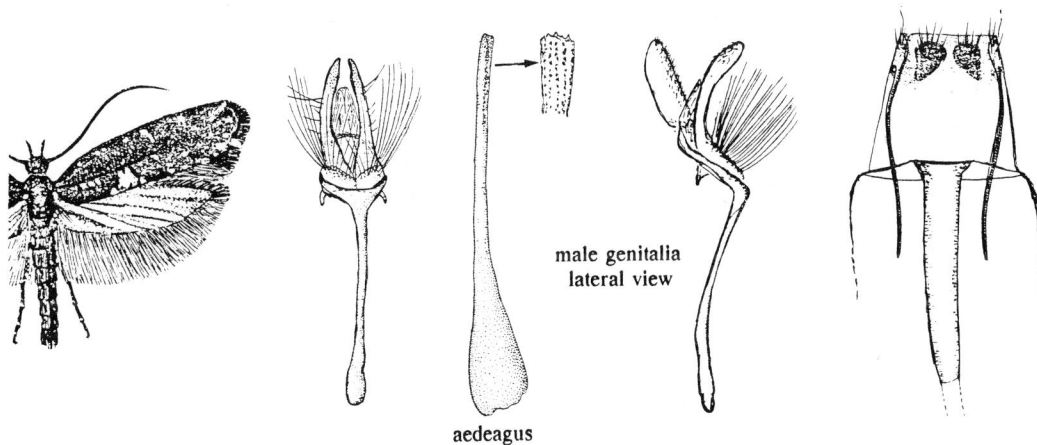

male genitalia
lateral view

aedeagus

Acrolepiopsis assectella

OCHSENHEIMERIIDAE

Monogeneric (*Ochsenheimeria*) family of about 20 known species of micro-moths inhabiting grassland and steppes of the Palaearctic region to the Himalayas, with one species (*O. vacculella*) introduced into North America. Although long regarded as tineoid, the group is now thought to show close affinities with the Yponomeutoidea, following the discovery (Kyrki, 1984) of a modification of the 8th abdominal segment similar to that found in yponomeutoids. Adults are diurnal, flying mostly at midday and are characterised by the coarsely scaled forewings and, in the male, the thinly scaled hindwing which is often almost transparent basally. Long scales on the crown of the head are usually furcate at the tip; second segment of labial palpus without pecten bristles (present in Tineidae); antenna are usually thickened with shaggy scales in basal half, especially in male.

Larvae feed chiefly on Gramineae, but also on Cyperaceae and Juncaceae, mining the leaves, stems and seeds.

Male genitalia: tegumen elongate, with pleural lobes from anterior margin; uncus bifid, forming two setose pads; gnathos short, fused medially, pointed; valva simple; sacculus small, with peg-like spines; vinculum a narrow rim, with slender saccus; juxta a small scobinate plate; aedeagus simple, with spine-like cornutus or cornuti.

Female genitalia: ovipositor extensile; ostium simple, cup-shaped, sometimes asymmetrical; signum a scobinate thickening of the bursa.

The species illustrated is *Ochsenheimeria vacculella* Fischer von Roeslerstamm, the cereal stem moth of the Palaearctic, recently introduced into North America.

References: Davis, 1975b; Karsholt & Nielsen, 1984; Kyrki, 1984.

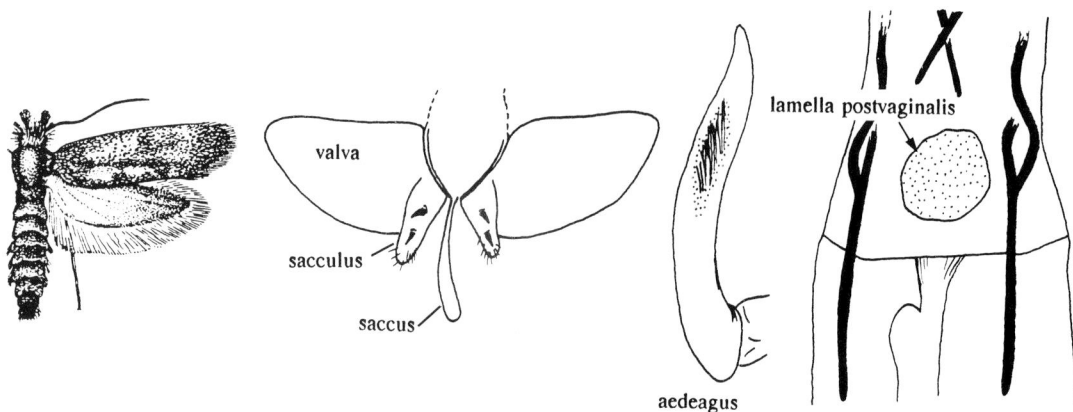

Ochsenheimeria vaculella

EPERMENIIDAE

Widely distributed family but contains relatively few species and is of little economic importance. Adults are mostly small (wingspan 9-15 mm), narrow-winged and with inconspicuous coloration, but are generally recognisable by the presence of tufts of raised scales on the dorsum (hind margin) of the forewing and the presence of stiff bristly hairs of the hind tibia. Tongue naked.

Larvae mine in leaves in early instars, often on Umbelliferae, later feeding externally on the leaves, usually living in a loosely spun web. Pupa in a slight open network cocoon amongst detritus; not extruded on eclosion.

94

Male genitalia: tegumen comparatively broad; uncus usually long, slender and hinged to tegumen; gnathos and socii absent; valva small, complex, with costal part free; sacculus developed; aedeagus short, with or without cornutus.

Female genitalia: ovipositor extensile; ostium usually variable, sometimes strong and with lamella antevaginalis deeply indented medially; signum single, in the form of a plate or pocket, or absent.

Opinions vary as to the correct systematic position of this family. Heppner (1977) assigned it to the Alucitoidea (Copromorphoidea), while Minet (1983) proposed a new superfamily Epermenioidea.

Epermenia chaerophyllella Goeze is a minor pest of carrot (*Daucus*), parsnip (*Pastinaca*) and other Umbelliferae.

References: Common, 1977; Gaedike, 1966, 1977; Minet, 1983.

GLYPHIPTERIGIDAE

Widely distributed but mostly non-economic, with numerous species in Australia and New Zealand. Adults are small micro-moths, the forewings elongate and relatively narrow, often notched below the apex and usually with glittering metallic markings. Many fly in sunshine and frequent flowers, sitting on them and fanning the wings. Larvae feed on seeds or bore into shoots or stems, especially of Juncaceae and Cyperaceae. Pupa usually in the larval feeding place; protruded on eclosion.

Main genus: *Glyphipterix**.

The Palaearctic cocksfoot moth, *Glyphipterix simpliciella* (Stephens), is illustrated. The larva feeds on grass (*Dactylis*, *Festuca*) seeds, pupating in the stem.

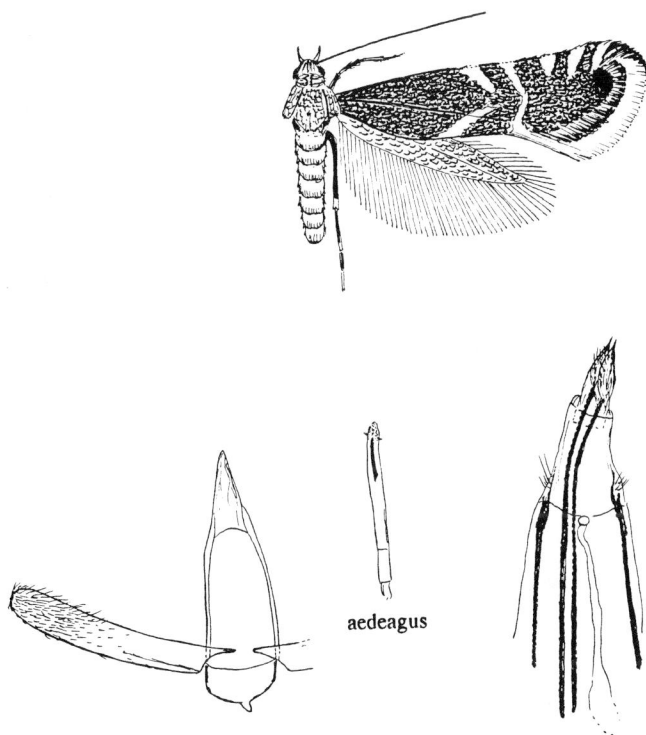

aedeagus

Glyphipterix simplicella

SESIOIDEA

Minet (1986) placed the Castniidae (see p.172) in the Sesioidea on the basis of several derived characters of adults and larvae.

SESIIDAE

Sesiids (clearwing moths) include numerous borers harmful to fruit trees and shrubs. The adults are remarkable for their mimicry of Hymenoptera, which they resemble not only in coloration and markings but also in form, the wings being narrowed, the abdomen often constricted basally (wasp-like), and the hind tibiae clothed with long bushy scales resembling a pollen sac. Most have a wing expanse under 35 mm, but some larger tropical species have a span of 50mm or more.

Larvae are borers in bark and living wood of a wide range of trees and shrubs, sometimes causing serious damage to fruit trees, or in the stems of vines and herbaceous and horticultural plants; a few are inquilines in galls, or are predaceous on Coccoidea (Bradley, 1956; Duckworth, 1969). Pupation usually occurs in the feeding place, with or without a cocoon of wood chippings or debris woven together with silk; herb boring species sometimes pupate in the earth, inside a silken tube which leads to the surface and through which the pharate adult moves. The sesiid pupa has rows of dorsal spines which aid movement along the tunnel to the exit hole, through which the exuviae protrude on eclosion.

Male genitalia: tegumen long and slender to short and moderately broad, with uncus distinctly demarcated or fused; tuba analis often prominent, with ventral sclerotisation (subscaphium); scopula androconalis usually conspicuous (i.e. an elongated setaceous sac situated at apex of tegumino-uncal complex); gnathos usually present, sometimes with paired flap-like lateral processes (Sesiinae), or absent (Tinthiinae); socii or lateral setaceous pads present, or absent (Tinthiinae); with dense specialised setae and a saccular ridge, or simple; vinculum usually with distinct, sometimes long saccus; aedeagus slender, usually with bulbous base; internal cornuti often present but external spines, spurs, etc. in apical portion are usually of greater taxonomic significance.

Female genitalia: ostium bursae on sternite 8 or on intersegmental membrane between sterna 7 and 8, or on posterior margin of sternite 7; signum usually absent, but sometimes present or represented by folds or bands.

Subfamilies: Sesiinae*, Parenthreninae, Tinthiinae.

References: Bradley, 1956, 1957, 1968; Duckworth, 1969; Duckworth & Eichlin, 1974; Engelhardt, 1946; Fibiger & Kristensen, 1974; Heppner & Duckworth, 1981.

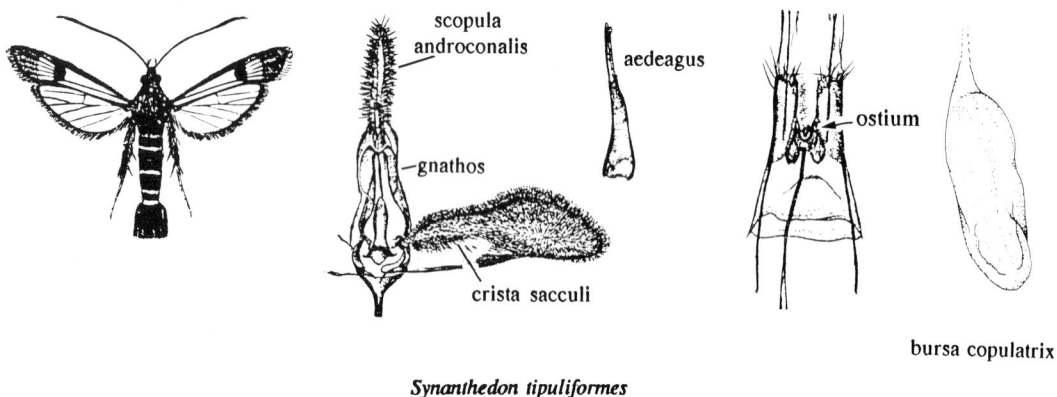

Synanthedon tipuliformes

SESIINAE

Worldwide and the largest and economically most important subfamily, with pest species in all regions. Sesiine adults have a hair-tuft present at the tip of the antenna (a tuft is also present in parenthrenines but not in tinthiines). Male genitalia with tegumen and uncus usually fused; gnathos usually well developed; socius and/or scopula androconalis present; subscaphium usually thin, strap-like; valva with dense specialized scales, and with crista sacculi usually well developed.

Main genera: *Bembecia, Carmenita, Melittia, Sesia, Synanthedon*.

Synanthedon tipuliformis Clerck, the currant clearwing, is illustrated above. It is a Palaearctic pest of soft fruit crops in the genus *Ribes*, boring in the stems, and has been introduced into N. America and temperate Australia.

PARENTHRENINAE

Worldwide, about 140 known species, chiefly Old World. Parenthrenine adults have a small tuft at the tip of the antenna. Male genitalia with tegumen reduced, short; uncus greatly elongated, several times longer than tegumen, heavily setose apico-laterally; gnathos small; valva with multifurcate scales dorsally, setaceous apically and ventrally, with a bare or sparsely setose field.

Main genera: *Parenthrene, Vitacea*

TINTHIINAE

Worldwide, about 60 species, mostly Oriental. Tinthiine adults lack the hair-tuft at the tip of the antenna. Male genitalia with tegumen unmodified; uncus simple or reduced; gnathos and socii absent; valva elongate-quadrate, without specialised setae, crista sacculi absent.

Main genera: *Ficivora, Tinthia*.

CHOREUTIDAE

The adults of this widely distributed but little known family somewhat resemble tortricids in general appearance, with the wings broad, the forewings rather rectangular (wingspan 10 to 35mm) and with fasciate markings. They may be distinguished by the presence of scales at the base of the tongue (not scaled in tortricids); ocelli are present in both of these families. Adults are generally diurnal and keep to the vicinity of the foodplant. The larvae are mainly leaf-feeders and of little economic importance.

Male genitalia: uncus and gnathos rarely present; socius-like setaceous area present; tuba analis often prominent; valva relatively simple, with ventral setal fields, or complex, occasionally bifurcate or fused along costal margin; vinculum developed; saccus often large, or sometimes absent; aedeagus usually with cornutus present.

Female genitalia: ostium usually on sternite 7, sometimes on intersegmental membrane between sternites 7 and 8; bursa copulatrix usually with single signum.

Subfamilies: Choreutinae*, Brenthiinae, Millieriinae.

References: Heppner & Duckworth, 1981; Heppner, 1982; Kuznetsov, 1981.

CHOREUTINAE

Including a few species of relatively minor economic importance, e.g. foliage feeders on *Ficus* and *Malus*.

Main genera: *Anthophila*, *Choreutis**, *Prochoreutis, Simaethis, Tebenna, Tortyra*.

The species illustrated below, *Choreutis pariana* (Clerck), is the Palaearctic apple leaf skeletoniser, introduced into N.America.

  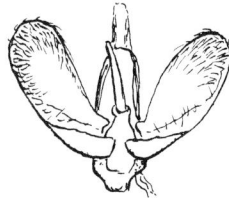 

head

signum

Choreutis pariana

BRACHODIDAE

Mainly Oriental and Indo-Australian, with a few species of minor economic importance. Brachodids are comparatively robust, elongate-winged, small to medium-sized moths (expanse 10-30mm), and are mostly diurnal. They are characterised by the prominent head and large ocelli, absence of chaetosemata, naked tongue, elongate forewings without pterostigma, and relatively broad hindwings with the basal area usually thinly scaled and often whitish or yellowish.

Larvae are either borers, usually in roots (Brachodinae), or external feeders on leaves (Phycodinae). Hostplants include Bromeliaceae, Gramineae, Leguminosae, Melastomataceae, Moraceae and Palmae.

Subfamilies Brachodinae and Phycodinae are keyed by Heppner & Duckworth (1981):
Labial palpus and thorax with normal sized scales; larva a borer with reduced prolegs ....
....Brachodinae
Labial palpus and thorax with large, plate-like scales; larva an external feeder, with relatively long prolegs ....Phycodinae
Reference: Heppner & Duckworth, 1981.

BRACHODINAE

Palaearctic and Pantropical, comprising over 70 known species. Forewing termen rather rounded; labial palpus and thorax with normal sized scales. Larvae are borers, usually in roots; pupa in subterranean chamber.

Main genera: *Brachodes* (on Gramineae), *Sagalassa** (on Palmae), the illustrated species being the Neotropical *Sagalassa valida* Walker.

Saraglassa valida

98

PHYCODINAE

Pantropical, the majority of the 20 or more known species occurring in the Old World tropics. Forewing termen usually somewhat truncated; labial palpus and thorax with large plate-like scales. Larvae feed externally on leaves. Pupa in a silk cocoon on host leaf surface.

Main genus: *Phycodes** (on Moraceae).

The illustrated species, *P.minor* Moore, has a web-dwelling larva that feeds on leaves of fig (*Ficus*).

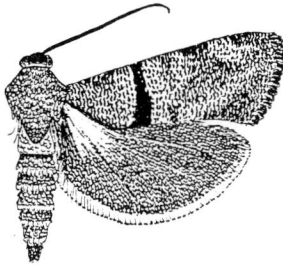

Phycodes minor

IMMOIDEA

IMMIDAE

Restricted to the tropics and subtropics, the majority of the 250 or so described species being distributed in the Indo-Australian region, Oceania and the Neotropics, and only a few in Africa. Although sometimes common they are not known to be of economic importance. Immids are mostly small to moderate-sized moths (wingspan 15-30 mm, but sometimes larger), broad-winged and with usually dull brown and yellow coloration, and are reminiscent of tortricids in general appearance. The tongue is naked and chaetosemata are present, as in tortricids, but ocelli are usually absent.

Little is known about the life histories of species in this family, but larvae are apparently external feeders, living on the foliage of various tropical trees and shrubs. Pupa is not spined dorsally (spined in tortricids); in a flimsy network cocoon spun on a leaf of the foodplant; not protruded on eclosion.

The male genitalia are varied: uncus simple, sometimes bifid, or absent; socii sometimes present; gnathos present or absent; tuba analis sometimes prominent; valva simple or complex, sometimes bifurcate; vinculum strong; aedeagus curved or straight, usually without cornutus.

Female genitalia: ovipositor lobes setose; ostium broad, usually on intersegmental membrane between segment 7 and 8; ductus bursae often spiralled or coiled; bursa copulatrix usually globular, with or without signum, or with paired signa.

Main genus: *Imma*.

References: Clarke, 1955-70 (vol.6); Common, 1979; Heppner, 1982a, 1984.

99

GELECHIOIDEA

OECOPHORIDAE

This large and heterogeneous family is well developed in most regions, especially in Australia and Central America, but contains comparatively few species of significant economic importance. Adults vary from small micro-moths to large robust noctuid-like moths, a few with wingspans exceeding 60 mm, e.g. Australian xyloryctines. The family-subfamily delimitations within the oecophorid group are not always well defined and opinions as to the proper status of some groups differ, e.g. Stathmopodinae, Stenomatinae and Xyloryctinae are sometimes given family status. Also the distinction between Oecophoridae and Gelechiidae is not always clear-cut, though the characters used for their separation in the Manual key should suffice for economic species.

Larval feeding habits of oecophorids are various and include foliage-feeders, stem, trunk, bark and root borers, and general feeders and scavengers on plant or animal detritus, decaying bark and wood, fungi, lichens, mosses, seeds and fruit. In Australia numerous species have evolved in the sclerophyllous leaf litter of eucalypt forests and xerophilous shrubby regions, and form an integral part of the local ecosystem. Some of the largest larvae in the family are those of the woodland bark-tunnelling and bark-ringing Australian Xyloryctinae belonging to the genus Cryptophasa.

Male genitalia are various and usually symmetrical: uncus-socii variously developed; gnathos usually present, single or double, spined or unspined; tegumen and vinculum strong; saccus not usually developed; valva simple or with some combination of costal, medial and/or saccular lobes; juxta usually present and often with a pair of lateral lobes; transtilla present or absent; aedeagus cylindrical, often bulbous basally, loosely attached by membrane (easily dissected out), often with group of minute thorn-like cornuti.

Female genitalia: ostium on sternite 8 or on membrane between sternites 7 and 8; walls of ductus bursae membranous or spiculose; signum usually stellate/dentate, sometimes double or multiple, or absent.

The following selected subfamilies contain species of economic importance: Oecophorinae, Depressariinae, Stathmopodinae, Autostichinae, Stenomatinae, Xyloryctinae.

OECOPHORINAE

Main genera: *Borkhausenia, Endrosis*, Hofmannophila*, Philobota, Psorosticha*.*

Hofmannophila pseudospretella Stainton, the brown house moth of temperate regions, is illustrated.

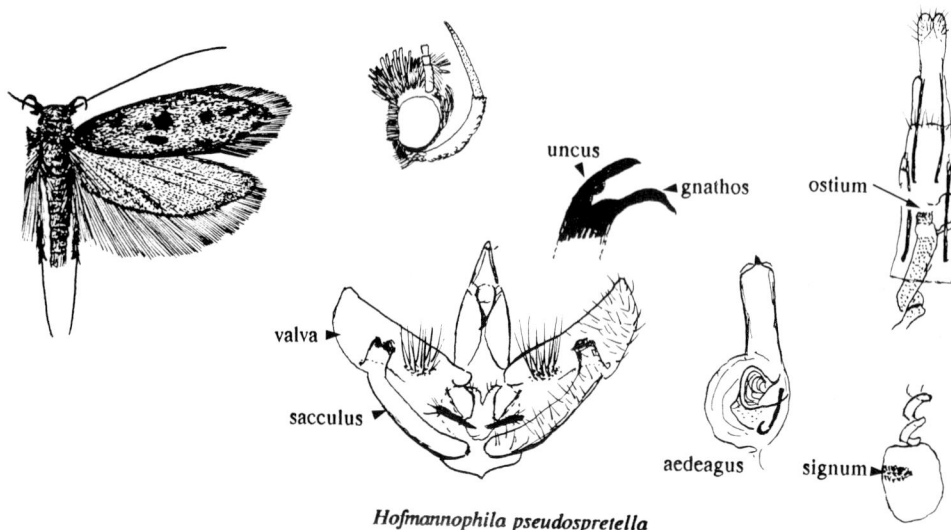

Hofmannophila pseudospretella

100

DEPRESSARIINAE

Mainly Holarctic, with a few representatives in the Neotropical and southern Afrotropical regions. Adults are known as flat bodies because of their peculiarly flattened or depressed form. In temperate regions with cold winters the adults hibernate, their flat bodies enabling them to slip into crevices which they use for shelter; they are frequently found in roof thatch and haystacks.

Main genera: *Agonopterix*, *Depressaria**, *Deloryctis**.

Depressaria pastinacella Duponchel, the Palearctic parsnip moth, is illustrated below. The larva feeds on the flowers and seeds of various Umbelliferae.

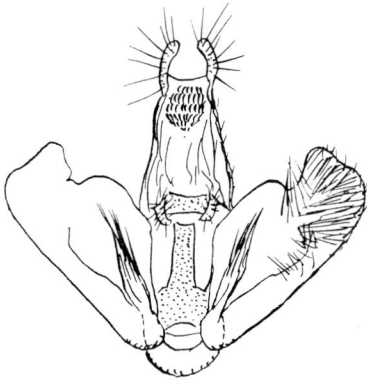

aedeagus

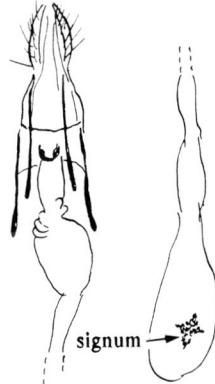

bursa copulatrix

signum

Depressaria pastinacella

STATHMOPODINAE

Mainly Afrotropical and Indo-Australian. Only one species, *Stathmopoda pedella* Linnaeus, occurs in Europe. Adults are small micro-moths with wingspan 10-15 mm, head smooth-scaled and shining, forewings narrow and attenuated, hindwings narrow-lanceolate. Many have a peculiar resting postion, holding the hind legs, the tibia of which are ornamented with whorls of stiff bristles, semi-erect between the fore and middle pairs.

Larvae bore in fruits, flowers, shoots, bark and galls, or are predaceous on scale insects (Coccoidea), etc. Pupa in a silk cocoon in the larval feeding place or amongst debris.

Main genera: *Eretmocera**, *Hieromantis*, *Stathmopoda**.

The illustrated species, *Stathmopoda ficipastica* Bradley, feeds on the fruits of *Ficus* species in Nigeria.

101

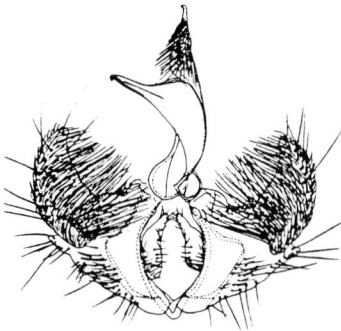

tip of aedeagus

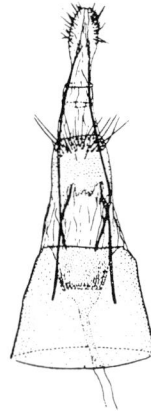

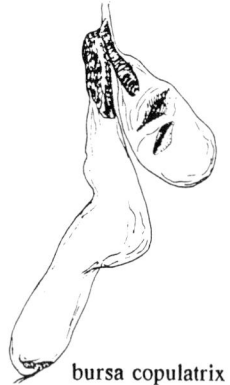

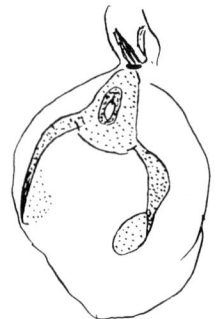

bursa copulatrix

Stathmopoda ficipastica

AUTOSTICHINAE

Asian-Oceanic distribution.

Main genera: *Autosticha**, *Stoeberhinus*.

Autosticha pelodes Meyrick (illustrated below) ranges from Malaysia to Polynesia and Hawaii, the larva feeding on decaying vegetation.

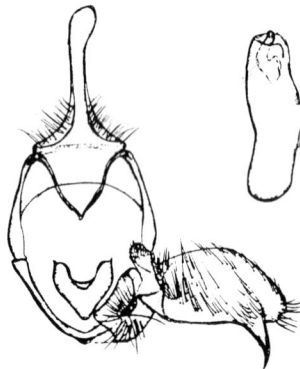

aedeagus

bursa copulatrix

Autosticha pelodes

102

STENOMATINAE

Predominantly Neotropical, with a few Old World genera, e.g. *Agriophara* (Australia, Papua New Guinea), *Synchalara* (India) and *Herbulotiana* (Madagascar). Stenomatine adults are mostly small to medium-sized, with comparatively broad forewings, the termen typically rounded or truncate. Speciation in this group has been greatest in the tropics of South America, and the genera *Stenoma* and *Antaeotricha* together contain several hundred species, but few are important economically. The subfamily status is to some extent uncertain, and the group has sometimes been treated as a distinct family and at other times amalgamated with the Xyloryctinae (Xyloryctidae).

Little information is available about the life histories of the majority of the Stenomatinae. Larvae of some species are known to mine in leaves or tunnel in the living wood of trees and shrubs. Australian *Agriophara* feed on eucalyptus, at first mining in a leaf, later feeding between spun leaves, and pupate amongst leaf litter.

Main genera: *Agriophara, Antaeotricha, Cerconota*, Gonioterma, Stenoma, Synchalara.*

Cerconota anonella Sepp is illustrated. It is a Neotropical pest of soursop or custard-apple (*Annona*), the larva feeding on the fruit.

References. Clarke, 1955-70 (vol. 2); Duckworth, 1973.

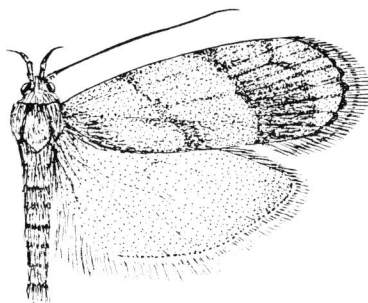

Cerconota anonella

XYLORYCTINAE

Predominantly Australian with a few outliers in the Oriental and Afrotropical regions, and represented in the Hawaiian islands by the endemic genus *Thyrocopa*. Adults range in size from small to moderately large and robust. Among the largest are the females of some of the sexually dimorphic Australian *Cryptophasa* with wingspans over 60 mm and the appearance of a heavy-bodied noctuid moth, and like the noctuids they are night-fliers and are attracted to light. Adults of some of the Australian *Xylorycta* species are almost as large and are shining white and very distinctive. Like the Stenomatinae, this subfamily is sometimes classified as a separate family, and at other times is amalgamated with the Stenomatinae.

Larvae are mostly borers in flower spikes, fruits, seedheads, bark, and living wood of various trees and shrubs, including *Acacia*, *Citrus* and *Ficus*, or feed on the foliage, often living in silken webs or tubes.

Main genera: *Cryptophasa, Deloryctis*, Neospastis*, Odites*, Opisina*, Pansepta, Ptochoryctis*, Xylorycta.*

Odites semibrunnea Bradley, the illustrated species, is an East African coffee pest, the larva feeding in the berries.

References: Becker, 1981; Bradley, 1958, 1959, 1967; Clarke, 1955-70 (vol.2); Common, 1970; Zimmerman, 1978.

103

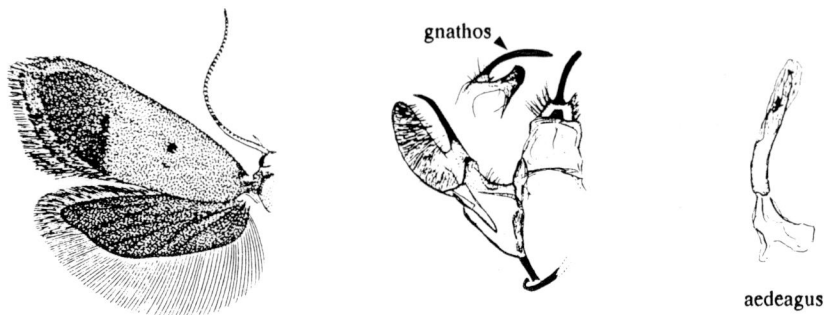

gnathos

aedeagus

Odites semibrunnea

ELACHISTIDAE

Mainly temperate and subtropical in distribution, and most numerous in the northern hemisphere. Elachistids are associated chiefly with monocotyledonous plants and are sometimes common on cultivated grassland and occasionally cereal (*Triticum*) but are not normally pests. Adults are inconspicuous micro-moths with a wing expanse around 8-10mm, the wings narrow but appearing broad because of the long fringes; the forewing is typically flat, but exceptionally the apex is upturned (*Dicranoctetes*, *Eupneusta*). Forewing coloration typically either white or ochreous-white with or without black or brown speckling and stigmata or black-brown with conspicuous whitish or yellowish and sometimes silvery transverse fasciae or patches. Tongue scaled at base.

Larvae are usually miners in leaves and stems of grasses (Gramineae), sedges (Cyperaceae) and rushes (Juncaceae), but some of the primitive genera are on dicotyledonous plants, e.g. Boraginaceae, Cistaceae and Caprifoliaceae). Pupa outside mine, either exposed or in a silken cocoon on the epiderm of the foodplant; not protruded on eclosion.

Male genitalia: valva simple, costa often thickened, cucullus sometimes enlarged and with a small spine (*Cosmiotes*, *Biselachista*); uncus well developed, bilobed in apomorphic forms; gnathos a spinose knob; digitate process arising from base of valva on inner side.

Female genitalia: ostium with distinct antrum; colliculum distinct; signum usually present in the form of single or multiple dentate plates(s), or absent.

Main genera: *Eupneusta**, *Cosmiotes*, *Dicranoctetes*, *Elachista*.

The species illustrated below, *Eupneusta solena* Bradley, is a sugar-cane leaf miner from Papua New Guinea.

References: Bradley, 1974; Braun, 1948; Traugott-Olsen & Nielsen, 1977.

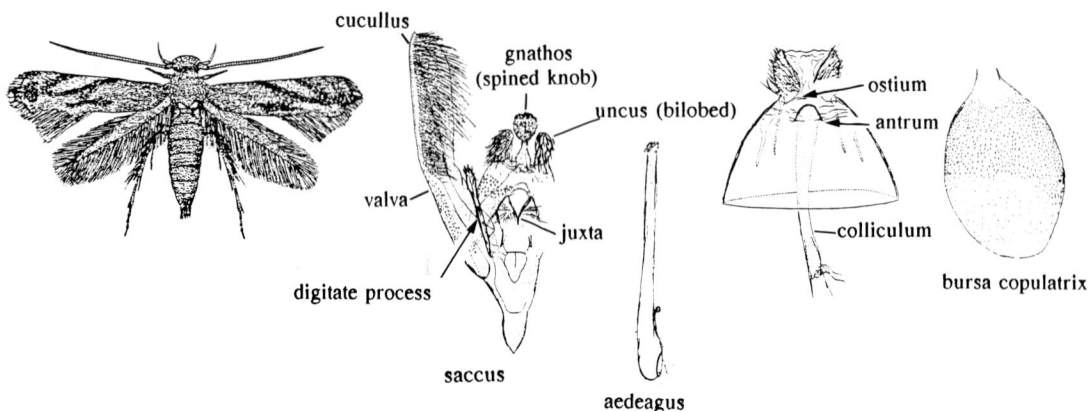

cucullus

gnathos (spined knob)

uncus (bilobed)

ostium

antrum

valva

juxta

colliculum

digitate process

bursa copulatrix

saccus

aedeagus

Eupneusta solena

BLASTOBASIDAE

Mostly tropical and subtropical, including oceanic islands. Adults are typically drab micro-moths with narrow wings (wingspan around 10-25mm), the forewing generally grey or brownish with an indication of the gelechioid triad of dark brown stigmata, and usually with a strongly developed pecten of long hair-like scales, male usually with notch at base of flagellum; labial palpus long and recurved, second segment often showing sexually dimorphic specific characters. Abdominal tergites with transverse caudal bands of spines, as in some oecophorids.

Larvae are mostly general scavengers on matured decaying vegetable matter, boring into twigs, berries, fruits, seeds, etc. Certain species of the genus *Holcocera* are predaceous on scale insects (Coccoidea).

Main genera: *Auximobasis, Blastobasis*, Holcocera, Pigritia*

Blastobasis inana Butler from Hawaii is illustrated. This widespread Indo-Australian scavenger of vegetable matter, including stored products, may well be distributed by commerce.

References: Powell, 1976; Zimmerman, 1978.

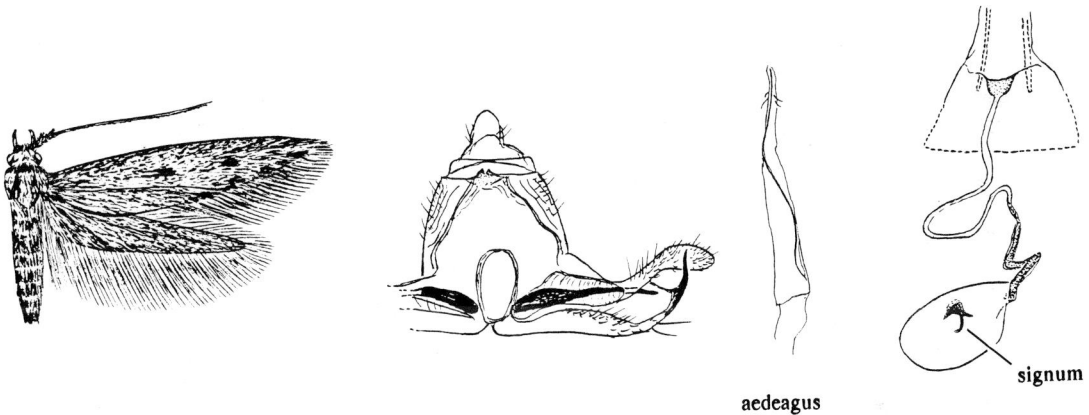

aedeagus

signum

Blastobasis inana

COLEOPHORIDAE

Mainly Holarctic, the great majority of the 1200 or more known case-bearer species are assigned to the genus *Coleophora* and are non-economic. Most are host specific or oligophagous, and feed on leaves, flowers and seeds, and occasionally on the stems of herbaceous plants. A few are harmful to seed production and others occasionally cause foliar damage to orchard trees, shrubs and ornamentals.

Adults are small (wingspan rarely over 20mm), with slender long-fringed wings, the forewing acuminate or weakly falcate, the hindwing linear-lanceolate. The forewing pattern is either linear, in the form of longitudinal vein-like markings, or is virtually non-existent, the wing being almost unicolorous; none of the adults show any transverse fasciate markings and few show any sign of stigmata. In repose the wings are folded along the body, which is pressed to the substrate, and the antennae are held straight forward (porrected). Head is smooth-scaled; ocelli absent; antennae often thickened with scales near base, the scape often with scale-tuft; tongue vestigial or absent. Abdominal tergites 1-7 have paired spinose patches which are often specific.

Larvae are concealed feeders, constructing portable cases of silk and pieces of

105

foodplant. The case may either be formed early in the larval stage or not until the final instar, the larva mining prior to its construction. The lip or mouth of the case is fixed to the leaf, seed or other part of the plant on which the larva is feeding, and the larva bores into the epidermis. Samples of the case should be preserved (either glued on card or pinned with the adult) as the structure is often characteristic of the species.

Male genitalia: uncus reduced and functionally replaced by the distally knobbed and spinose gnathos; valva short and deeply bifid, the digitate costal part being free distally.

Female genitalia: ovipositor moderately extensible; ostium usually strongly formed and ductus bursae distinctly spiculate, both these structures being highly specific; signum usually thorn-like, sometimes double.

The mass of species comprising the genus *Coleophora* can be divided into species-groups according to host plant(s), genitalia structure, forewing pattern, etc. Some authorities treat these groups as distinct genera, but such division can serve no practical purpose and the 40-50 described genera have been amalgamated into one genus (Sattler & Tremewan, 1974, 1978).

The illustrated *Coleophora frischella* Linnaeus (=*alcyonipennella* sensu auctt.) is a Palaearctic species that has been introduced into temperate Australia. The larva bores into green seeds of white clover (*Trifolium repens*).

References: Baldizzone, G. and Falkovitsh, G., numerous contributions by both authors in current literature; Sattler & Tremewan, 1974, 1978; Patzak, 1974; Toll, 1953.

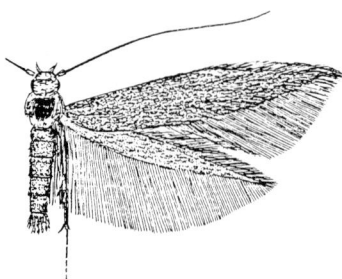

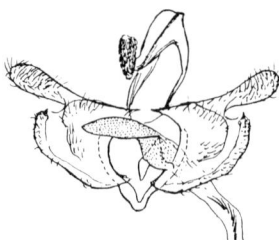

bursa copulatrix

Coleophora frischella

AGONOXENIDAE

Monogeneric (*Agonoxena*; coconut flat moths) comprising four described species of micro-moths (expanse 9-16mm) associated with palms and distributed from Indonesia to Queensland and the S.W. Pacific to Hawaii. An undescribed fifth species is recorded from Lakeba I., Fiji, by Dugdale (1978). One species, *A. argaula* Meyrick ~~which is illustrated~~, rates as a pest, attacking the leaves of coconut (*Cocos nucifera*).

Agonoxenid adults are characterized by the flattened head and abdomen, explanate (flattened and divergent) labial palpi and their generally yellowish brown coloration and forewing markings, sometimes with obscure whitish or reddish longitudinal markings. The flattened form of the moths is reminiscent of certain depressariine oecophorids, and may have evolved from the habit of concealment between folded leaflets of the palms. The venation differs between the species; the forewing lacks a pterostigma; hindwing venation

106

is reduced, with discal cell open. Abdominal terga are not spined (spined in Coleophoridae).

Male genitalia: uncus reduced; gnathos a large spinose knob; valva short, oval, semimembranous.

Female genitalia: ovipositor short, setose; signa a pair of serrate-dentate plates.

References: Bradley, 1966; Dugdale, 1978; Zimmerman; 1978.

MOMPHIDAE

A composite family - Momphinae, Batrachedrinae and Blastodacninae - of mostly dull-coloured micro-moths, the forewing lanceolate, without a pterostigma, often with raised scales, the apex rounded or falcate, the hindwing lanceolate-linear. Head smooth-scaled; tongue developed, scaled at base; labial palpus recurved, smooth-scaled, apex acute; antenna from two-thirds to full length of forewing, simple or ciliate, pecten present or absent. Abdominal tergites 2-7 with paired longitudinal patches of spines as in Coleophoridae (except Blastodacninae).

The status and relationship of the three "subfamilies" mentioned above is debatable. Both Batrachedrinae and Blastodacninae are sometimes separated as distinct families, or the Batrachedrinae are placed as a subfamily of the Coleophoridae and Blastodacninae as a subfamily of the Agonoxenidae. Economically, the most important are the Batrachedrinae and Blastodacninae.

Male genitalia: symmetrical; valva deeply furcate, with costal part free distally (Momphinae), or entire, with costal part united (Batrachedrinae, Blastodacninae); uncus and gnathos absent (Momphinae), or uncus absent and gnathos present, specialised, with arms either fused or separate and forming slender single or double recurved process(es) spined at tip(s) (Blastodacninae); uncus when present articulated with tegumen; socii absent; anellus and aedeagus not anchylosed; aedeagus short, with heavy cornuti (Momphinae), or moderately long, with a patch of cornutal spines (Blastodacninae), or very long and slender (Batrachedrinae).

Female genitalia: ovipositor slightly extensile; ostium deeply cleft, sometimes projecting (Momphinae), or indefinite; ductus bursa sometimes with accessory diverticula; signum usually sickle-shaped or curved, single or double, sometimes absent.

References: Hodges, 1962,1966,1978; Riedl, 1969.

MOMPHINAE

Non-economic; about 80 known species, mainly Holarctic. Larvae feed mostly on herbaceous plants, especially Onagraceae (*Epilobium, Circaea, Ludwigia*), mining leaves and

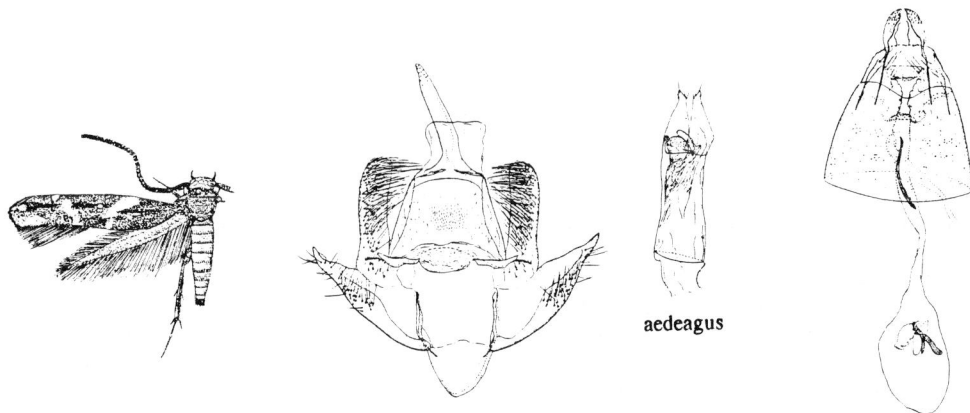

Mompha ludwigiae

aedeagus

forming blotches, or in shoots, or forming galls in stems. Pupa in a silken cocoon, usually outside the mine.

Main genus: *Mompha**.

M. ludwigiae Bradley, illustrated above, acts as control agent for the water weed *Ludwigia* in N.E. India.

References: Bradley et al., 1973; Reidl, 1969.

BATRACHEDRINAE

About 130 species in 5 genera, the largest being *Batrachedra* which is cosmopolitan and contains some economically important tropical species associated with palms. Adults are slender brownish coloured micro-moths, the forewings smooth-scaled and usually without markings excepting an elongate blackish stigma at the end of the cell.

Larvae mine leaves or bore in flowers, developing seeds and fruits, e.g. in the flowers of palms (reducing set of seed), or are scavengers, feeding on dry vegetable matter, e.g. at the base of pineapples, or animal matter, e.g. on insect remains in the nests of spiders, or are predaceous on scale insects, usually feeding beneath silk webbing. Pupa in a flat, elliptical cocoon outside the mine, or in the feeding place.

Main genera: *Batrachedra**, *Batrachedrodes*, *Homaledra*.

Batrachedra arenosella Walker is illustrated below, but the name probably embraces a species complex distributed throughout the Old World tropics. The larvae attack developing flowers of palms, including coconut.

References: Hodges, 1966; Lever, 1969; Zimmerman, 1978.

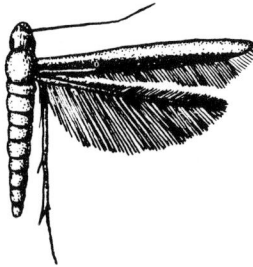

Batrachedra arenosella

BLASTODACNINAE

Mainly Holarctic; minor economic importance. Blastodacnine adults are narrow-winged micro-moths (wingspan averaging 12-13mm), the forewing often with scale-tufts on the upper surface, coloration generally fuscous, often with white admixture, or sometimes bright orange. Head smooth-scaled; labial palpus with second segment loosely scaled beneath, terminal segment shorter, acute. Abdominal tergites not spined.

Larvae feed in fruit, twigs or shoots of various trees and shrubs. Pupa in feeding place.

Main genera: *Blastodacna, Parametriotes**.

Parametriotes theae Kuznetsov, the species illustrated below, is found from Central Europe to China and is a pest of tea in part of its range.

References: Hodges, 1978 (in Agonoxenidae).

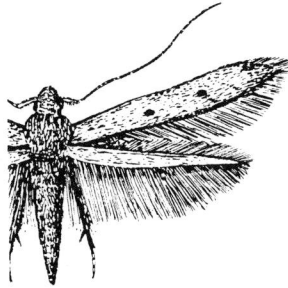

Parametriotes theae

COSMOPTERIGIDAE

Worldwide, with a number of economically important species. Cosmopterigids typically have long-pointed, long-fringed wings (wingspan 6-12mm), the forewing with bright metallic coloration. Head smooth-scaled; labial palpi slender and recurved; tongue developed, scaled at base; abdominal tergites not spined. The genitalia are often strongly asymmetrical in both sexes; in the male abdominal segments 7 and 8 are modified to form a pair of flap-like structures (epiptygmata).

Larvae are mostly miners in leaves, buds, seeds, stems (sometimes causing galls) and roots, or scavengers on dried or decaying plant material. Pupa not protruded on eclosion.

Male genitalia: valva strongly asymmetrical to nearly symmetrical, simple or complex, sometimes with basal or anellar lobes; aedeagus anchylosed to diaphragma by sclerotised manica; uncus absent and arms (brachia) of gnathos asymmetrical (Cosmopteriginae), or uncal brachia asymmetrical and aedeagus not anchylosed but free (Antiquerinae).

Female genitalia: ostium on sternite 7, with lamellae often forming a projecting tube; bursa copulatrix often double or with an accessory sac; signa usually double but sometimes single or absent.

Subfamilies: Cosmopteriginae*, Chrysopeleiinae, Antiquerinae.

Reference: Hodges, 1978.

COSMOPTERIGINAE

Includes a number of mainly tropical and subtropical species which mine in leaves and seeds, especially of Gramineae, Leguminosae, Cyperaceae, Pandanaceae, etc., or are scavengers in cotton bolls, sorghum heads and similar matured or decaying plant material.

The genus *Pyroderces* contains several important scavenger-pests, most prevalent in the dry tropics. The adults are often very similar externally, and dissection of the genitalia is usually a prerequisite for identification. *Pyroderces badia* Hodges is illustrated below. It is a scavenger of vegetable matter and is native to the Neotropics. The closely related *P. rileyi* Walsingham is sympatric but has been introduced to Hawaii (Zimmerman, 1978).

Main genera: *Cosmopterix, Pyroderces*, *Labdia, Limnaecia, Trissodoris*.

109

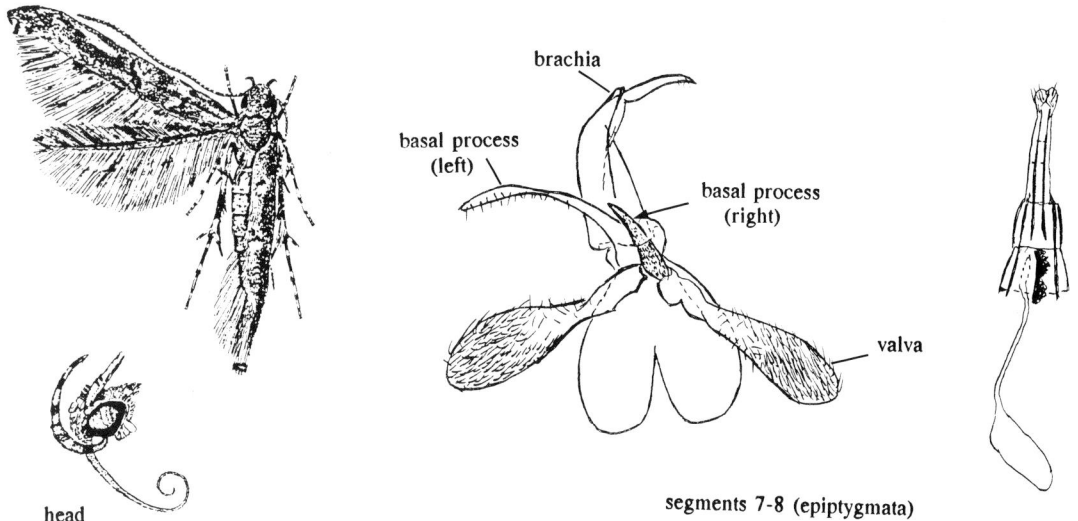

brachia

basal process
(left)

basal process
(right)

valva

segments 7-8 (epiptygmata)

head

Pyroderces badia

CHRYSOPELEIINAE

Mainly Holarctic and of minor economic importance. Adults are generally inconspicuous micro-moths with obscure markings, and dissection of the genitalia is usually
necessary for identification of the species. Forewing often with groups of raised scales.

Larvae feed on a wide variety of host plants, including Leguminosae, Rhamnaceae,
Fagaceae and Betulaceae, living internally in stems, roots, bark, thorns, etc. Pupa in the
larval feeding place or amongst debris.

The male genitalia do not usually show such a strong tendency to asymmetry as most
cosmopterigines; gnathos and socii absent; aedeagus anchylosed; valva simple, or complex
with three well defined lobes (*Sorhagenia*); uncus present, either simple or bifid.

Main genera: *Ascalenia, Sorhagenia, Stilbosis, Walshia.*

The *Ascalenia* species-complex includes numerous very similar dark-coloured species
distributed in the Mediterranean region through Eurasia to the Himalayas. Many are
associated with *Acacia* and *Tamarix*, feeding in the inflorescence (Kasy, various papers).

GELECHIIDAE

Includes a number of important pest species in tropical and temperate regions.
Adults are mostly moderately broad-winged micro-moths, the wings well fringed and generally
with inconspicuous coloration. Head is smooth or with scales loosely appressed, sometimes
with long scales curved forward; tongue present and densely scaled at base; antenna usually
simple, scape seldom with pecten; labial palpus slender to stout, usually upturned or
recurved, rarely porrect, second segment rough-scaled or tufted beneath, third segment
usually slender and acute; ocelli present or absent; chaetosemata absent. Forewing with
veins R4 and R5 usually stalked, R5 to costa, vein CuP obsolescent (represented by a fold

110

or crease) or absent; hindwing generally trapezoidal, the apex prominent, the termen often sinuate or emarginate, veins Rs and Ml usually approximate at base or stalked, CuA sometimes with basal pecten, CuP present or absent.

Larvae are concealed feeders, living in rolled leaves, shoots, etc., mining or boring in leaves, stems, seeds, fruit, tubers, etc. Pupa usually in a silken cocoon in the larval feeding place or nearby; not protruded on eclosion.

The male genitalia are variable in structure and form, rarely asymmetrical, varying from simple and reduced to modified and partially anchylosed; 8th sternite sometimes divided into distal lobes covering genitalia; tegumen elongated with uncus usually short and squat; gnathos well developed, usually hook-like; valva generally narrow or rod-like, costa sometimes separate and sacculi developed as basal, saccular processes, sometimes fused medio-ventrally; aedeagus usually strong and loosely attached by membrane, cornuti usually present.

Female genitalia: ostium varying from membranous and not well defined to strongly sclerotised, sometimes with elaborate foam-like cellular structure (gnorimoschemine complex), or sometimes with projecting ostial plate; signum/signa single or double, of various shapes and often of generic significance.

Subfamilies: Gelechiinae, Chelariinae, Anacampsinae, Dichomeridinae, Symmocinae.

References: Bartiloni, 1951; Povolný, 1964 et seq.

GELECHIINAE

Main genera: *Aristotelia, Chionodes, Ephysteris, Exoteleia, Filatima, Gelechia, Gnorimoschema, Keiferia, Lita, Megalocypha, Phthorimaea*, Recurvaria, Scrobipalpa, Ptycerata, Stegasta, Symmetrischema, Tildenia.*

The species illustrated below is the South American potato tuber moth *Phthorimaea operculella* (Zeller), now widely distributed by commerce.

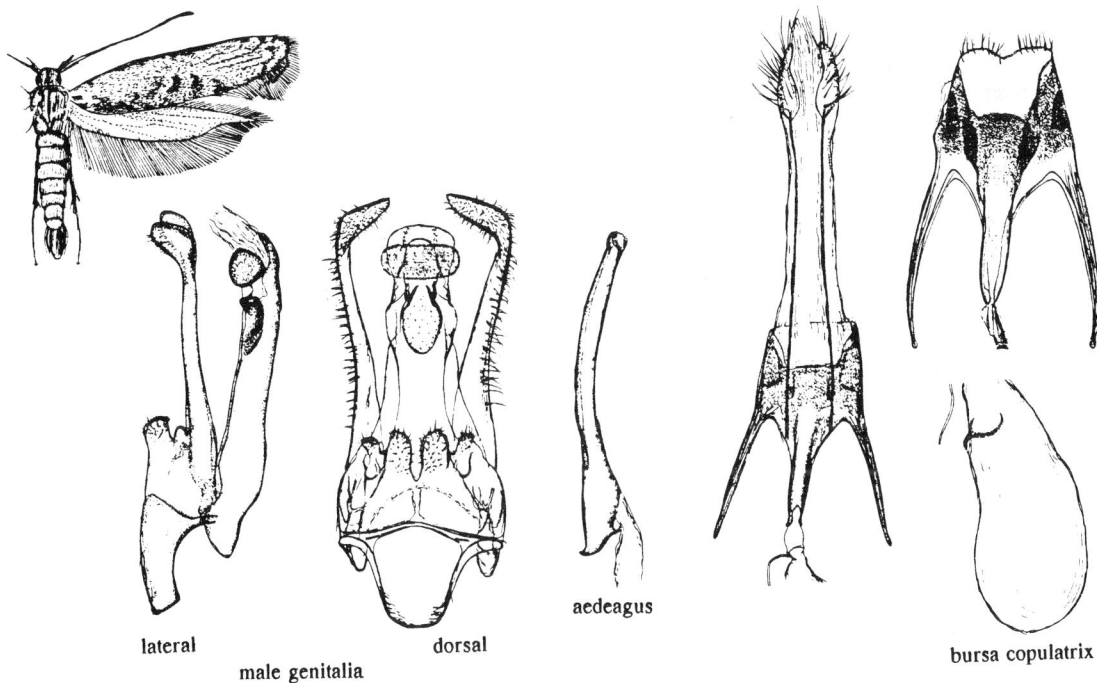

lateral   dorsal   aedeagus   bursa copulatrix

male genitalia

Phthorimaea operculella

CHELARIINAE

Main genera: *Anarsia**, *Hypatima**, *Pectinophora**, *Platyedra*, *Sitotroga**.

The grain moth *Sitotroga cerealella* Olivier, illustrated, is a pest of stored grain particularly in warm moist regions; widely distributed by commerce.

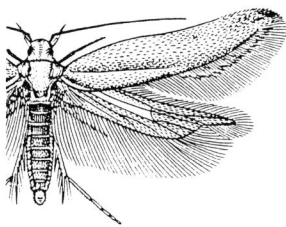

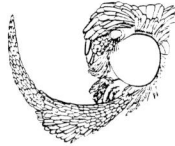

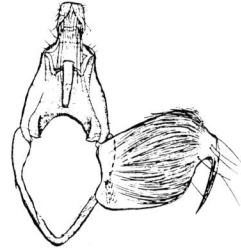

aedeagus

Sitotroga cerealella

ANACAMPSINAE

Main genera: *Anacampsis*, *Aproaerema**.

Aproaerema modicella Deventer is illustrated below, a groundnut leaf miner of the Oriental tropics.

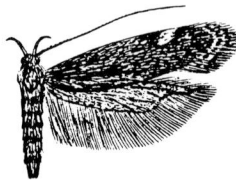

Aproaerema modicella

DICHOMERIDINAE

Main genera: *Brachmia**, *Brachyacma*, *Dichomeris*, *Trichotaphe*.

Brachmia convolvuli Walsingham (illustrated above) is a sweet potato leaf folder found throughout the African and Oriental tropics. below

Brachmia convolvuli

SYMMOCINAE

Chiefly scavengers and of little economic importance.

Main genera: *Oecia**, *Oegoconia*, *Symmoca*.

The illustrated *Oecia oecophila* Staudinger is a detritus feeder that has been distributed throughout the tropics and into temperate latitudes by commerce.

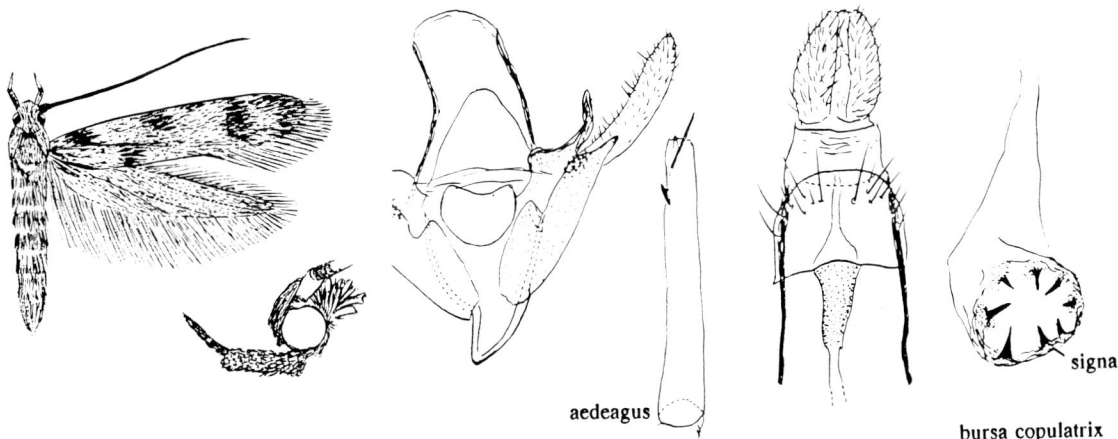

aedeagus

signa

bursa copulatrix

Oecia oecophila

PYRALOIDEA

PYRALIDAE

Worldwide and numerous in all regions and containing some of the economically most important species of Lepidoptera, pyralids are found on most agricultural, horticultural, forest and garden crops and plants, and some are pests of stored food. Adults are generally small to medium-sized, e.g. *Myelobia* (Crambinae), *Melitara* (Phycitinae) and *Rhamphidium* (Chrysauginae) exceeding 50 or 60mm. The majority are slender, often with spindly legs, but the various subfamilies show considerable diversity in form and structure, particularly in wing shape and pattern and specialisation of the labial palpus. The principal characters that distinguish a pyralid moth from other Lepidoptera are the presence of paired tympanal organs directed anteroventral of the basal sternum of the abdomen; and the fusing in the hindwing of veins Sc+R1 and Rs for a short distance beyond the cell, except in the subfamily Pyralinae in which it is sometimes only approximated. The tongue is usually developed and strongly scaled at the base, but is sometimes obsolescent. The maxillary palpi are usually well developed and prominent, but are sometimes obsolescent or absent, and the labial palpi often modified and sexually dimorphic.

Male genitalia: tegumen simple, generally broad, but sometimes narrow (Crambinae) or elongated (Phycitinae), articulating pleurally with the vinculum; vinculum U- or V-shaped; saccus weakly developed (shallow) or absent; uncus present and usually simple and articulating with gnathos, or absent (Galleriinae); gnathos usually present, strong and hook-like, but sometimes weak or absent; valva simple or complex; anellus usually well developed; juxta present; coremata often present on 7th-8th abdominal segments, sometimes strongly developed and complex (Phycitinae).

Female genitalia: ovipositor varying from short and simple to long and extensile, sometimes specialized and with blade-like or serrate lobes for cutting; ostium varying from membranous and simple to sclerotized and complex (Crambinae); antrum often strongly developed; signum or signa usually present, either single, double or multiple, varying from simple and thorn-like, plate-like, scobinate, serrate, stellate, etc. (e.g. Pyralinae) to highly complex bands or groups of spines or denticulations (Phycitinae).

Larvae are usually concealed feeders, living under webbing, or sheltering in spun shoots or leaves, or boring into stems and fruits, etc.; some are scavengers, and a few (Phycitinae) are predaceous (e.g. *Laetilia coccidivora* (Comstock) in N. America, *Euzophera coccidiphaga* Hampson in India and *Cryptoblabes proleucella* Hampson in the Oriental region, all on coccids, and *Isauria aphidivora* (Meyrick) (= *Cryptoblabes aphidivora* Yoshiyasu & Ohara) in the Malay-Papuan region on aphids, while *Dicymolomia julianalis* (Walker) (Glaphyriinae) lives parasitically on psychid larvae in N. America. Larvae of many Nymphulinae and some Pyraustinae and Schoenobiinae are aquatic or semi-aquatic. Pupation is usually in a silken cocoon in the larval feeding place or nearby; pupa not protruded on eclosion.

Pyralid subfamilies may be divided into two main groups depending on the presence or absence of a praecinctorium in the tympanal organs. This is a small membranous or sclerotised structure situated medially in front of (anteriad) the pair of opposed tympanic bullae on the sternum of the first abdominal segment, and hangs down into the cavity between the thorax and abdomen. It is usually fringed or tufted with scales distally, and may be transversely flattened near the tip, and sometimes is enlarged and bilobed (Pyraustinae). A full description of the structure is given by Munroe (1972: 10-12), and according to its presence or absence the principal subfamilies may be grouped as follows:

Praecinctorium present; tympanic bullae fused medially: Crambinae*, Schoenobiinae*, Scopariinae, Nymphulinae*, Evergestinae*, Odontiinae*, Glaphyriinae*, Pyraustinae*, Cybalomiinae.

Praecinctorium absent; tympanic bullae separate or approximated: Pyralinae*, Galleriinae*, Chrysauginae, Epipaschiinae*, Midilinae, Phycitinae*, Peoriinae*

References: Balachowsky, 1972; Carter, 1984; Corbet & Tams, 1943; Fletcher & Nye, 1984 (generic names); Heinrich, 1956 (Phycitinae); Minet, 1982, 1983; Munroe, 1972 et seq.; Zimmerman, 1958.

CRAMBINAE

This subfamily includes some stem-borers of major importance on graminaceous crops, particularly rice, maize, sorghum and sugar-cane. Adults have characteristically long narrow forewings which are somewhat truncated distally, and broad hindwings which are much folded. Wings always smooth-scaled, the hindwing usually with a distinct pecten of hairs on the upperside. Labial palpi typically very long and porrected straight forward, often exceeding 3 times eye width; maxillary palpi triangularly scaled; chaetosemata absent.

114

Larvae are mostly borers in stems and rootstocks of grasses and sedges, or live beneath a silken web low down on the rootstock. Pupa in a silken cocoon, usually in the feeding place.

Main genera: *Acigona**, *Ancylolomia**, *Calamotropha**, *Chilo**, *Diatraea**.

References: Bleszynski, 1961, 1965, 1970; Bleszynski & Collins, 1962; Box, 1931, 1935, 1952.

The genus *Chilo*

The 30 or so described species belonging to this important genus of stem-borers are distributed mainly in the southern Palaearctic and Oriental regions, with outliers in the Australian and Neotropical regions. Adults are very alike superficially, and accurate identification usually requires genitalia dissection; for figures see Bleszynski (1970). The Afrotropical and Oriental *C. partellus* (Swinhoe) is illustrated.

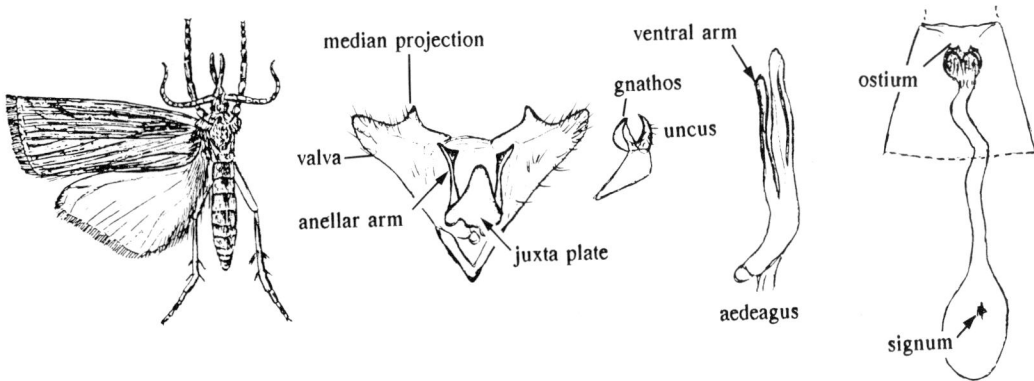

Chilo partellus

The genus *Diatraea*

Virtually the Neotropical counterpart of *Chilo*, with species whose larvae are stem-borers of Gramineae and Cyperaceae, and are major pests of sugar-cane, rice and maize. A check-list of described species has been published by Box (1948) and other papers by the same author figure genitalia and provide bibliographical references (Box, 1931, 1935, 1952).

D. saccharalis (Fabricius), widely distributed in C. & S. America, Florida and the W. Indies (C.I.E. map 5), is illustrated.

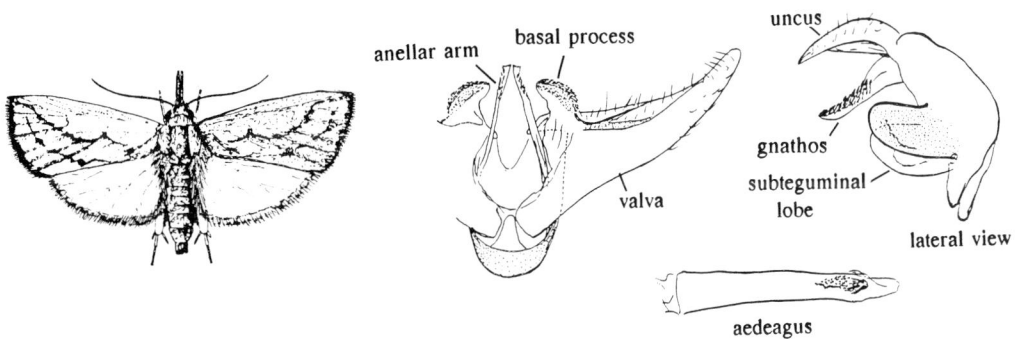

Diatraea saccharalis

SCHOENOBIINAE

A comparatively small group but including a number of stem boring pest species in both the Old and New World. Adults are generally white or yellowish white, sometimes with a blackish discal spot or spots in the central area of the forewing and neural dots along the termen, or with dark suffusion or diffuse oblique streaks, but seldom with other markings. The female is usually larger than the male and often has a bulky anal tuft of long deciduous hairs surrounding the lobes of the ovipositor, some of which stick to the eggs when the latter are deposited. Tongue obsolescent; ocelli and chaetosemata present; forewing with vein CuP developed at margin.

Main genera: *Donacaula, Patissa*, Rupela*, Schoenobius, Scirpophaga*, Topeutis.*
References: Common, 1960; Lewvanich, 1981.

Scirpophaga and *Rupela*

These are both stem borers of rice, sugar-cane and other Gramineae. The Old World genus *Scirpophaga* includes about 35 described species widely distributed from central Europe to Australia. In the forewing veins R2 and R5 are free and arise from the cell; in some species R1 is connected (anastomosed) to Sc, and the presence or absence of a connecting vein can be of specific importance; a scale tuft (corema) is present on the 7th abdominal sternite in the male and an anal tuft is present in the female. The New World genus *Rupela* includes about 20 known species and represents the Neotropical counterpart of *Scirpophaga* (Walker); about 20 species are known. *S. innotata* (Walker) is illustrated; a rice pest ranging from Malaysia to Australia.

References: Heinrich, 1937 (*Rupela*); Lewvanich, 1981 (*Scirpophaga*).

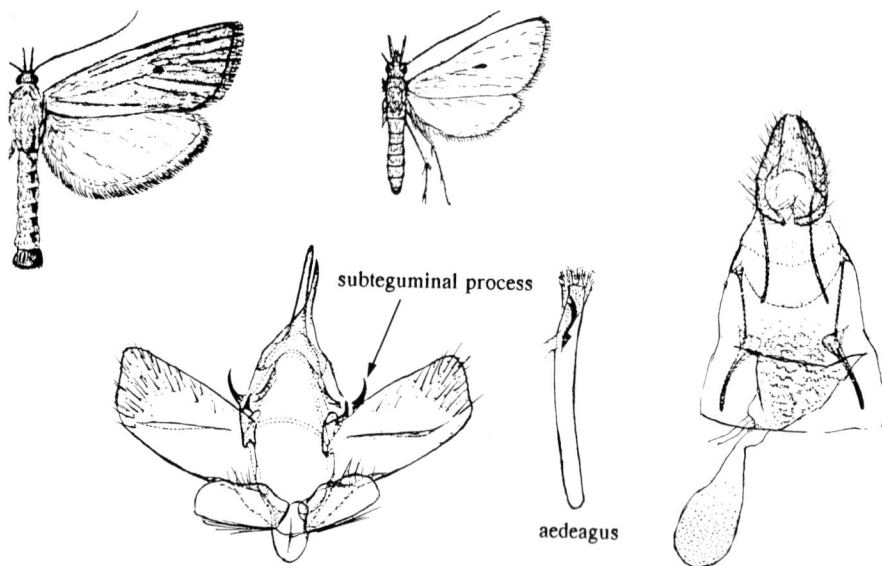

subteguminal process

aedeagus

Scirpophaga incertulas

116

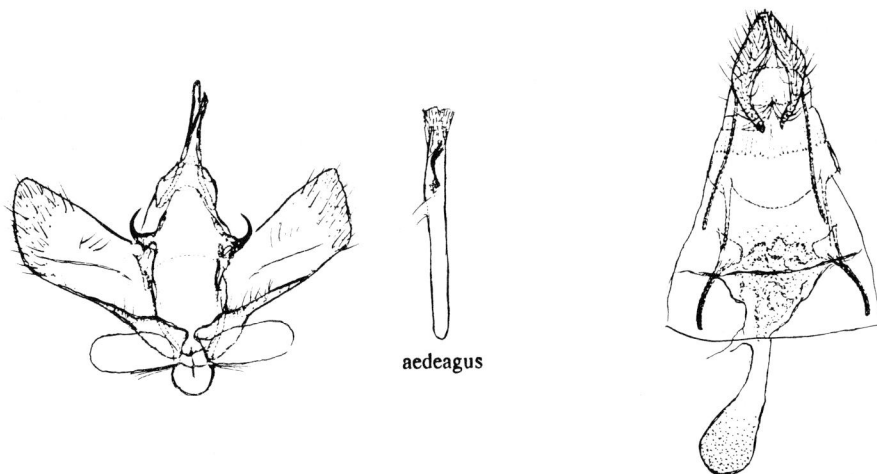

aedeagus

S. innotata

NYMPHULINAE

A small and widely distributed subfamily containing a number of specialised aquatic or semi-aquatic species of economic importance; the taxonomy of the group requires revision. Adults typically have an intricate pattern of either bold or fine transverse bands on a white or pale background which is duplicated on the fore and hindwings. In many species the hindwings have a series of black and metallic marginal spots. Tongue present; chaetosemata present.

Larvae mine in leaves and stems or live in a silk web spun amongst leaves, sometimes constructing flat, silk lined floating cases from leaf fragments. Those living submerged breathe air which is either entrapped in silken reservoirs or obtained from plant tissues, or directly from the water via their integument or by means of gills or branchiae.

Main genera: *Cataclysta**, *Elophila**, *Nymphula**, *Parapoynx**, *Piletocera*, *Synclita*.

The illustrated species, *Parapoynx fluctuosalis* Zeller, probably originated in Asia but is now almost cosmopolitan through introduction (Hubei, 1978; Zimmerman, 1958).

Reference: Speidel, 1984.

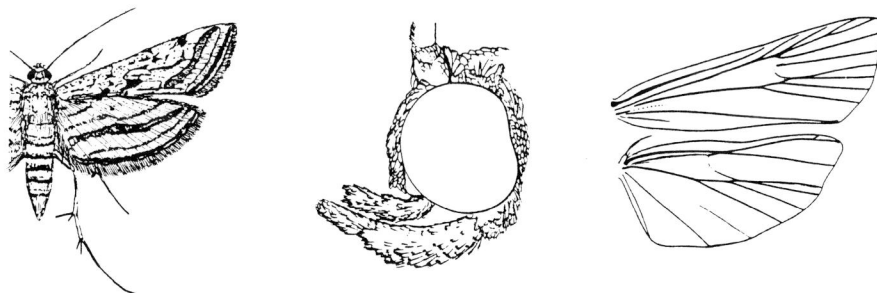

Paraponyx fluctuosalis

117

ODONTIINAE

Worldwide but chiefly in northern subtropical eremic regions. Adults often have the frons produced; hindwing often with broad band along outer margin. Larvae spin or mine leaves, or feed on the seeds mainly of shrubs.

Main genera: *Cynaeda, Noctuelia, Noorda*, Odontia, Tegostoma*.

The illustrated *Noorda blitealis* Walker is Afrotropical and extends to India and Sri Lanka; larva feeds on *Moringa*. Minet (1980) has recently proposed the subfamily Noordinae for this genus.

References: Minet, 1980; Munroe, 1961, 1972.

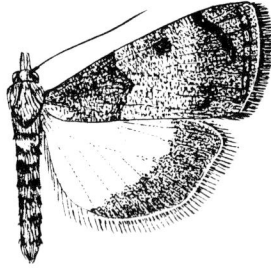

Noorda blitealis

GLAPHYRIINAE

Worldwide; chiefly northern hemisphere. Adults are mostly small and broad-winged; labial palpus short; chaetosema absent. In the male genitalia the uncus is usually well developed and slender, but sometimes modified or reduced; gnathos absent. Larval habits are various, the web-makers (*Hellula*) on Cruciferae are economically important.

Main genera: *Dicymolomia, Hellula**.

The Oriental cabbage webworm *Hellula undalis* (Fabricius) is illustrated. It ranges from S. Europe to Japan, Hawaii, Australia and New Zealand (CIE map 427).

References: Capps, 1953; Munroe, 1972.

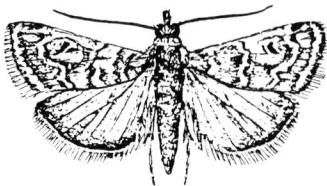

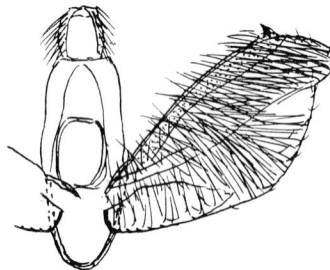

aedeagus

Hellula undalis

EVERGESTINAE

A small mainly Holarctic subfamily of about 100 species with a few tropical and subtropical and of minor economic importance. Adults are generally distinguished by the oblique or produced, sometimes conical frons.

Main genera: *Cornifrons, Crocidolomia*, Evergestis.*

The illustrated species, *Crocidolomia binotalis* Zeller, is a pest of cabbage and other Cruciferae that is found throughout the Old World tropics (CIE map 236).

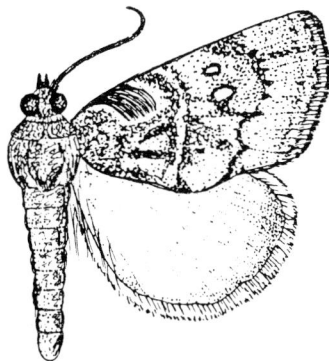

Crocidolomia binotalis

PYRAUSTINAE

Worldwide and numerous in most regions, with many important pest species. Forewing usually triangular, with scale-tufts in the cell area; vein 1A forming a loop, curved forward and joining 2A. Hindwing without defined pecten of hairs on median vein (but sometimes with scattered hairs); veins M3 and M2 closely approximated or stalked, Rs arising out of M1 near at base; chaetosemata absent. In the male genitalia the gnathos is sometimes absent.

Larvae characteristically slender and shining-translucent; usually living in a slight web spun amongst leaves, osscasionally in stems or roots, or on dead or decaying leaves. Pupation usually in the larval feeding place.

Main genera: *Agathodes, Antiercta*, Antigastra*, Bocchoris, Botyodes, Conogethes, Diaphania*, Dichocrocis*, Ercta*, Glyphodes*, Hedylepta*, Hendecasis, Herpetogramma*, Hymenia*, Lamprosema*, Leucinodes*. Loxostege, Marasmia*, Maruca*, Nacoleia*, Nausinoe, Nomophila, Omphisa, Ostrinia*, Palpita*, Polygrammodes, Psara, Pyncarmon, Pyrausta, Samea, Sameodes, Syllepte*, Sufetula, Syngamia, Terastia, Udea, Uresiphita.*

References: Munroe, 1972; Zimmerman, 1958.

The genus *Ostrinia*

Worldwide, containing about 20 described species (Mutuura & Munroe, 1970) four of which are recognized pests, among them the European corn borer *Ostrinia nubilalis* (Hubner) which is illustrated below. Males are characterized generically by the oval, broad-based valva with a complex, spinose clasper or mesal pad, the often strongly spined sacculus, and the short uncus which is distally either bluntly triangular or bifid or trifid.

119

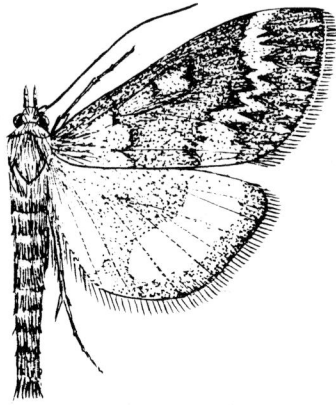

Ostrinia nubilalis

Marasmia trapezalis

The genus *Marasmia*

The relationships, identities and distribution of the several species assigned to this genus, which includes some important leaf-eating pests of rice and other Gramineae, are not fully known and require taxonomic revision. The Old World tropical species *M. trapezalis* Guenee is illustrated above.

PYRALINAE

Represented in most regions, with a small number of pest species some of which, notably *Pyralis* and *Aglossa*, have become less common as a result of the use of machines instead of horses in farming, and modern storage methods. Adults are typically pyralid-like, often with a well developed wing-pattern. Forewing without scale-tufts on upper side; hindwing without defined median pecten, with veins Sc+R1 and Rs usually only approximated, and Rs out of M1 near base, free or sometimes anastomosing with Sc+R1. Maxillary palpi developed; tongue developed or obsolescent; chaetosemata usually present; ocelli absent.

Larvae mostly on dry or decaying vegetable matter, e.g. seeds, cereals, hay. Pupa in a silken cocoon near larval feeding place.

Main genera: *Aglossa**, *Endotricha**, *Hypsopygia**, *Herculia*, *Pyralis**, *Synaphe*, *Tyndis**.

References: Balachowsky, 1972; Carter, 1984; Corbet & Tams, 1943; Whalley, 1963 (*Endotricha*).

The genus *Pyralis*

The several species of this genus recorded in the literature as warehouse and store pests appear in recent years to have become less common, probably as a result of improved storage methods. *P. farinalis* Linnaeus, the cosmopolitan (possibly originally Asiatic) meal moth, is illustrated.

Pyralis farinalis

GALLERIINAE

A comparatively small subfamily, cosmopolitan though nowhere very prominent, but containing a number of species of economic importance. Adults are typically rather sombre-coloured and with elongate wings, the forewings often broadly rounded distally and with subdued markings. Tongue rudimentary or obsolete; chaetosemata absent; ocelli absent; labial palpi often sexually dimorphic; maxillary palpi obsolete in male, often distinct but small in female.

Male genitalia: gnathos absent; uncus broad, rounded terminally.

Larvae mostly feed on dry or harvested vegetable material, some are scavengers or inquilines in bee and wasp nests, eating the comb, and a number are pests on palm trees and sugar-cane.

Main genera: *Achroia**, *Arenipses*, *Corcyra**, *Doloessa**, *Eldana**, *Galleria**, *Lamoria**, *Mampava*, *Melissoblaptes*, *Paralipsa*, *Patna**, *Tirathaba**, *Trachylepidia**

The illustrated species are *Corcyra cephalonica* Stainton, a cosmopolitan stored food pest, and *Tirathaba rufivena* Walker, the Indo-Australian coconut spike moth.

References: Whalley, 1964b; Zimmerman, 1958.

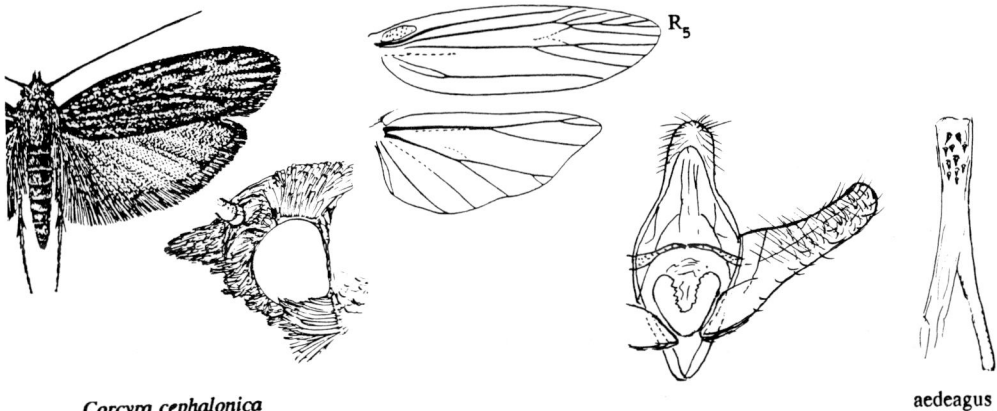

R_5

Corcyra cephalonica

aedeagus

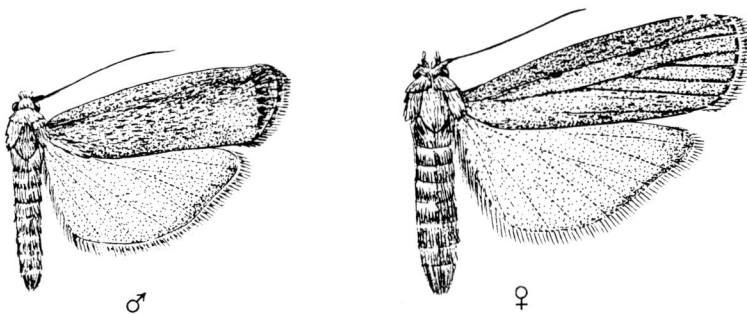

♂ ♀

Tirathaba rufivena

121

EPIPASCHIINAE

A small subfamily, mostly tropical or subtropical and of little economic importance. Adults are mostly moderate-sized pyralids with a somewhat noctuid-like appearance. They are usually distinguishable by the presence of scale-tufts on upper surface of the forewing and, in the males, the displaced venation and development of a fovea. Males of some species are especially distinctive in having the labial palpus or basal segment of the antenna enlarged and plume-like and curved upward and backward over the head vertex. Antenna usually fasciculate in male. Tongue developed; chaetosemata present but weak; ocelli present. Hindwing usually with veins Sc and Rs anastomosed; frenulum simple in both sexes; pecten present or absent.

Main genera: *Epipaschia, Jocara, Lamida, Macalla, Orthaga**.

The illustrated *Orthaga euadrusalis* Walker is a widespread Oriental tropical species, also found in Japan.

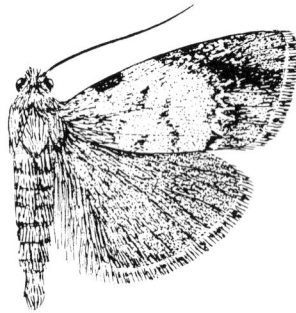

Orthaga euadrusalis

PHYCITINAE

Numerous in most regions, including many pest species, some highly injurious to stored food produce. Also belonging to this subfamily is the beneficial cactus moth, *Cacotoblastis cactorum* Berg, which was introduced into Australia from South America and effectively used as biological control agent for the eradication of *Opuntia* cactus from vast areas of grazing land.

Phycitine adults generally have elongate forewings with grey-black-brown-white coloration, sometimes tinged or mixed with reddish, yellowish or greenish; vein R5 is always absent; hindwing with cubital pecten present. Labial palpi usually curved upwards, often sexually dimorphic, and well developed but occasionally reduced or absent (*Anerastia* and related genera); chaetosemata present.

Larval feeding various, but generally living in silken gallery amongst its food plant, or boring into stems, seeds and fruits, etc., often living in a tube of silk mixed with grass.

Main genera: *Acrobasis, Ancylostomia, Anerastia, Assara, Cryptoblabes*, Dioryctria, Ectomyelois*, Elasmopalpus, Ephestia*, Etiella*, Euzophera*, Fundella*, Hypsipyla*, Mussidia*, Phycita*, Plodia*, Salebria, Sciota.*

The species illustrated below are: the Mediterranean honeydew moth, *Cryptoblabes gnidiella* Milliere, a fruit pest distributed by commerce and now found widely in the tropics and subtropics (Carter, 1984); *Etiella zinckenella* Treitschke, a tropicopolitan pod borer of Leguminosae; *Plodia interpunctella* Hübner, the widely distributed 'Indian meal moth' of stored products. Distinguishing characters of the important *Ephestia* group of stored product pests are also figured.

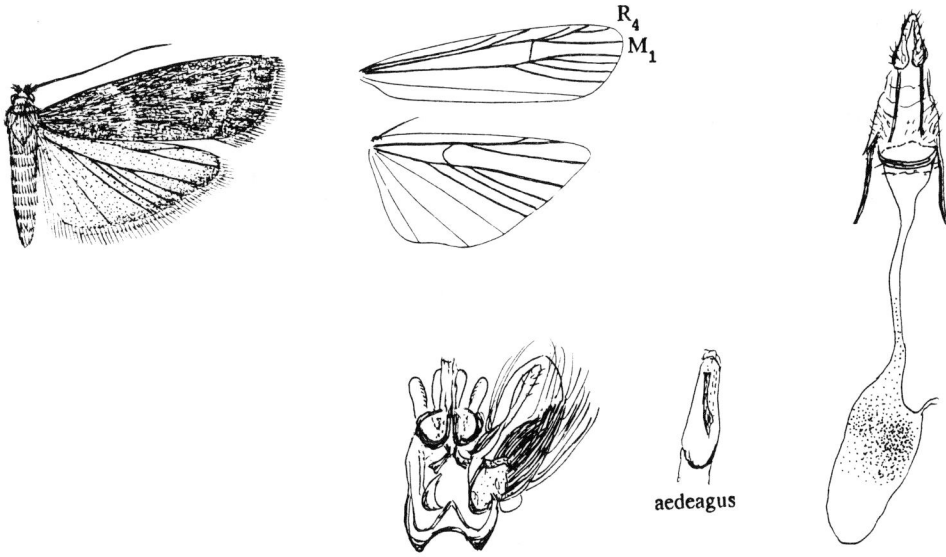

R_4
M_1

aedeagus

Cryptoblabes gnidiella

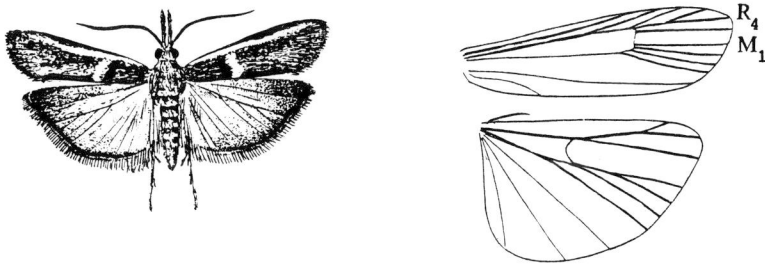

R_4
M_1

Etiella zinckenella

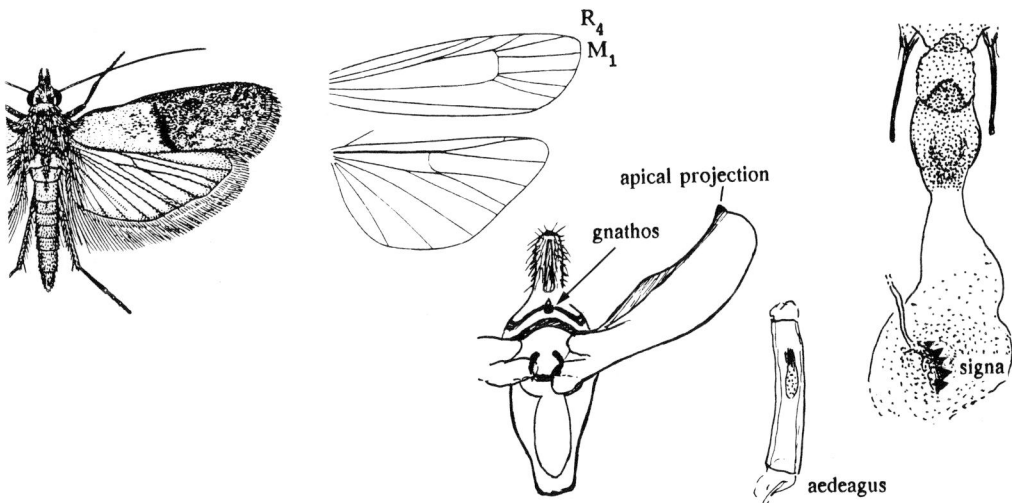

R_4
M_1

apical projection

gnathos

aedeagus

signa

Plodia interpunctella

123

Ephestia elutella

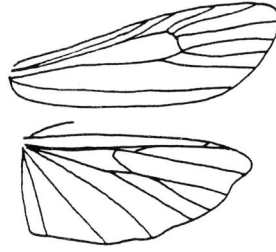

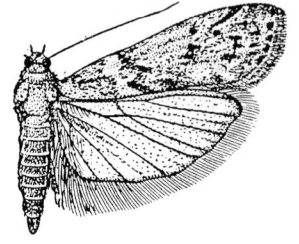

Ephestia kuehniella

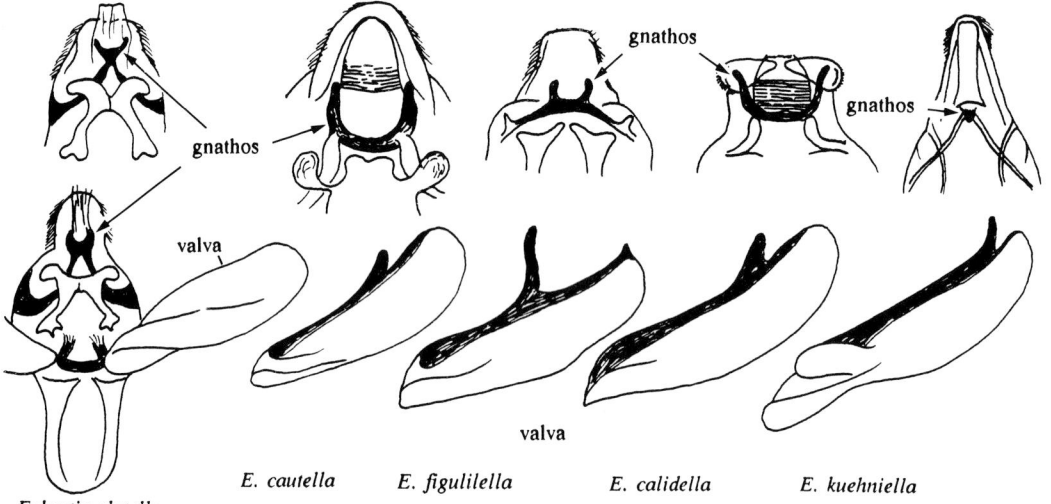

gnathos

gnathos

gnathos

valva

valva

Ephestia elutella

E. cautella

E. figulilella

E. calidella

E. kuehniella

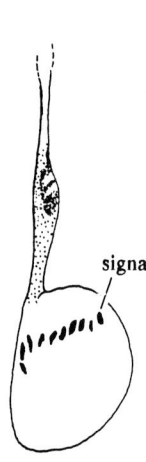

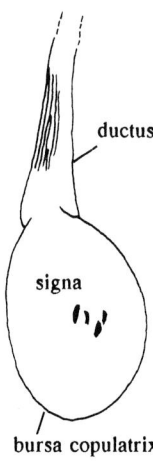

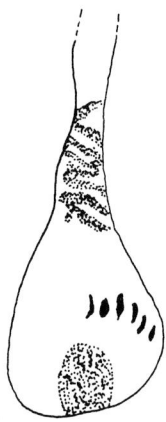

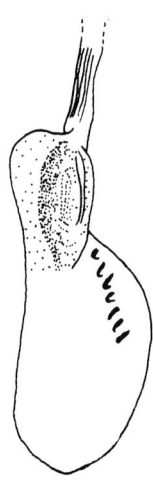

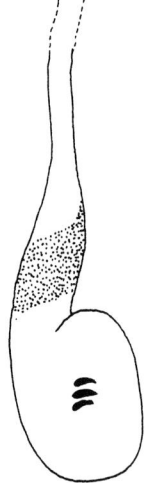

ductus bursae

signa

signa

bursa copulatrix

E. elutella

E. cautella

E. figulilella

E. calidella

E. kuehniella

Ephestia species

124

THYRIDIDAE

The typical thyridid wing pattern is reticulate and is duplicated on fore- and hind-wings; it varies from subdued to bold and silvery translucent, sometimes with hyaline patches or 'windows', and generally produces a leaf-like effect. Adults are mostly small to medium-sized, a few larger species having a wing expanse around 50mm. They differ from pyralids in lacking an abdominal tympanum and in having the tongue, when developed, unscaled at the base; in the hindwing Sc+R1 is approximate and not fused with Rs beyond the cell. Minet (1986) separated this family off in the Thyridoidea.

Generally the 600 or so known species in this exotic family are of little economic importance except as casual leaf rollers or stem borers, mainly of forest trees and shrubs in the tropics. In Africa *Cecidothyris pexa* Hampson causes galls in twigs of *Terminalia*.

Main genera: *Banisia**, *Rhodoneura*, *Sicula*, *Striglina**, *Thyris*.

Banisia myrsusalis Walker is illustrated. It is pantropical, though with genitalically distinct subspecies, the larva feeding on the leaves of sapodilla (*Achras zapota*).

References: Whalley, 1964a, 1971, 1974, 1976.

Banisia myrsusalis

HYBLAEIDAE

A small isolated family comprising about 20 species (*Hyblaea*) from the Indo-Australian region and one sexually dimorphic species (*Torone*) from tropical America. Adults somewhat resemble small catocaline noctuids in appearance, typically having brown forewings with subdued pattern and black hindwings with conspicuous yellow or orange patches, and in earlier classifications were placed in the Noctuidae. But hyblaeids are readily distinguished from noctuids by the presence of well developed maxillary palpi and the absence of a metathoracic tympanal organ; they also lack the abdominal tympanal organ present in Pyralidae, but they have pyralid-like larvae and pupae. Minet (1986) separated this family from the Pyraloidea as a monobasic superfamily, Hyblaeoidea.

Larvae are foliage-feeders, spinning the leaves, and pupate in a cocoon near the feeding place.

Hyblaea puera Cramer, illustrated below, is widespread in the Indo-Australian tropics, the larva attacking teak (*Tectona*) and *Vitex* (both Verbenaceae), and also *Catalpa* (Bignoniaceae)

References: Common, 1970; Singh, 1955.

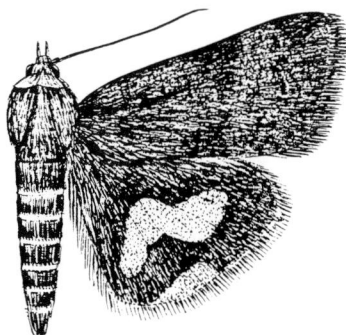

Hyblaea puera

ALUCITOIDEA

Minet (1986) separated the Copromorphidae and the Carposinidae from the Alucitoidea as a separate superfamily, Copromorphoidea.

ALUCITIDAE

A worldwide, small family of little economic importance but often attracting attention because of the curious 'multi-plumed' wings of the adults. The fore- and hind-wings are each divided usually into six feathery plumes or lobes (a few species have seven lobes in the hindwing). Most species are small to medium-sized (wingspan 10-20mm) and delicate, but a few comparatively robust giants occur in the mountain regions of Africa. Tongue not scaled.

Larvae tunnel in buds, seeds, leaves, shoots and stalks, of mainly herbaceous plants and shrubs, sometimes causing gall-like excrescences. Pupa in a silken cocoon in the feeding place or nearby; not protruded on eclosion.

Main genus: *Alucita.*

CARPOSINIDAE

The species of this rather distinctive family are thinly spread outside the tropics and subtropics, and are most numerous in the Indo-Australian and Pacific region, with more than 40 endemic species in the Hawaiian islands. A few species are occasional pests on soft fruits such as peach and plum or on various berries, but the majority are non-economic. Adults are typically small to medium-sized moderately broad-winged micro-moths, the forewings typically whitish with cryptic black or brown markings (reminiscent of pyralid scopariines) and scattered scale-tufts; some have greenish coloration but this tends to fade quickly. Head with erect scales on vertex and appressed scales on frons; ocelli and chaetosomata are absent; antenna shorter than forewing, densely ciliated ventrally in male, pubescent in female, scape without pecten; tongue short, not scaled; labial palpus sexually dimorphic, short in male and recurved with apical segment porrect, long in female and porrect. The loss of one and often two medial veins in the hindwing is characteristic.

Larvae bore in berries, fruits, bark, twigs, buds, flowers and shoots of various trees and shrubs, some start as leaf miners, some cause swellings or galls in stems. Pupa in a silk cocoon, usually in the ground.

Main genera: *Bondia, Carposina*, Heterocrossa, Heterogymna, Meridarchis.*

126

Carposina berberidella Herrich-Schäffer is illustrated. The larva feeds on *Berberis* berries, and the species ranges from the meditterranean and S. Europe to the U.S.S.R.

References: Common, 1970; Davis, 1968; Zimmerman, 1978.

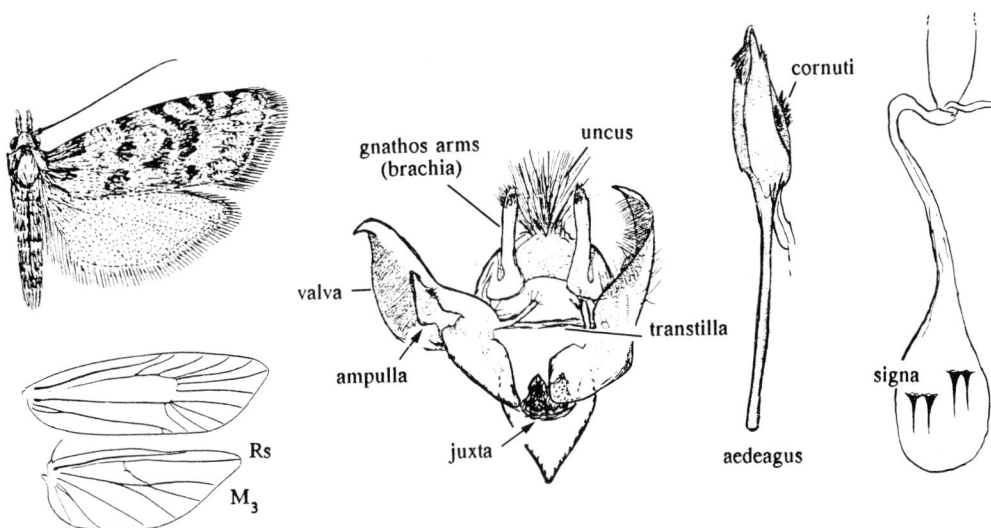

Carposina berberidella

COPROMORPHIDAE

Mainly Indo-Australian, the 60 or so known species in this family are of little economic importance. Copromorphids are comparatively broad-winged, robust micro-moths, and the forewings generally have a roughened appearance due to the loosely appressed or raised scales, sometimes forming ridges, and in some species the roughened appearance is enhanced by thin scaling on the interneural areas. Forewing coloration is generally brownish, but some rain-forest species are a rich mossy green.

A few species are associated with cultivated fig (*Ficus*), the larva boring in the fruits and stems, but nothing is known about the biology of the majority.

Main genera: *Copromorpha**, *Phycomorpha*; the Indian *Copromorpha myrmecias* Meyrick is illustrated below.

Reference: Common, 1970.

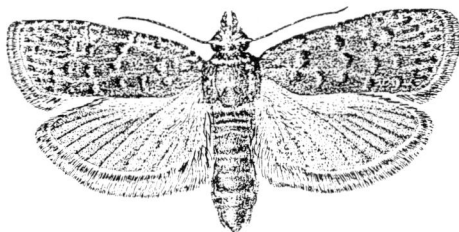

head - lateral view

Copromorpha myrmecias

127

PTEROPHOROIDEA

PTEROPHORIDAE

This distinctive family of 'plume-winged' moths is well represented in most regions but is generally of minor economic importance. The narrow plume-like wings of the adults, slender body, spindly long-spurred legs and habit of resting with the wings rigidly outstretched at right angles to the body, either tightly rolled or flat, with the hindwings usually tucked under the forewings, gives them a characteristically unmoth-like appearance. In most species the forewing is cleft into two plumes and the hindwing into three plumes, and only those in the small subfamily Agdistinae have the wings entire. The fragile structure of the plume moths makes them weak fliers, and most species are sedentary and take only short flights, usually at sunset. Maxillary palpi, ocelli and chaetosemata are absent.

Larvae are unusually hairy or bristly for Microlepidoptera, since they have numerous secondary setae. They live externally on flowers and leaves or tunnel in stem, buds, seed capsules, etc., generally of herbaceous plants and shrubs, including Compositae, Vitaceae, Convolvulaceae and Cucurbitaceae.

Three principal subfamily groups are recognized: the Agdistinae, which are not of economic importance and can be readily distinguished by their non-cleft (entire) wings, and the Platyptiliinae and the Pterophorinae, both of which have the wings cleft or 'plumed' but can be separated by differences in the hindwing, which in Platyptiliinae usually has dark scales mixed with the cilia (fringe) and lobe 2 with 3 veins and lobe 3 with 1 vein, while in Pterophorinae the cilia are usually not mixed with scales and lobes 2 and 3 each have 2 veins.

References: Fletcher, 1909; Lange, 1950; Zimmerman, 1958.

PLATYPTILIINAE

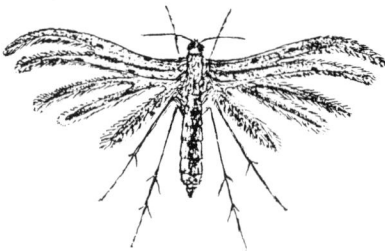

Main genera: *Marasmarcha** (= *Exelastis*), *Oxyptilus**, *Platyptilia**, *Sphenarches**, *Stenoptilia**, *Stenoptilodes**.

Marasmarcha atomosa Walsingham, the Indian pigeon pea plume moth, is illustrated below.

Platyptiliinae

Pterophorinae

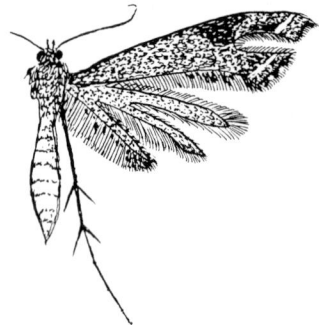

Marasmarcha atomosa

Lantanophaga pusillidactyla

PTEROPHORINAE

Main genera: *Lantanophaga**, *Pterophorus*, *Trichoptilus*.

Lantanophaga pusillidactyla Walker is illustrated above, a lantana-feeding plume of Caribbean origin now widely introduced for control of its host.

TORTRICOIDEA

TORTRICIDAE

The simplified classification proposed by Razowski (1977) of the family-subfamily-tribe hierarchy within the superfamily Tortricoidea advocates the single family Tortricidae, with the subfamilies Tortricinae, Olethreutinae, and Chlidanotinae. Under this system the former family/subfamily Cochylidae or Cochylinae is relegated to tribal status (Cochylini) in the Tortricinae.

Tortricids are well represented in all vegetated zones and include a number of important agricultural, horticultural and forest pest species. Adults are generally moderately robust micro-moths, with a wing expanse seldom over 40mm. Diagnostic characters are unscaled (naked) tongue; presence of ocelli (except in Australian Schoenotenini) and chaetosemata; the usually beak-like form of the labial palpus (except some Olethreutinae); frequent presence of a costal fold on the forewing of the male and, in many olethreutines, an ocellus-like or spectral marking in the tornal area of the wing.

The feeding habits of tortricid larvae are diverse. Many species are external feeders, constructing shelters by rolling or tying leaves, and often are polyphagous or oligophagous, others are borers in stems, rootstocks, seeds, fruits, etc. Some are gall makers and others mine in early instars but feed externally later. Pupation is usually in a cocoon spun in the larval feeding place or nearby, sometimes amongst surface debris or in the earth. In temperate regions with icy winters the larvae of many species which become fully grown as the vegetation period ends and low temperatures prevail, spin a cocoon but overwinter in it before actually pupating, and those of others still in an early instar (usually 2nd instar) construct a silken hibernaculum on the food plant. Pupa usually with characteristic rows of dorsal spines; protruded from the cocoon on eclosion.

Male genitalia: uncus usually developed, long and hood-shaped or narrow (Tortricinae), deeply bifid (Tortricinae, Cochylini), sometimes bifid at the tip or weak (Olethreutinae, Grapholitini), or obsolescent (Olethreutinae, Grapholitini); gnathos well developed or weak and obsolescent (some Olethreutinae); socii often large and prominent, erect or drooping (pendulous), but sometimes small (e.g. *Adoxophyes*, *Archips*); valva relatively simple, but usually with sacculus and costa differentiated (especially in Cochylini) and cucullus often demarcated (Olethreutinae); transtilla usually present, often sclerotised and spined, or membranous; vinculum U- or V-shaped, saccus seldom developed; aedeagus very characteristic, usually short and curved, loosely attached by membrane and easily dissected out (Tortricinae) or hinged to sclerotised juxta and usually best left *in situ* (Olethreutinae); vesica often with deciduous internal spine-like cornuti (these are sometimes found in the bursa of a dissected female), or with fixed spines or barbs; coremata often present on 7th-8th segment(s).

Female genitalia: ovipositor usually short, with flattened setose lobes; ostium often cup- or funnel-shaped, usually with sclerotised sterigma; signum or signa usually present and often diagnostic for tribe or genus, e.g. single horn with prominent capitulum (Tortricinae, Archipini), single dentate band (Tortricinae, e.g. *Cnephasia*), a pair of thorn-like signa (Olethreutinae, e.g. *Cydia*), a pair of flattened horn-like signa, e.g. two triangular, hollow plates (Olethreutinae, *Ancylis*), single or double scobinate circles, discs, or a fold or pocket (Olethreutinae), extensive dentate patch or patches covering large area of bursa (Cochylini).

References: Bradley et. al, 1973, 1979; Common, 1970; Horak, 1984; Kuznetsov, 1978; Razowski, 1977; Yasuda, 1972, 1975.

TORTRICINAE

Main genera: *Acleris, Adoxophyes*, Archips*, Argyrotaenia, Cacoecimorpha*, Choristo-neura*, Clepsis*, Cochylis, Epiphyas*, Cnephasia, Epichoristodes*, Eupoecilia, Homona*, Pandemis*, Platynota*, Sparganothis*, Tortrix**

Archips-Homona complex

This includes a number of polyphagous Old World tropical species. Adults are sexually dimorphic and show considerable individual variation. The genera are closely related but adults can usually be distinguished by the position of veins R4 and R5 in the forewing, which are separated in *Archips* and stalked in *Homona*; a more reliable difference has been described by Diakonoff (1982) in the structure of the sacculus of the male genitalia. Identification of *Archips* species can be made from the monograph by Razowski (1977). *A. seminubila* Meyrick is illustrated. The larva is polyphagous, attacking *Citrus*, litchi, apple, mango, tea and legumes among other crops.

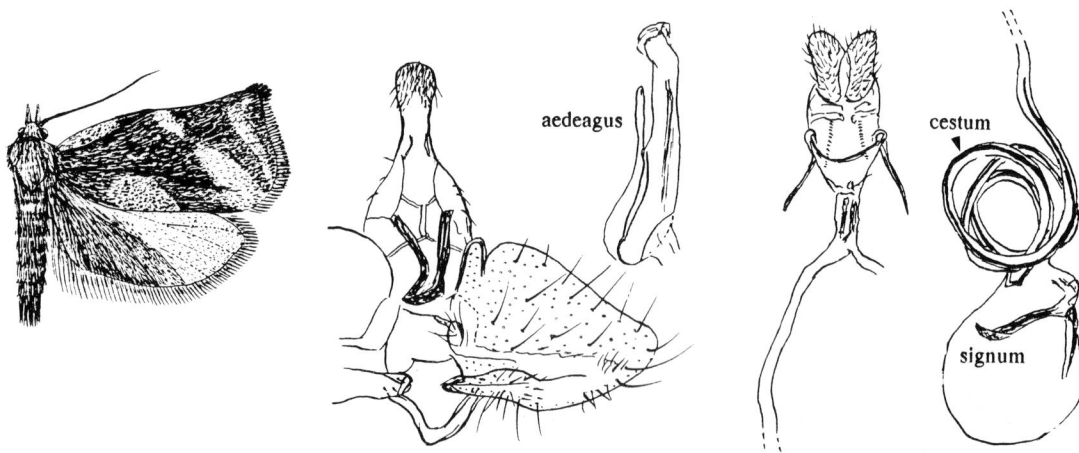

Archips seminubila
Archips species complex

OLETHREUTINAE

Main genera: *Acroclita*, Ancylis*, Bactra*, Crocidosema*, Cryptophlebia*, Cydia*, Dudua*, Eccopsis, Epinotia*, Eucosma*, Fulcrifera*, Lasiognatha*, Leguminivora*, Lobesia*, Olethreutes*, Pammene*, Rhopobota*, Rhyacionia*, Spilonota*, Statherotis*, Strepsicrates*, Tetramoera*, Zeiraphera**.

The genus *Bactra*

A worldwide genus of over 50 known species whose larvae are mainly stem-borers in sedges (Cyperaceae), grasses (Gramineae) and rushes (Juncaceae). The similarity of the adults and the intraspecific variation shown by some often makes dissection of the genitalia necessary for accurate identification. *Bactra venosana* Zeller, the Old World nutgrass (*Cyperus rotundus*) borer is illustrated.

References: Clarke, 1955-70,vol. 3, 1976; Diakonoff, 1956, 1963, 1964.

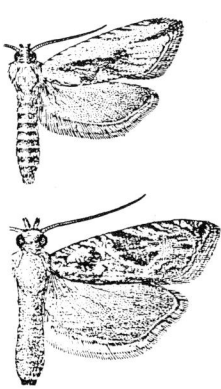

pattern variants

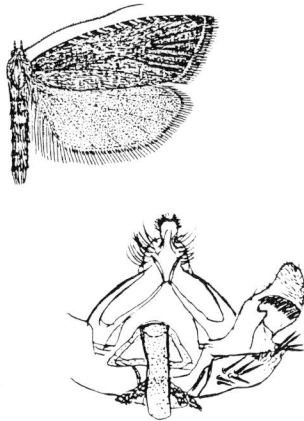

Bactra venosana

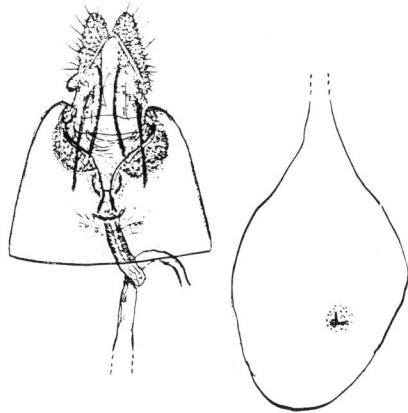

bursa copulatrix

The genus *Cryptophlebia*

Several of the 40 or more described species of this pantropical genus are important pests. Most of the species occur in the Afrotropical and Indo-Australian regions and are fruit- and seed-borers of trees and shrubs. *C. ombrodelta* Meyrick, the African false codling moth, is illustrated below. The larva is polyphagous but attacks *Citrus* in particular.

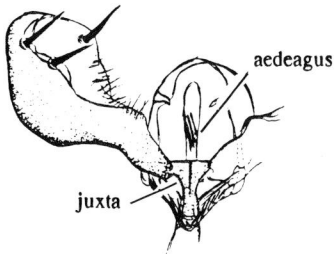

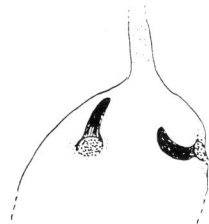

aedeagus

juxta

bursa copulatrix

Cryptophlebia ombrodelta

Cydia (= *Laspeyresia*) -*Grapholita* complex

The relationship of many of the 100 or more species in this worldwide economically important complex are not yet clear taxonomically. For practical reasons the name *Cydia* is at present used for the complex as a whole. Those species in which the male has a pair of coremata on abdominal segment(s) 7/8 are sometimes assigned to *Grapholita* (= *Grapholitha*) and treated as either a subgenus or genus. *C. pomonella* Linnaeus, the Palaearctic codling moth, is illustrated below with similar congeners. The species has been widely introduced and occurs wherever apples are grown.

131

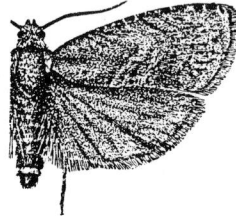

C. pomonella   *C. molesta*

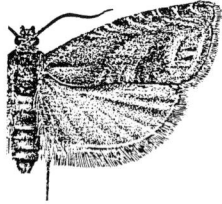
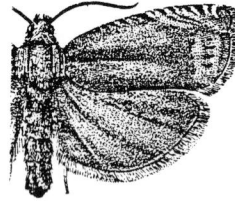

C. funebrana   *C. nigricana*

Cydia **species**

THE DITRYSIAN 'MACROLEPIDOPTERA' SUPERFAMILIES

(JDH)

The 'macrolepidoptera' are generally considered to include the larger moths and the butterflies, though they are far from being a natural grouping more an assemblage of historical usage. Thus the first two superfamilies discussed, Cossoidea and Zygaenoidea, perhaps have more in common with 'microlepidopterous' groups, and their definition is currently unclear and disputed in the literature (Brock, 1971; Common, 1970, 1975). However, the Limacodidae, Megalopygidae, Dalceridae and Chrysopolomidae share larval modifications and may therefore be a natural grouping in the Zygaenoidea (S.E.Miller, unpublished Ph.D. thesis). Here, the Castnioidea are placed near the Papilionoidea but this is probably a convergent association due to the day-flying habit. They have features in common with Cossoidea and Sesioidea (Common & Edwards, 1981; Minet, 1986) but there may also be morphological convergence here due the stem-boring habit.

No satisfactory characters have yet been found to pull the remaining 'macro' superfamilies together as a natural group: loss of the M vein stem in the cell is the sort of character that could have evolved several times, like the development of a clubbed antenna in day-flying groups. However, Scott (1986) purports to demonstrate the monophyly of the Bombycoidea, Noctuoidea, Geometroidea and Papilionoidea.

The major superfamilies are the Noctuoidea, Geometroidea, Bombycoidea and Papilionoidea. The Noctuoidea are defined by the presence of a thoracic tympanum and would appear to be a natural group, though the definition of families and subfamilies is less satisfactory. The Geometroidea families mostly share the feature of possession of an abdominal tympanal organ, though this occurs also in Pyraloidea, and Minet (1983) has suggested that this organ may have evolved independently on several occasions, indicating that the Geometroidea *sensu lato* are an unnatural assemblage.

The Bombycoidea are again defined primarily by unsatisfactory 'loss' characters, but about two thirds of included families share larval characteristics that may indicate they are a natural group

The Papilionoidea *sensu lato* (Brock, 1971) include all the butterflies, but are sometimes divided into 'true butterflies' (Papilionoidea *sensu stricto*) and 'skippers' (Hesperioidea). To these must be added the Hedylidae, a Neotropical 'moth' family that possesses a number of papilionoid apomorphies (Scoble, 1986).

Literature giving accounts of global, regional and more local butterfly faunas is profuse, as listed at the end of the account of the Papilionoidea. Coverage of 'macro' and 'micro' moths is more patchy. Important literature on various families is referred to in relevant descriptions. The only attempt at a global coverage for macrolepidoptera is the series edited by Seitz entitled *Macrolepidoptera of the World*. This gives extensive coverage of butterflies and the more 'popular' moth groups (Bombycoidea, Noctuoidea, Geometroidea, Cossidae and Limacodidae) although the treatment of the Noctuidae and Geometridae was not completed and the series was not resumed after World War II. Volumes of Seitz are now rare and therefore unavailable to most workers. Illustrated accounts of the Order in Dictionary or Encyclopaedia style have been published by Watson & Whalley (1975) and Sbordoni & Forestiero (1985).

The Neotropics are the subject of an ambitious project, the *Atlas of Neoptropical Lepidoptera*, of which one volume has so far been published (Heppner, 1984). North America will be covered by the *Moths of America North of Mexico* series, and several volumes of this are referred to in discussions of the Bombycoidea and Lymantriidae. A checklist has also been published (Hodges et al, 1983), and there is a field guide to the moths of eastern North America (Covell, 1984). A useful general work on the Florida fauna is by Kimball (1965). Hayes (1975) covered many widespread Neotropical species in his account of the Galapagos fauna.

The Afrotropical Region is poorly served apart from Southern Africa where the works of Janse (1932-1964) and Pinhey (1975) provide a useful but incomplete guide to the fauna. Pinhey's work, however, does include a high proportion of those pest species that are widespread in Africa. Fletcher (1963, 1958-1968) gave broad coverage of Lepidoptera in two restricted localities. A guide to the fauna of the island of Reunion has been provided by Guillermet & Guillermet (1986).

The Middle East fauna has been covered to a large extent by the works of Wiltshire

(1948-9, 1957, 1977, 1980-6, and other smaller works referred to therein), though a comprehensively illustrated account is still needed.

Local faunas within Europe are being or have been covered by the *Moths and Butterflies of Great Britain and Ireland* series, by Forster & Wohlfahrt (1954-1974), and Rougeot & Viette (1978), although further sources are referred to by Carter (1984), who described the pest Lepidoptera in detail, with particular emphasis on the early stages. In the eastern Palaearctic the Japanese fauna is the subject of a comprehensive account by Inoue et al. (1982).

For the Indian Subregion the work of Hampson (1893-1896) is still the most comprehensive treatment, though nomenclaturally much out of date. In South-east Asia Barlow (1982) has provided an introduction to a wide selection of the commoner species. The *Heterocera Sumatrana* series (e.g. Diehl (1980), Kobes (1985), Bender (1985), and Roesler (1983)) and the Moths of Borneo series (Holloway, 1983, 1985, 1986) will in time provide a more comprehensive account. The male genitalia of many species were illustrated by Holloway (1976).

Australasia lacks comprehensive faunistic treatments except in eastern Melanesia where the works of Tams (1935), Robinson (1975) and Holloway (1977, 1979) cover the 'macro' moths. For Australia the reader may consult Common (1963, 1970), D'Abrera (1974) and McQuillan (1985) for popular accounts. The New Zealand fauna is covered by Hudson (1928) and that of small islands to the south by Dugdale (1971). Both Common and Dugdale have more extensive works in preparation.

The Pacific coverage is also patchy, with works on Micronesia and Rapa by Clarke (1971, 1976), Hawaii by Zimmerman (1958, 1978) and a general review of Polynesian 'macros' by Holloway (1983b).

Generic names of the macrolepidoptera of the world have been catalogued by Nye (1975), Watson, Fletcher & Nye (1980), Fletcher (1979) and Fletcher & Nye (1982).

COSSOIDEA

COSSIDAE

The Cossidae, with their destructive shoot-, bark-, wood- and root-boring larvae, are of considerable economic significance. The male antennae in the adult are mostly bipectinate, broadly so only over the basal half in the subfamily Zeuzerinae. The female antennae are more likely to be filiform. The forewings are elongate, usually with a fine, irregular, reticulate pattern. The hindwings are very much shorter, and the abdomen usually extends well beyond them. Vein M branches within the cell in both wings. Though there is a frenulum, wing coupling is enhanced by expansion of the costal area of the hindwing; this expansion is relatively more distal in the Zeuzerinae. In collections the accumulation of fats in the body often breaks down and seeps out, investing the wings and body with grease.

The male genitalia are relatively simple, with a broad, well developed uncus. In the female genitalia the ovipositor lobes, eighth segment and apodemes of both are extremely elongated and slender.

In the larva the head is broad and has large mandibles as might be expected from the tough, fibrous diet. The prothorax bears a distinctive plate or shield, posteriorly rough in the Zeuzerinae but smooth in *Cossus* Fabricius.

The family is most diverse in the tropics and subtropics. The Oriental fauna has been described by Roepke (1955, 1957), Arora (1976) and Holloway (1986), and the classification of the family as a whole is being studied by Dr. J.W. Schoorl.

The majority of the Old World pest species are in the subfamily Zeuzerinae, e.g. in the genera *Azygophleps* Hampson (Carter & Deeming, 1980), *Zeuzera* Latreille and *Xyleutes* Hubner. Two species widespread in the Indo-Australian tropics are illustrated below,

Zeuzera coffeae Nietner, a stem borer of coffee, tea and other shrubs, and *Xyleutes ceramica* Walker, a borer of economic timber.

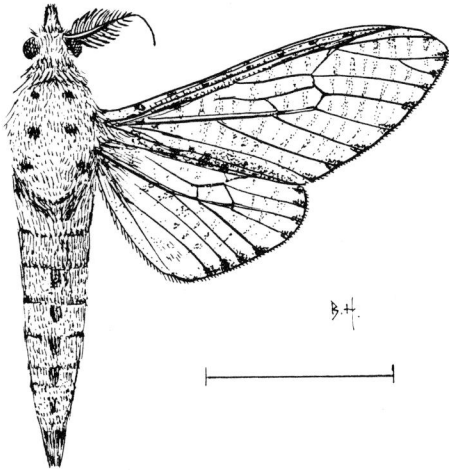

Zeuzera coffeae

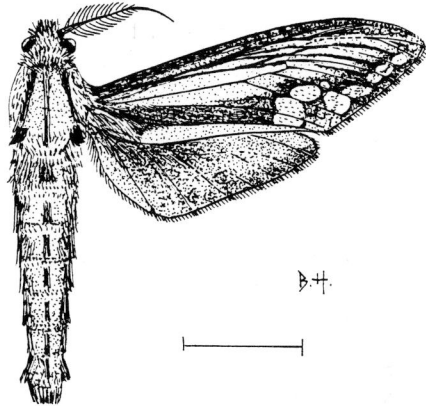

Xyleutes ceramica

METARBELIDAE and RATARDIDAE

These families are closely allied, having only one strong anal vein in the forewing and sharing several characters of the female genitalia (Holloway, 1986). The Ratardidae are extremely rare in collections and are restricted to the Oriental tropics. There are only about ten species known, and these only from females. The rounded wings are distinctive; the antennae are bipectinate. The family has been monographed by Hering (1925) and reviewed by Holloway (1986).

The Metarbelidae include species of economic importance, the larvae boring the bark of citrus trees, cocoa, mango, tea, guava and mangosteen in South-east Asia (Holloway, 1986) and Africa. The patterning of the forewings is cossid-like, but the frenulum is absent. The antennae are bipectinate in both sexes. The male genitalia have a long, broad, flattened uncus with rather drumstick-like appendages that may be gnathi; the valves are small, rather rounded, with a short process from the sacculus; the aedeagus is relatively small also. In the female genitalia the small ductus and corpus bursae, the deep, narrow ovipositor lobes that are dorsally expanded, and the extensive membrane between tergites 7 and 8 are features shared with the Ratardidae.

The larvae of Metarbelidae have a rugulose head with an irregular arc of ocelli (regular in Cossidae); the head is not overlapped by the prothorax (overlap occurs in Cossidae). The larvae shelter in a tunnel in the bark by day, abrading the cambial zone by night, covering the damage with a mass of frass held together with silk (Roepke, 1957).

The species illustrated below is *Squamura disciplaga* Swinhoe, a bark borer of avocado and citrus in Borneo and the Philippines.

137

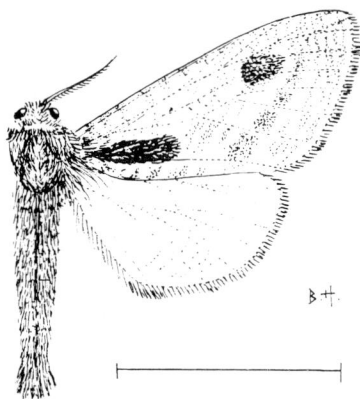

Squamura disciplaga

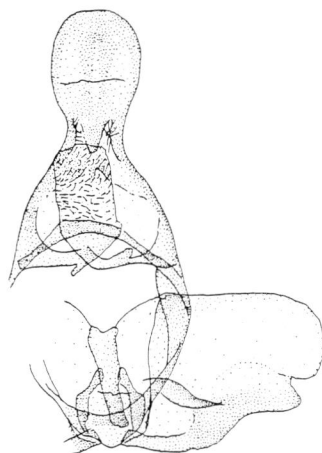

Metarbelid male genitalia (*Squamura*)

DUDGEONEIDAE

This family consists of a small, uniform assemblage of cossoid-like species with abdominal tympanal organs (described by Minet (1983)) and a forewing pattern of silvery spots on reddish brown. It is known from Madagascar, Africa, India, Peninsular Malaysia, New Guinea and Australia. Species have been illustrated by Roepke (1955) and Holloway (1986). The family is not of economic importance. Minet (1986) placed the family in Pyraloidea.

ZYGAENOIDEA
(sensu Common, 1970 and Minet, 1986)

EPIPYROPIDAE

Species of this small, mainly tropical family are probably all ectoparasitic on Homoptera as larvae. The ovate, flattened larva is covered dorsally by a waxy secretion. The hooked thoracic legs and crochets of the prolegs enable the larva to cling to silken threads it has spun on the surface of its host.

The adult moths are small, blackish brown or dark grey. The forewings are usually very deeply triangular or with a rather rounded margin. In most genera all the forewing veins arise directly from the cell (see illustration at couplet 21 of key).

The World fauna has been reviewed by Krampl & Dlabola (1983), and a key to the genera based on wing venation was presented by Kato (1940).

HETEROGYNIDAE

This family consists of two genera, the western European *Heterogynis* Rambur and the South African *Janseola* Hopp. The adult moths are small with rather elongate, dull brown, semitransparent wings. Pupation in European species is in a double cocoon. The family is not of economic importance.

138

CHRYSOPOLOMIDAE

This small, African family consists of rather round-winged, coarsely scaled, moderate sized species. The frenulum is absent. Pinhey (1975) noted similarities of the smooth, sluglike larva to those of Limacodidae, but did not refer to the condition of the abdominal prolegs. The family was monographed by Hering (1937).

DALCERIDAE

The Dalceridae are a small Neotropical family, the species all having rather deep forewings. They are usually shades of yellow and rather lightly built, reminiscent of some Lymantriidae. However, the forewing venation is characteristic with the stem of vein M present in the cell and often an accessory cell or areole formed by the radial veins (subfamily Acraginae). There are accessory glands in the female genitalia that may be diagnostic (Miller, unpublished Ph.D. thesis). The family is currently being studied by S.E. Miller. several species are of minor economic importance on oil palm, cocoa, coffee, and citrus, with sporadic outbreaks as in Limacodidae.

The larvae are rather gelatinous or glossy in appearance with gelatinous tubercles in rows; the tubercles are easily dislodged. Prolegs are absent and motion is snail-like. Details of the larvae and a review of the family have been presented by Sick (1939). The larvae would indicate a close relationship to the Limacodidae, the two families sharing reduction of the abdominal prolegs to produce a peristaltic surface. Crochets are present on abdominal segments 2 and 7 in addition to segments 3 - 6 (Miller, unpublished Ph.D. thesis).

MEGALOPYGIDAE

The adults of Megalopygidae have triangular forewings, white, grey or occasionally pink, with vein R5 arising most distally from Rs such that all other radial veins branch off anteriorly. Exceptions to this are the genera *Aidos* Hubner and Brachycodella Dyar of the subfamily Aidinae and some species of *Podalia* Walker in the Megalopyginae. The male antennae are bipectinate to the apex in most genera, but the Aidinae have them only partially bipectinate, a character state that, like the forewing venation of the subfamily, is shared with the Limacodidae. The Aidinae also have more limacodid-like male and female genitalia, hence their assignment to the Megalopygidae is in question (Holloway, 1986).

The male genitalia lack a gnathos and typically have the valves divided entirely in two, the upper portions giving the impression of setose socii on the tegumen. The female genitalia have the ovipositor lobes opposed (cf. coplanar, disc-like in the Limacodidae) and a mass of dense, fine pilosity on an expanded membrane between segments 7 and 8.

The larva has three rows of spined scoli (dorsolateral, lateral and subspiracular), six pairs of prolegs, only four of which bear crochets, as well as the anal claspers, and are variably pilose (more so in Megalopyginae than in Trosiinae as a rule). The spines are stinging as in the Limacodidae. The pupa has a peculiar eye-plate that is shared with the Limacodidae (Mosher, 1916) and thus, with the larval characters just mentioned (Fracker, 1915; Peterson, 1962), the relationship between the two families appears to be strong. Indeed, modifications to the larvae point to the Chrysopolomidae, Dalceridae, Megalopygidae and Limacodidae forming a natural group (Miller, unpublished Ph.D. thesis).

Pupation is in an extensive cocoon of irregularly matted silk.

The species are not of great economic importance. The subfamily Trosiinae, consisting mainly of white species best distinguished on genitalic characters, has been monographed by Hopp (1927).

The genera *Somabrachys* (N. Africa) and *Psycharium* (S. Africa) are sometimes placed in the family Somabrachyidae but are assigned to the Megalopygidae by Fletcher & Nye (1982). The larva is zygaenid-like but with two extra pairs of crochetless prolegs as in

139

Megalopygidae, though these are not tending towards a peristaltic surface. Females of *Somabrachys* are wingless.

LIMACODIDAE

This is a primarily tropical family but is found in all zoogeographic regions of the world. It is perhaps best known for its highly ornamented larvae. These often bear stinging spines (nettle-caterpillars), are sometimes smooth (gelatine-caterpillars, though this term is perhaps more strictly applied to dalcerid larvae), and all progress by a peristaltic method of locomotion, sometimes leaving a viscous trail (slug-caterpillars).

The moths are small to medium-sized, usually stout, coarsely scaled, often brightly coloured. The forewings are triangular. The antennae of the male are usually bipectinate, most often only partially so. The male genitalia have a basic ground plan of a simple, robust uncus, usually slightly downturned and more sclerotised apically, opposed by a fused gnathus. The valves are simple, elongate, and the aedeagus is often simply flexed at right angles; cornuti in the vesica are uncommon. The female genitalia often have the ductus bursae long and spiralled. The ovipositor lobes are setose, coplanar to form a disc, a reflection of the flat, disc-like eggs. The eighth segment has unusual lateral lugs that may be definitive for the family.

The 'nettle' larvae have urticating spines on rows of scoli or tubercles as in the Megalopygidae, but only the dorsolateral and lateral rows persist. The subspiracular area is modified into a flange to assist with adhesion of the abdominal undersurface to the substrate. This lacks prolegs and is relatively smooth, enabling progress by a slug-like peristalsis. The head and first thoracic segment are retracted beneath the rest of the thorax at rest, but are protruded when feeding occurs.

Pupation is in a tough, spherical cocoon, usually attached to the substrate by silk. Emergence of the adult is through a circular cap at one end.

The adults have a distinctive resting posture with the body axis held at an angle to the substrate, supported by the extended legs, protected by the wings held closely against the flanks.

The larvae are often polyphagous and attack a wide range of trees and shrubs, including many of economic importance such as coconut and oil palm (Wood, 1968; Cock, Godfray & Holloway, in press), tea (Austin, 1931-1932), coffee, cocoa and banana. The South-east Asian fauna has been reviewed by Holloway (1986) who suggested various groupings of genera as a first step to a higher classification, and by Cock, Godfray & Holloway (in press). The African fauna has been reviewed by Hering (1955) and Janse (1964).

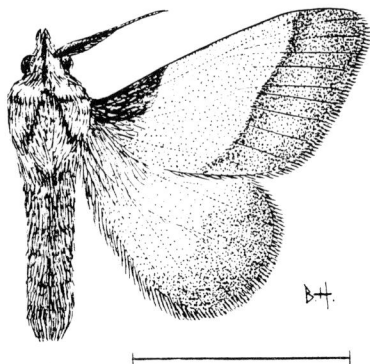

Parasa lepida

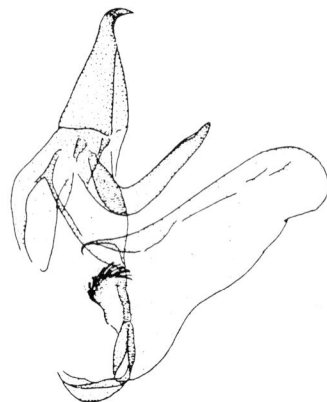

Limacodid male genitalia (Praesetora)

140

The species illustrated above is *Parasa lepida* Cramer, an Oriental pest of a number of crops. It belongs to a complex of species with similar green-banded forewings that is discussed in detail by Cock, Godfray & Holloway (in press). Other important genera are *Thosea* Walker, *Setora* Walker, *Darna* Walker and *Sibine* Herrich-Schäffer.

ZYGAENIDAE

This family is a large assemblage of mainly day-flying, brightly coloured, often iridescent species that in some instances have rather dilated antennae. The species are mainly tropical though the subfamilies Zygaeninae and Procridinae are most diverse in the Palaearctic Region. The species are often involved in mimicry complexes and may be generally distasteful; the Palaearctic *Zygaena* Fabricius species contain cyanide.

In the adults both ocelli and chaetosemata are present, the latter consisting of distinctive fields of bristles set in a honeycomb of rather short, broad scales in a deep zone posterior to the antennae and compound eyes. The frons is frequently rather domed, produced, and covered in very fine scales, particularly in the subfamily Chalcosiinae. The male antennae are bipectinate in most genera but prominently dilated in *Zygaena*.

The larva has tufts of setae, sometimes urticating, on verrucae.

Few of the species are of economic importance. *Artona catoxantha* Hampson, illustrated below,is a drab brown species that damages coconut and oil palm fronds in South-east Asia. The species *Levuana iridescens* Bethune-Baker was a major defoliator of coconut in Fiji until wiped out following the deliberate introduction of a tachinid (Diptera) parasite of *catoxantha* from South-east Asia (Tothill, Taylor & Paine, 1930; Robinson, 1975). This is one of the most spectacular examples of the success of classical biological control methods; the introduction also caused the demise in Fiji of another native zygaenid, *Heteropan dolens* Druce.

Tarmann (1984) has monographed the New World fauna, where only the subfamily Procridinae is present, and is currently studying the Indo-Australian fauna, dominated by the Chalcosiinae.

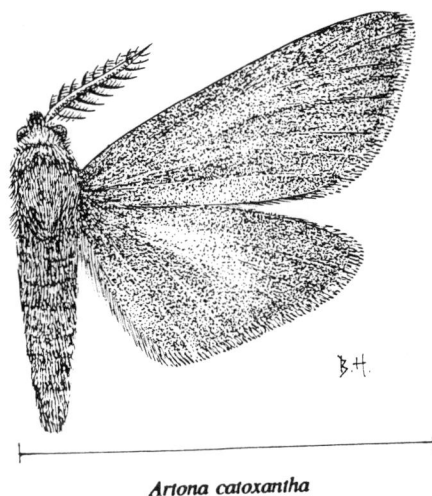

Artona catoxantha

CYCLOTORNIDAE

This small Australian family has been discussed by Common (1970). The larvae are external parasites of Homoptera in early instars. Later instars are associated with ants much as in some lycaenid butterflies.

BOMBYCOIDEA

This superfamily groups together many families of large moths mainly on characters of absence (Common, 1970; Franclemont, 1973): of the M vein stem of the cells of both wings; of tympanal organs; of the tongue and frenulum (present in some families. e.g. Sphingidae); of ocelli and chaetosemata. It is thus questionable whether the superfamily is a natural group. Characters such as bipectination of male and, usually, female antennae and complete patterning of the hindwings are not unique to the group, albeit perhaps more frequent within it than outside. A fine silken cocoon is also a widespread characteristic and in many cases this has been exploited commercially (Dusuzeau, Sonthonnax & Conte, 1897-1911).

Characters of the early stages indicate that two groupings within the superfamily could be recognised. The larvae of Lasiocampidae, Lemoniidae, Apatelodidae, Eupterotidae and Anthelidae are densely clothed in secondary setae giving them a uniformly hairy appearance. The larvae of most of the other families have a caudal horn on abdominal segment 8, at least in early instars, and often paired horns or filaments on some or all of the thoracic and/or abdominal segments. Thoracic and caudal processes are seen in Brahmaeidae, subfamilies Rhescyntinae (Arsenurinae), Citheroniinae (Ceratocampinae) and Agliinae of the Saturniidae, the South African genus *Spiramiopsis* Hampson placed in the Bombycidae by Fletcher & Nye (1982), and some Sphingidae such as the genus *Ceratomia* Harris. A caudal horn only is seen in the Oxytenidae, most Sphingidae, many Bombycidae and early instars of the Carthaeidae. The Cercophanidae have caudal and thoracic areas produced into tapering processes. It is not clear whether the caudal hump of the Endromidae is homologous. The larvae of other Saturniidae subfamilies do not have conspicuous horns but have setose or spinose scoli on each segment; in early instars the thoracic ones, possibly homologous with the horns of other subfamilies, can be longer than the rest (Michener, 1952). Some Saturniidae (e.g. Saturniinae) have three rows of verrucae or scoli on each side as in other Lepidoptera groups such as Zygaenidae.

Franclemont (1973) noted a character of first instar chaetotaxy that placed the Bombycidae with the lasiocampid group rather than the saturniid one.

Kuznetzov & Stekolnikov (1985) have studied genital musculature in the group, together with other characters. Their studies support the monophyly of the horned larva group of families, with the Sphingidae placed in a sister-relationship with the rest and sharing a secondary type of muscle insertion with them. The Lasiocampidae are sister-group to the Bombycoidea and Sphingidae but the other 'hairy larva' families are not included in the phylogenetic diagram. The authors concluded the Bombycoidea *sensu lato* were all inter-related, but referred them to three superfamilies: Lasiocampoidea; Sphingoidea; Bombyc-oidea, *sensu stricto*. The Endromidae were assigned to the last group.

Further consideration of the Bombycoidea as a natural group may be found in Holloway (in press).

A number of families are small and of no economic importance, so will only be touched on briefly here. The Endromidae, Mirinidae and Carthaeidae are very small, the first two consisting of single, monobasic Palaearctic genera and the last of an ocellate saturniid-like species from Australia (Common, 1970: 856).

The Oxytenidae and Cercophanidae are small Neotropical families, the former with some species resembling microniine Uraniidae and the latter containing species that resemble small Saturniidae. Both were monographed by Jordan (1924).

The Brahmaeidae consist of a fairly homogeneous group of large species with striking multifasciate wing patterns. They extend from eastern Europe to the Far East and to Sulawesi in the Oriental tropics; they also occur in Africa. Non-African taxa are mainly on Oleaceae as larvae, whereas African ones are on Asclepiadaceae (see Holloway, in press) Sauter (1986) divided the Brahmaeidae into two subfamilies and provided a key to the genera. The Lemoniidae have a similar distribution outside Africa, though excluding the Oriental tropics and being more extensive in Europe.

The Anthelidae are exclusively Australasian, the species generally resembling

Eupterotidae. The major genus is *Anthela* Walker, several species of which have been recorded as damaging *Pinus* plantations; most taxa defoliate native *Acacia* and *Eucalyptus* (Common, 1970: 853-4).

The Apatelodidae are a small New World group. The frenulum is well developed but the tongue is reduced or absent. The N. American fauna was described by Franclemont (1973).

LASIOCAMPIDAE

Lasiocampidae are generally large species with deep forewings and rather rounded hindwings, though several Palaeotropical genera have the forewings rather narrow, strongly produced apically, and relatively small hindwings. The hindwings tend to repeat the pattern of the forewings. The origin of R5 and M1 from a common stalk on the forewing (sometimes these veins are merely connate at the cell) would appear to be a definitive feature for the family though there are a few exceptions in the Oriental fauna. The frenulum is absent, and the hindwing humeral area is expanded and sometimes contains humeral veins. The tongue is vestigial or absent. The body is densely clad with long scales.

In the male genitalia the uncus is usually vestigial or lost, also the gnathi, leaving the tegumen as a thin arc of chitin. The valves also may be reduced, their function taken over by modifications of the vinculum and saccus.

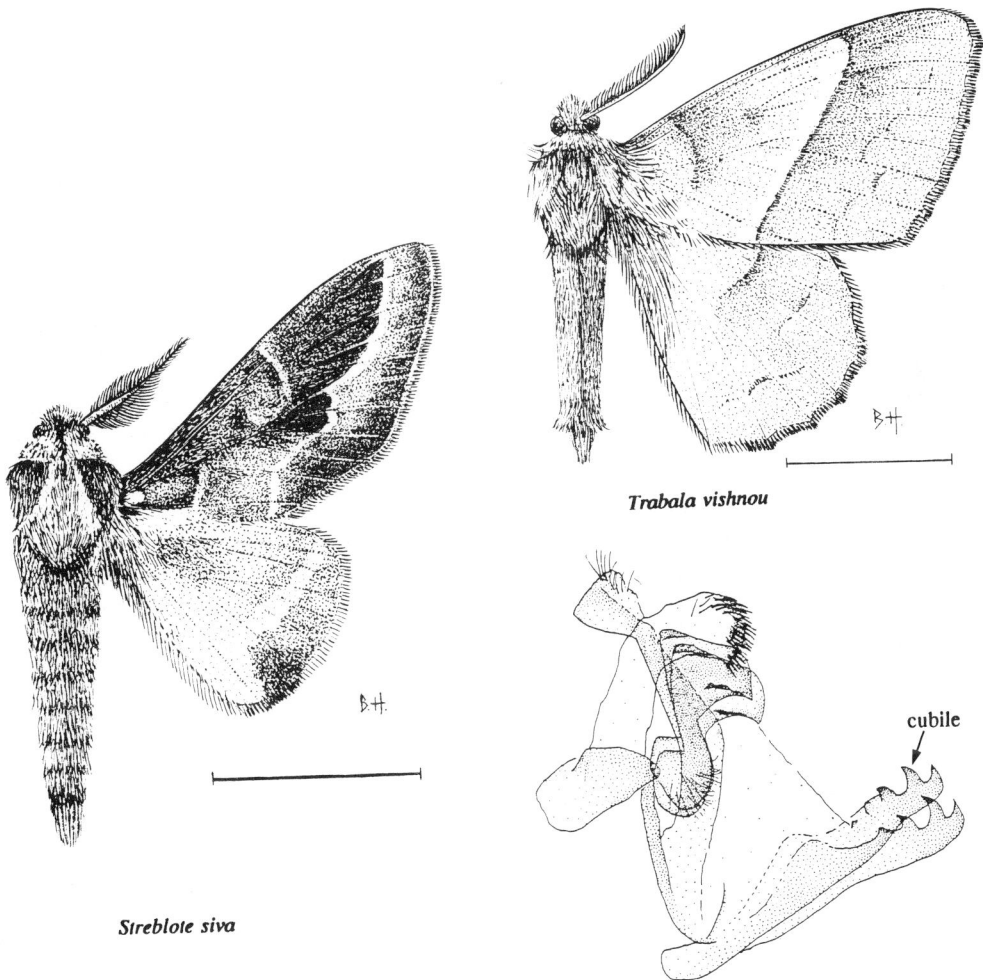

Trabala vishnou

Streblote siva

cubile

lasiocampid (*Kunugia*) male genitalia lateral view

143

Franclemont (1973) reviewed the subfamilial classification of the family, referring to Aurivillius' (1927) treatment of the African fauna. He recognised three in the New World: the Macromphaliinae, with male genitalia of more generalised type, the uncus, gnathos or socii variably present; the Gastropachinae with a much expanded and modified hindwing humeral area; the Lasiocampinae, with a specialised ventral clasping structure in the male genitalia, modified from the vinculum (the cubile of Lajonquiere (1968)). Both the second and third subfamilies have highly modified male genitalia with structures on the tegumen vestigial. The Macromphaliinae are mainly Neotropical and comprise the majority of the fauna there. The other two are worldwide. Aurivillius' Chondrosteginae, Chionopsychinae and Archaeopachinae are small African groups. His Gonometinae may prove to be best combined with the Gastropachinae.

The larvae are very hairy, cylindrical or flattened, usually the latter with lateral lappets on each segment (e.g. in Gastropachinae) and sometimes with transverse bands of setae on the mesothorax and metathorax. The diet tends to be polyphagous. The cocoon is well developed, spun of silk; this has been used for textiles.

The family is most diverse in the tropics. The larvae of many species defoliate plants of economic importance. The two species illustrated above are *Trabala vishnou* Lefebvre, a green species that is widespread in the Oriental Region with many congeners (Roepke, 1951), and *Streblote siva* Lefebvre a brown species from the Indian Subregion. In addition to the literature already mentioned, important Old World genera have been reviewed by Lajonquière (1973, 1974, 1976, 1977, 1978).

EUPTEROTIDAE

This family is virtually restricted to the Old World tropics. The species are mainly large, with deep, rather coarsely scaled wings, often with a multifasciate pattern. The male antennae are bipectinate, those of the female usually not so. In the male genitalia the uncus tends to be broadly bifid.

Little detailed work has been done on the family but some of the Oriental fauna was reviewed by Holloway (1982). The larvae may be catholic in host-plant selection but too little is known of their biology. However, the illustrated species, *Nisaga simplex* Walker, is a pest of rice in India.

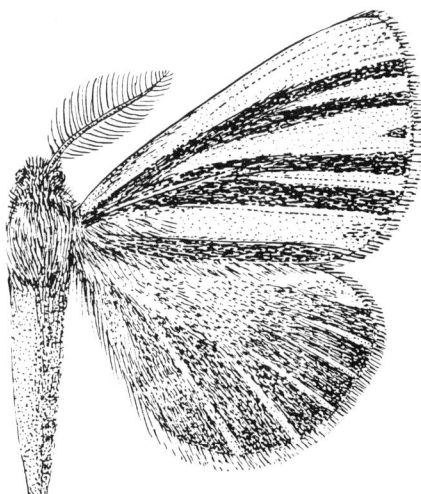

Nisaga simplex

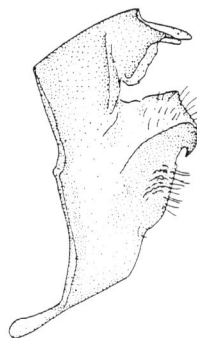

lateral view

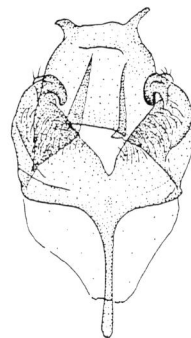

ventral view

Eupterotid male genitalia (*Eupterote*)

144

BOMBYCIDAE

This small, mainly tropical Old World family is best known through *Bombyx mori* Linnaeus (illustrated below), the most important silk producing moth, now unknown outside cultivation though it may have originated from *B. mandarina* Moore, a species that ranges from the Himalaya to Japan (Franclemont, 1973).

The most easily recognisable feature in the adults of the family is the way the dorsum of the hindwing is pleated, somewhat concave and the most strongly patterned part of the wing. This is correlated with the resting posture of the adults. The forewings are held out sideways with the dorsal margin of the hindwing folded up round the forewing dorsum. This posture is also seen in *Spiramiopsis* (N.Duke, pers. comm.) and some smerinthine Sphingidae. There is some variety in genitalic structure. The uncus can be single or bilobed, the gnathos can be two separate arms, a fused structure as in ennomine geometrids, or absent, and the saccus well developed or absent. The valves are usually reduced in the *Ocinara* Walker group and the eighth abdominal segment modified.

The larva is smooth, often with a caudal horn and usually with the thoracic segments laterally expanded. There is some evidence in the Oriental Region for two lineages restricted to the plant families Moraceae (*Bombyx* Linnaeus, *Ocinara* and allies; Bombycinae) and Theaceae and Symplocaceae (*Andraca* Walker, *Mustilia* Walker; Oberthuerinae). The *Ocinara* alliance has been reviewed by Dierl (1978, 1979).

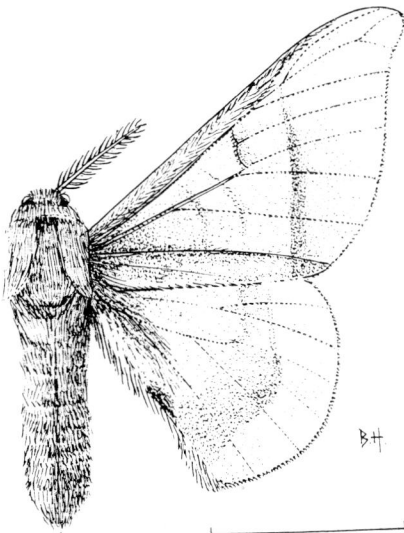

Bombyx mori

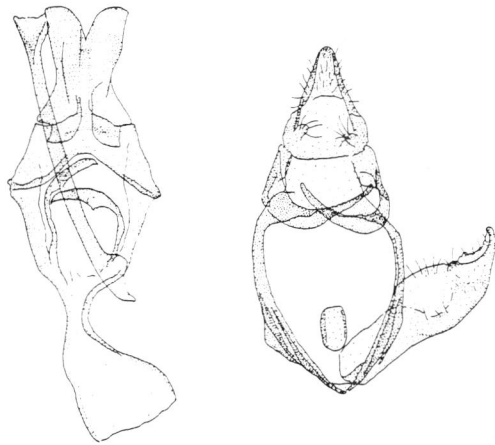

Bombycid male genitalia (*Penicillifera*, *Andraca*)

SATURNIIDAE

Species of this family include the world's largest in terms of wing area, though the bodies of the males at least are comparatively small. Generally the patterning of fore- and hindwings is approximately equivalent, though this is not the case in most of the New World family Citheroniinae. In most species the discal area of one or both wings is conspicuously marked, most characteristically with a complex ocellus, but often with only one or more transparent 'windows'. The frenulum is absent, the humeral zone of the hindwing being correspondingly expanded. The mouthparts are reduced. The male antennae are usually quadripectinate, a diagnostic feature for the family, though in *Hemileuca* Walker the antennae are only bipectinate (Ferguson, 1971).

In the male genitalia the uncus is often bifid (cf Brahmaeidae) and the gnathos reduced or lost, its position usurped by development of the transtilla (Ferguson, 1971).

The larva has setose scoli in early instars. In some groups these develop into the thoracic and caudal horns discussed in the superfamily account. In other groups the scoli remain evenly developed over the length of the larva, being reduced in later instars to spined tubercles or becoming evenly expanded and branched as in the Hemileucinae.

Many subfamilies form silken cocoons, some of which have been used for textiles (e.g. the species discussed by Arora & Gupta (1979)). Others, such as the Citheroniinae, pupate in the ground like the Brahmaeidae. The pupa differs from that in other Bombycoidea families in lacking setae; a cremaster is present in Citheroniinae and Hemileucinae but not in other subfamilies and other Bombycoidea (Ferguson, 1971).

Ferguson recognised seven subfamilies. The Saturniinae are represented in both Old and New Worlds. The Citheroniinae, Hemileucinae and Rhescyntinae (Arsenurinae) are confined to the New World, the Agliinae to the Palaearctic, the Salassinae to Asia and the Ludiinae to Africa.

The species illustrated below is the African *Gonimbrasia belina* Westwood, the Mopane Worm. The larva feeds on a wide range of trees, sometimes causing serious defoliation; however, in some parts of Africa they are used by man for food (Pinhey, 1975).

Treatments of the New World fauna may be found in Michener (1952), Ferguson (1971, 1972), Lemaire (1971-1974), and Riotte & Peigler (1980). The African fauna is best studied by reference to Pinhey (1956, 1972), Rougeot (1955, 1962) and Griveaud (1961). There is no comprehensive treatment of the Oriental fauna though regional works of relevance are by Arora & Gupta (1979), Allen (1981) and Lampe (1985).

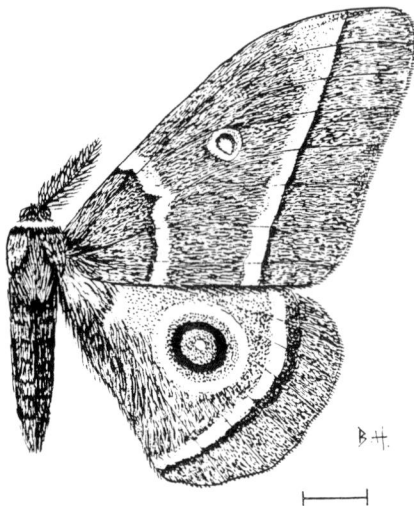

Gonimbrasia belina

SPHINGIDAE

This family is sometimes (e.g. Hodges, 1971) placed in its own superfamily, Sphingoidea, but it seems more informative, considering the characteristics of the larva mentioned in the superfamily description, to include the family in the Bombycoidea until such a time as the classification of that superfamily can be studied thoroughly.

The species are mainly robust, powerfully flighted insects, as indicated by the possession of the narrow wings usually associated with rapid, direct flight. The antennae are usually filiform to clavate (dilated), sometimes with setae, and apically somewhat hooked in many cases. Ocelli and chaetosemata are absent. The tongue shows great variation from being vestigial to having the greatest length and development seen in Lepidoptera, associated with the pollination of long, tubular flowers. The frenulum is present.

The male genitalia mostly have the uncus and gnathos robust. The valve is usually simple, with a spur-like (or other shape) harpe arising distally from the sacculus.

The caudally horned larva is characteristic, though the horn is absent in a few cases. Pairs of horns are present on the thoracic segments in genera such as *Ceratomia* Harris, a combination of characters shared with other bombycoid groups. Pupation is in the soil, more rarely in a cocoon on the ground or above ground; there is a pupal cremaster as in the Saturniidae: Citheroniinae, where pupation is also in the soil.

Hodges (1971) recognised two subfamilies, the Sphinginae and Macroglossinae, rather than the several included by Rothschild & Jordan (1903); Carcasson (1968) also recognised only two subfamilies, as has Derzhavets (1984) who presented a good synopsis of the higher classification.

The Sphinginae lack a patch of short sensory setae that is present on the interior naked area of the first segment of the labial palp in the Macroglossinae. The male eighth abdominal sternite is unmodified in the Sphinginae, but usually unevenly sclerotised or modified in the Macroglossinae. The larvae of Sphinginae tend to have oblique lateral bands on each segment; these slope down from the back to the front. In the Macroglossinae the markings are more linear, and the anterior, mainly thoracic part of the body is often swollen, bearing eye-like markings; these are the larvae that appear to mimic snakes; the caudal horn is often reduced to a knob in the final instar.

In the Oriental tropics the Sphinginae: Smerinthini tend to be more associated with natural forest whereas the Macroglossinae and Sphingidae: Sphingini are encountered more frequently in open, disturbed habitats with secondary growth, and in agricultural eco-systems. The host-plants favoured by these groupings show little overlap. The member of the Sphinginae illustrated here, *Agrius convolvuli* Linnaeus, is encountered widely in open habitats and is a pest of crops in the Convolvulaceae such as sweet potato (*Ipomoea batatas*). It is widespread in the Old World tropics and subtropics, migrating to more temperate latitudes. Similarly widespread and migratory is the macroglossine illustrated, *Hippotion celerio* Linnaeus. This is also a pest of sweet potato but also attacks other crops such as taro (*Colocasia*). The Sphingidae, particularly the Macroglossinae and Sphinginae: Sphingini, have a high proportion of geographically widespread, mobile, often migratory species.

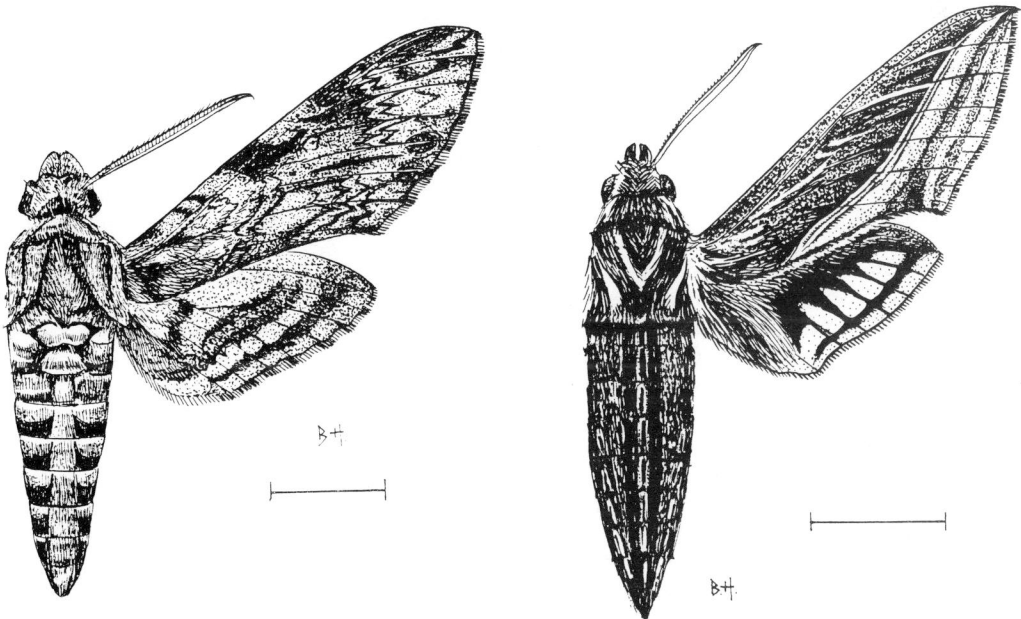

Agrius convolvuli

Hippotion celerio

147

The North American fauna has been treated by Hodges (1971), and Schreiber (1978) has published a conspectus of the South American genera. Carcasson (1968) and Pinhey (1962) have covered parts of the African fauna and Griveaud (1959) that of Madagascar. The Western Palaearctic fauna will be discussed by Pittaway (in preparation), and the fauna of the USSR is listed and discussed by Derzhavets (1984). More localised treatments of the Oriental fauna may be found in Bell & Scott (1937), Inoue (1973) and Diehl (1980). Moulds (1981, 1984) has described the larvae of economically important Australian species. The monograph of the world fauna by Rothschild & Jordan (1903) remains the most comprehensive treatment, though an illustrated guide to most species (D'Abrera & Hayes, 1986) provides a most valuable complement to it.

MIMALLONOIDEA

The sole included family, Mimallonidae was placed in its own superfamily by Franclemont (1973) because he considered it anomalous within the Bombycoidea and therefore to merit exclusion. Scott (in press) suggested a relationship with the Pyraloidea.

The family is small and restricted to the New World, being most diverse in tropical latitudes.

The forewings are often falcate, the radial vein system having veins R2 and R3 arising from a common stalk independently from that of R4 and R5. Ocelli, tongue and palps are absent or vestigial.

The larva is striking in that later instars it builds a portable case from silk, fragments of vegetation and sometimes frass.

NOCTUOIDEA

The families included in the Noctuoidea all share the feature of a thoracic tympanum. This is situated laterally on the metathorax and is opposed in most families by an ear-like structure on the first abdominal segment: the counter-tympanal hood.

NOTODONTIDAE

The Notodontidae are of moderate diversity in most regions of the world except high latitudes. They are probably a natural group though the characters used to define them in the key - the anterior position of forewing vein M2 and the ventrally oriented tympanum - are not wholly reliable. Other characters that are widely distributed through the family and may be defining apomorphies, lost in some lineages, are: deciduous stellate spicules in the aedeagus vesica; a corrugate, membranous area to the valve sacculus; a pair of often elaborate socii or gnathal processes on the uncus; a pair of crested spurs (cteniophores) on the lateral margin of the fourth sternite. All these are characters of the male abdomen.

The family shows great diversity in wing pattern. The abdomen is normally very much longer than the hindwings.

There is similar diversity in the form of the larvae, some being hairy with secondary setae (*Phalera* Hübner, *Clostera* Samouelle), some being smooth with all prolegs normal, often with various dorsal humps and prominences; a large group of smooth larvae has the anal prolegs modified into slender stematopodiform processes that are held clear of the substrate, often connected (e.g. *Stauropus* Germar, *Teleclita* Turner) with a modification of shape into a 'crumpled leaf' effect. In *Cerura* Schrank and the *Gargetta* Walker group of genera the processes have protrusible whiplash organs that are produced on alarm. In the

148

Gargetta group the larva is elongate, slender, of semi-looper type, with reduction of the anterior two pairs of prolegs.

The majority of larvae are arboreal defoliators, though an Oriental group including the genera *Ambadra* Moore and *Eupydna* Fletcher (see monograph by Kiriakoff (1962)) is restricted to monocotyledons. Further discussion of host specialisation iṅ the Oriental fauna may be found in Holloway (1983a).

The species illustrated below are *Epicerura pulverulenta* Hampson, a defoliator of Combretaceae in West Africa that has a number of very similar congeners, and *Neostauropus alternus* Walker, a polyphagous defoliator from the Oriental Region that occasionally damages trees of economic importance.

Recent works with extensive coverage of the Oriental Notodontidae are by Bender (1985), Cai (1979) and Holloway (1983a). The Old World genera and species have been illustrated and catalogued by Kiriakoff (1964, 1967, 1968) who has also published extensively on the family. Some of his larger works are listed in the bibliography. However, the reader should be aware that, whilst establishing a usefully broad foundation treatment of the family, Kiriakoff's works are prone to numerous errors such as redescription of named species as new, and indiscriminate establishment of new genera. Thus Holloway (1983a) identified twelve generic synonyms and as many specific ones created by Kiriakoff and has subsequently noted several more.

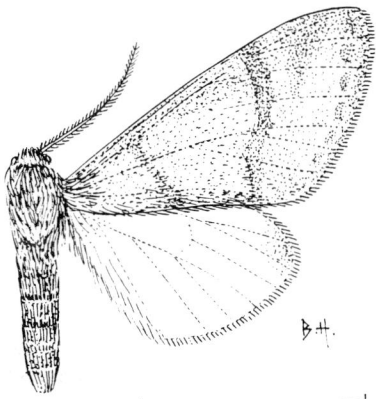

Epicerura pulverulenta

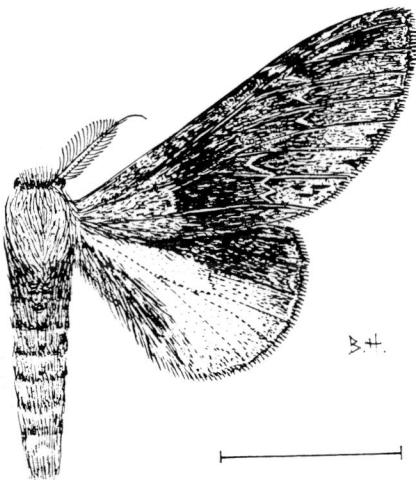

Neostauropus alternus

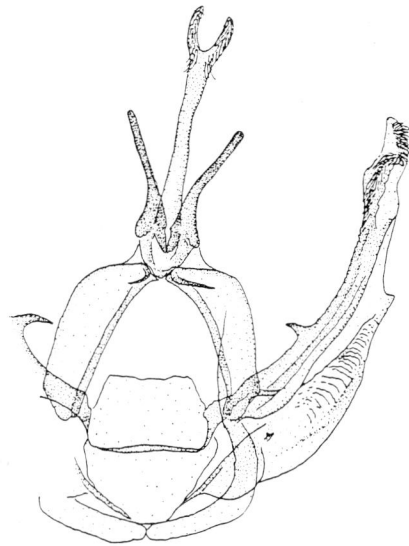

Notodontid male genitalia (*Hexafrenum*)

149

The Notodontidae, together with the Thaumetopoeidae (treated next) and Dioptidae (New World, mainly tropical), may represent the sister-group to the rest of the Noctuoidea. Minet (1982) has subordinated all these families and the Thyretidae (Afrotropical; Kiriakoff, 1960) to a single family, Notodontidae. The Thyretidae are probably Arctiidae as will be discussed later.

THAUMETOPOEIDAE

This is a very small group, geographically disjunct, with representation in Africa, the Mediterranean and S. Europe to N. India, and Australasia. The colonial, processionary, hairy caterpillars form siken nests for shelter and may cause serious defoliation of the trees they feed on. The larval hairs are irritant, and species of the Mediterranean genus *Thaumetopoea* Hubner can often attain sufficiently large populations to cause a minor health hazard in populated areas when numerous hairs are shed into the atmosphere (Carter, 1984: 241-244). The larvae provide a character distinguishing the family from the Notodontidae (see "Intorduction to larvae" section).

The species illustrated below is *Anaphe reticulata* Walker, one of an African genus; the species are often referred to as bagnest moths because of the silken shelter built by the larvae. The family has been reviewed by Kiriakoff (1970).

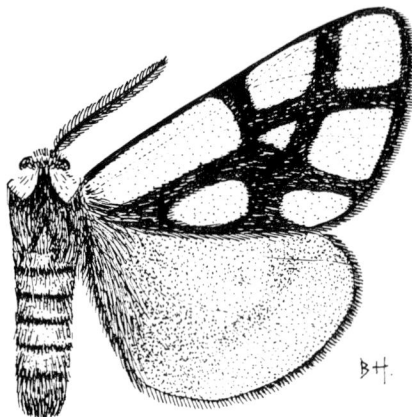

Anaphe reticulata

THYRETIDAE and DIOPTIDAE

These two families key out as Notodontidae on the basis of their ventrally directed tympanum. Minet (1982) placed the Thyretidae as subfamily of the Notodontidae but it has been shown recently that Thyretidae have a tymbal organ (A. Watson, pers. comm.) and a typically arctiid larva (J.E. Rawlins *in litt.* to A. Watson) and hence are best subordinated to that family (see p.143). The group consists of small to moderate, brightly coloured (mainly red and orange) Afrotropical species.

The Dioptidae are a moderate-sized Neotropical group, much more delicately built than the Notodontidae and with abdomens roughly equal in length to the hindwings. Both wings are deep and more equal in size than those of most Notodontidae. The patterns consist of striking designs in black, white, yellow, occasionally red or yellow, or transparent areas. The species are likely to be involved in the numerous mimicry rings shown by Neotropical Lepidoptera. Minet (1982) included the group as tribe of the Notodontinae on the basis of reduction or absence of the larval anal claspers on the grounds that this character was shared with several notodontid groups.

LYMANTRIIDAE

Species of Lymantriidae resemble Arctiidae in having a prespiracular counter-tympanal hood, but the tongue is vestigial or absent. The pair of pockets, possibly sound-producing, on the fourth abdominal sternite (see p. 50) may be definitive for the family or a large part of it. The male antennae are strongly bipectinate and may have a few long, divergent spinules at the end of each pectination (Ferguson, 1978).

The resting posture of the adult was considered by Ferguson to be diagnostic. The wings are held flattened against the substrate, the forewings meeting at their dorsal margins, tending to form a triangle. The densely scaled forelegs are extended forward in front of the head.

The male genitalia have no features that could be said to characterise the family, though members of the tribe Orgyiini usually have the valve divided into dorsal and ventral portions. The gnathos is weak or absent.

The larvae are hairy or tufted, spines or hairs arising in clumps from verrucae. The tribe Orgyiini of Ferguson (1978) has larvae with four conspicuous dorsal brushes on the first four abdominal segments, often in association with a pair of longer pencils flanking the head and a similar trio directed backwards at the caudal end. The tribe Lymantriini has shorter, more even tufts of hair and conspicuous yellow or red dorsal glands on abdominal segments 6 and 7. These glands are present, but less conspicuous, in the Orgyiini and might thus present a useful definitive character for the family.

The two tribes recognised by Ferguson may merit subfamily status when the world lymantriid fauna has been more satisfactorily studied. They would appear to embrace most South-east Asian genera.

The larvae tend to be polyphagous, primarily arboreal feeders and include many pest species. Some larvae have urticating hairs and in places are sufficiently numerous to represent a human health hazard. In the Lymantriini genera such as *Lymantria* Hübner and *Euproctis* Hübner include many pest species. *Euproctis xanthorrhoea* Kollar, illustrated below, is a pest of rice and other crops in India. The Orgyiini include genera such as *Calliteara* Butler, *Orgyia* Ochsenheimer and *Dasychira* Hübner. *Dasychira mendosa* Hübner, also illustrated here, is a defoliator of many economic plants in the Indo-Australian tropics. Females of *Orgyia* species have vestigial wings and tend to mate and oviposit on the cocoon from which they have emerged. The third species illustrated, *Psalis pennatula* Fabricius, is a pest of graminaceous crops in the Indo-Australian tropics. *P. africana* Kiriakoff is a similar species from Africa. The species also belong to the Orgyiini.

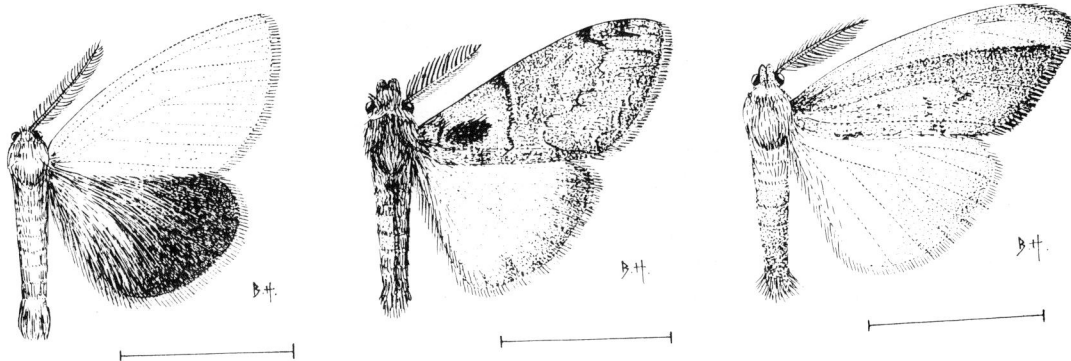

Euproctis xanthorrhoea   *Dasychira mendosa*   *Psalis pennatula*

151

Apart from Ferguson's (1978) work on the N. American fauna, regional accounts of the family are rare. For the Oriental Region one may cite Collenette's (1932, 1947, 1949a, b) reviews of the faunas of Peninsular Malaysia and the Indonesian islands of Java, Bali and Sulawesi. Dall'Asta (1981a, b, 1982) has commenced the task of revising the African species that have been assigned in the past to *Daschyira*, and Holloway (1982) made some preliminary observations on the same group in South-east Asia. Collenette (1955) published a key to African genera and Griveaud (1977) has monographed the Madagascar fauna. Inoue (1956-1957) revised the Japanese fauna. Maes (1984) has examined the generic status of Palaearctic species.

ARCTIIDAE

p.52

This family, like the Lymantriidae, has a prespiracular counter-tympanal hood and is most clearly defined by the presence of a tymbal organ, a bulbous sclerite found laterally on the metathorax (p.48). It is thought to be a sound-producing organ. Some wasp-mimicking species (e.g. in the Ctenuchinae) have the base of the abdomen and posterior of the thorax narrowed, with consequent absence or reduction in tymbal and tympanal organs. The family includes a high proportion of brightly coloured, aposematic species, some of which are day-flying.

The larvae have long and conspicuous tufts of secondary setae on verrucae giving them a hairy appearance and leading to the name of 'woolly bear'.

At present the division of the family into subfamilies is in a state of flux, and these cannot yet be defined satisfactorily in terms of uniquely derived characters (apomorphies). The male and female genitalia are unhelpful.

CTENUCHINAE

The Ctenuchinae in the New World consist of narrow-winged, often rather metallic species. In the Old World they have reduced hindwings, banded abdomens and a pattern consisting usually of white or yellowish more or less translucent patches on a black or brown ground. The antennae often have a white zone apically or subapically as is frequently found in insects mimicking Hymenoptera. The species illustrated below, *Syntomis sperbius* Hubner, is a pest of sugar and rice in the Indian Subregion.

Amata

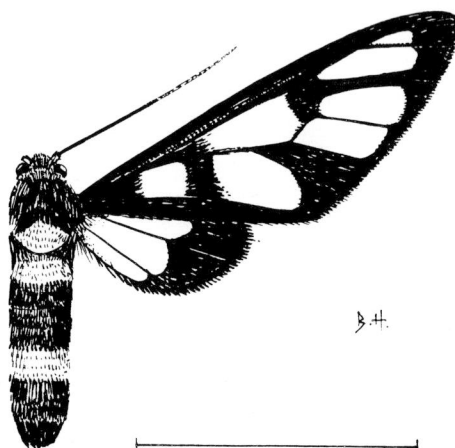

b.#.

Syntomis sperbius
Amata

Minet (1986) treated most of the Old World, wasp-mimicking taxa as a separate family, Syntomidae.

THYRETINAE

The Thyretinae (see p.143) are a small Afrotropical group mostly resembling rather robust versions of Old World Ctenuchinae. The abdomens are much longer than the rather reduced hindwings. The ground colour is usually red, orange or yellow, the equivalents of the ctenuchine transparent patches being more thickly scaled and therefore opaque. The male antennae tend to be long and bipectinate. The group was reviewed by Kiriakoff (1960).

LITHOSIINAE

The Lithosiinae generally have narrow, almost rectangular forewings, the hindwings being approximately of equal area, sometimes greater, and oval to rounded. The narrowness of the forewing has led to a variety of modifications of the venation: often one of the veins arising from the cell (R5 or M2) is lost so that only nine can be counted apart from subcostal and anal veins; R1 often converges with Sc and usually fuses with it over most of its length; M2 and M3 and CuA1 (*Eilema* Hübner) can have a common stalk. The antennae are usually filiform, serrate or finely pectinate. The abdomen is without dark maculation on each segment.

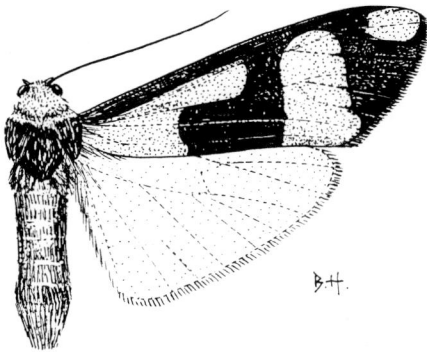

Oeonistis entella   **Lithosiine venation (*Eilema*)**

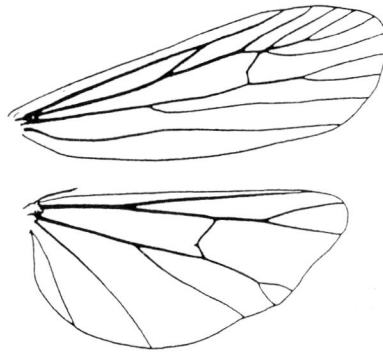

The larvae are thought to be mainly lichen, alga or moss feeders so care is needed when recording data on rearing to be sure that the larva is genuinely cutting holes in foliage rather than browsing on epiphytic growths. Only then can an angiosperm host be recorded.

The species illustrated above is *Oeonistis entella* Cramer, found widely in the Indo-Australian tropics.

The definition of lithosiine genera has in the past been mainly on venation characters. This has led to anomalies that become apparent when the genitalia are examined (Birket-Smith, 1965; Holloway, 1982). The subfamily is in need of revision.

ARCTIINAE

The Arctiinae are robust, deeper winged species, mostly with moderate sized hindwings, and generally much larger than species of the previous three subfamilies. A regular feature of Old World species at least is the presence of black maculation or transverse bands on each abdominal segment, dorsally, laterally, or both.

The larvae are often polyphagous defoliators, feeding on a wide variety of plants, particularly herbs, and often those with marked chemical defences. Such allelochemicals may be sequestered by the moth, rendering it distasteful and thus leading to the development of striking aposematic wing patterns that often involve yellows, reds and white. Other defences include secretion of strong-smelling froth from the thorax and, against bats, the production of sound from the tymbal organ.

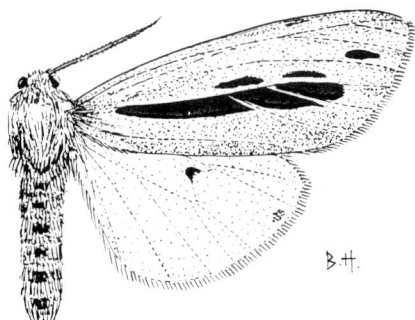

Amsacta lineola   *Pericallia* sp. of *ricini* complex

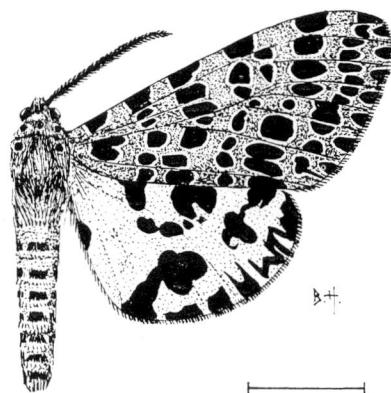

Two Arctiinae are illustrated above. *Amsacta lineola* Fabricius is an Indian member of a group of Old World tropical species, many of which are recorded as minor pests. *A. lineola* feeds on various Gramineae. The other species is a member of the *Pericallia ricini* Fabricius complex of species found also in India and currently being investigated by G. Orhant. The larvae are polyphagous. Some Neotropical groups have been reviewed by Watson (1971, 1973, 1980). The Indo-Australian *Nyctemera* Hübner were monographed by Roepke (1957b). The large Old World genus *Spilosoma* Hübner is currently being reviewed by W. Thomas.

PERICOPINAE

The Pericopinae are found only in the New World, predominantly in the tropics. They tend to be large, relatively delicately built, with wing proportions approximately as in the Arctiinae. They are also aposematically patterned, many with transparent areas and others with brightly coloured bands against a black ground. They, like the Dioptidae (p.142), are involved in mimicry rings.

AGANAINAE

The Aganainae have had a chequered history taxonomically. They include many of the old 'Hypsidae', others of which, such as *Nyctemera*, are in the Arctiinae. They have been included in the Arctiidae and, in colour pattern and abdominal maculation, resemble the Arctiinae strongly. However, there is no tymbal organ and the position of the first abdominal spiracle relative to the counter-tympanal hood is ambiguous as the latter appears to be double, with the spiracle in a central position (A. Watson, *pers. comm.*), and there is now a school of thought that would place the Aganainae with the quadrifine Noctuidae. Kitching (1984: 214-5) has reviewed the problem.

The Aganainae are predominantly an Old World tropical group. *Asota heliconia* Linnaeus, illustrated below, is one of the more abundant and widespread species of the group in the Indo-Australian tropics. Larvae of *Asota* species have been recorded mainly from *Ficus* (Moraceae).

Faunistic treatments of the family Arctiidae are uncommon, though Watson (1975) has covered many of the Neotropical species and Watson & Goodger (1986) have produced a checklist of the Arctiinae and Pericopinae. A similar work on the Afrotropical fauna is in progress. Toulgoet (1984) has published a checklist for Madagascar.

154

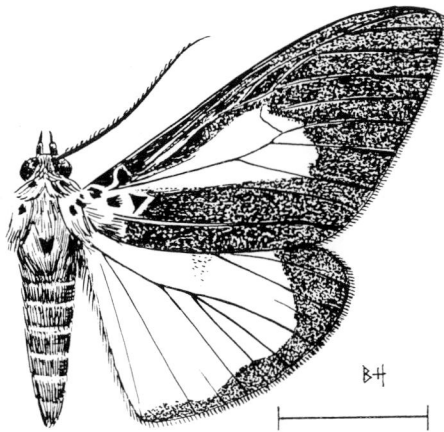

Asota heliconia

NOCTUIDAE (including Nolidae, Agaristidae, Cocytiidae and Dilobidae)

This is probably the largest macrolepidopteran family with well over 20,000 described species. It includes numerous species of economic importance such as all armyworms and cutworms, but also stem and shoot borers, root feeders, many defoliators, flower, fruit and seed feeders, detritus feeders and predators.

The family is best defined by the postspiracular position of the counter-tympanal hood (Kitching, 1984), a character that excludes the Herminiinae, a group that is now coming to be regarded as family in its own right, the Herminiidae. The presence of an orbicular stigma within the forewing cell may also define the Noctuidae and exclude the Herminiinae, though comparable features may be found in some Aganainae and Notodontidae (e.g. *Phalera* Hubner). For convenience they will be treated with the Noctuidae here.

The family Noctuidae is divided into numerous subfamilies, the definition of which leaves much to be desired, and the status of which is currently under review (Kitching, 1984). The subfamilies fall broadly into two groups: the Trifinae and Quadrifinae. The former have vein M2 of the hindwing weak or vestigial whereas in the Quadrifinae it is well developed. The trifines may well be a natural group and the quadrifines paraphyletic according to a tentative cladogram presented by Kitching. Within both groupings there are several subfamilies that can be defined clearly by derived characterisitcs, indicating they are likely to be monophyletic, and therefore natural, groupings: Noctuinae and Heliothinae amongst the trifine subfamilies; Euteliinae, Stictopterinae, Sarrothripinae + Chloephorinae, Plusiinae and Herminiinae amongst the quadrifines.

Local faunas covering Noctuidae are available for parts of Canada (Rockburne & Lafontaine, 1976), Finland (Mikkola & Jalas, 1977, 1979) and Poland (Kostrowicki, 1956, 1959 etc.). The last two contain useful illustrations of habitus and genitalia.

155

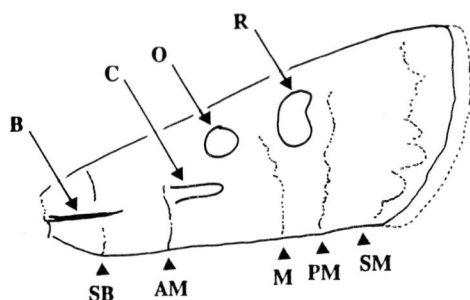

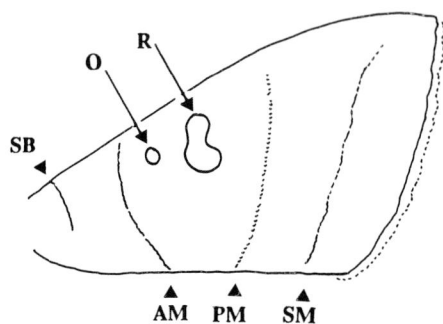

Agrotis   *Ophiusa*

B basal streak	**M** medial fascia
C claviform stigma	**PM** postmedial fascia
SB subbasal fascia	**SM** submarginal fascia
AM antemedial fascia	
O orbicular stigma	
R reniform stigma	

Nomenclature of pattern elements in Noctuidae

QUADRIFINE SUBFAMILIES

HERMINIINAE

This group is coming increasingly to be regarded as a full family, as mentioned above. The counter-tympanal hood is prespiracular and the orbicular stigma is absent. In the past it has tended to be associated with, or subordinated to, the Hypeninae. The species are usually rather lightly built brownish moths with an obtusely angled submarginal line to the hindwing. The foretibia is produced into a sheath alongside an elongate first tarsal segment, giving the species their English name of 'fan-foots'. The antennae of males of a number of species have a central swelling or other modification. The palps are usually prominent, upcurved.

There are a number of records of the larvae as detritus feeders (especially dead leaves); some may be fungus or root feeders. Characteristic genera are *Simplicia* Guenée, *Hydrillodes* Guenée, *Nodaria* Guenée, *Herminia* Latreille and *Progonia* Hampson.

HYPENINAE

This subfamily has been characterised by the course of vein M2 in the hindwing, arising from well above the lower angle of the cell and running parallel to M3: this character is also found in the Herminiinae (Kitching, 1984). Kitching mentioned that the subfamily *sensu stricto* has lashed eyes and diagnostic characters of the tympanal organ. The palps are usually long, slender, directed forward.

A number of *Hypena* Schrank species have been recorded as minor pests but the species illustrated below, *H. laceratalis* Walker (sometimes referred to erroneously as *strigatus*), has been introduced to parts of Australasia for lantana control; however, it is naturally widespread through the Indo-Australian tropics.

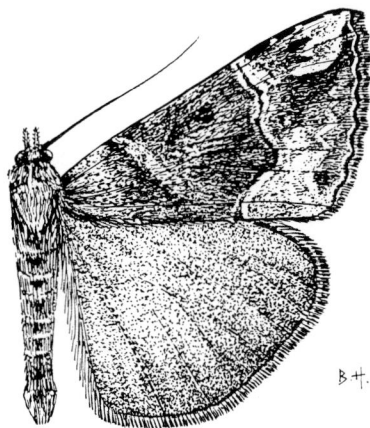

Hypena laceratalis

CATOCALINAE (including OPHIDERINAE)

This predominantly tropical subfamily as currently defined (Kitching, 1984) embraces perhaps half of the species in the Noctuidae. The old Catocalinae/Ophiderinae division was based on presence/absence of spines in the mid-tibia. Basing a taxonomic division on two states of a presence/absence character is unsatisfactory if one is aiming for a natural (cladistic) classification. In this particular instance the spining character has split between the old subfamilies a number of natural groups that can be defined by strong morphological characters. The following are good examples (Catocalinae/Ophiderinae):

Avatha Walker/*Pseudathyrma* Butler. There are several apomorphies in wing pattern and male genitalia. The complex is allied to *Serrodes* Guenée. Larval host-plants are in the Sapindaceae. The group was reviewed by Kobes (1985).

Cocytodes Guenee/*Arcte* Kollar. There are apomorphies in wing pattern, especially iridescent blue bands on the hindwing, and in possession of a striate, sclerotised organ on the dorsum of the abdomen. The larvae feed on members of the family Urticaceae.

Caranilla Moore/*Buzara* Walker. Despite striking modification of the latter genus (black with chrome yellow bands; reduced hindwings), consisting of two species from New Guinea, the male genitalia show similar complex asymmetry to those in *Caranilla*, a genus consisting mainly of dull brown '*Parallelia*'-like species.

No satisfactory definitive characters for the group as a whole have yet been identified. The moths are mainly robust, with deep forewings. The palps are usually long, upcurved, the third segment long, slender. The larvae are mostly of the 'semi-looper' type, with reduction or loss of the first pair of abdominal prolegs and reduction of the second, though this characteristic is found in other quadrifine subfamilies (e.g. Plusiinae and Acontiinae) and is also seen in the *Gargetta* group of the Notodontidae (p. 140).

However, various groupings of genera can be recognised withtin the huge assemblage, some of which include important pest species. Two illustrated below, *Ophiusa coronata* Fabricius and *Mocis frugalis* Fabricius, belong to a large group of 'catocalines' with coremata from the valves of the male genitalia, cruciform sclerotisation of the juxta, and a sclerotised gnathos opposing a massive uncus. The former species, as an adult, is a fruit-piercer, attacking citrus fruit in the Old World. The latter is a grass feeder and a minor pest of pastures and graminaceous crops in the Old World. The group contains a large assemblage of euphorbiaceous feeders, and others attacking legumes, myrtaceous shrubs and

157

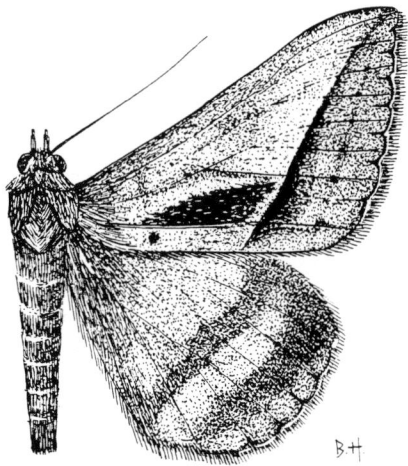

Mocis frugalis

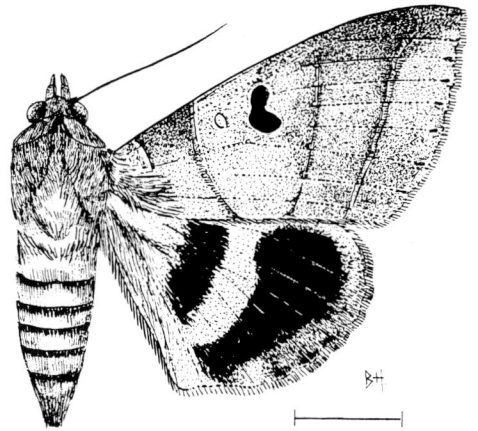

Ophiusa coronata

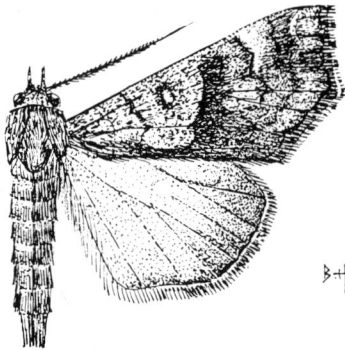

Anomis flava

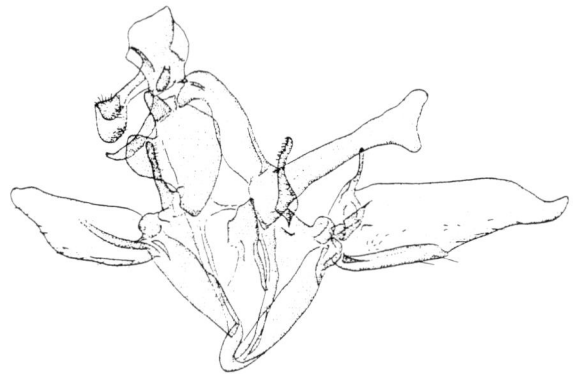

Ophiusa male genitalia

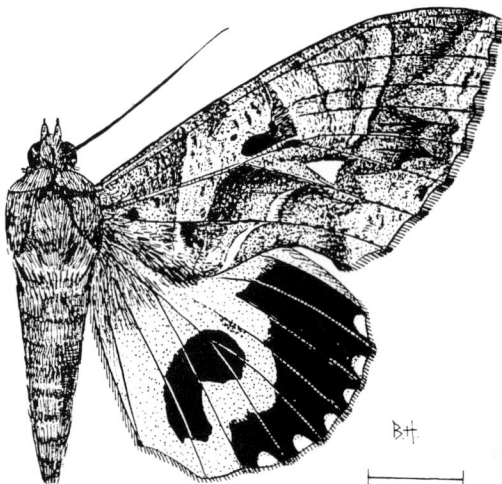

Othreis fullonia

Anticarsia gemmatalis

trees, as well as further graminaceous feeders. Other included genera are *Anua* Walker, *Phyllodes* Boisduval, *Trigonodes* Guenée, *Achaea* Hübner, *Parallelia* Hübner sensu lato, and *Grammodes* Guenée. Berio (1965) reviewed many of the African species, and discussions of Oriental groups may be found in Holloway (1979, 1982) and Kobes (1985).

Kobes also illustrated many species in the *Sypna* Guenée complex of genera, a group revised by Berio & Fletcher (1958).

Another important group contains further genera of fruit-piercing species such as *Othreis fullonia* Clerck, illustrated above, and congeners, and the genera *Calyptra* Ochsenheimer, *Oraesia* Guenée, *Plusiodonta* Guenée and *Gonodonta* Hübner. There is some concentration of larval host-plant records in the family Menispermaceae, and the proboscis of the adults has been modified into a system of ratcheted barbs for fruit piercing and, in the case of *Calyptra eustrigata* Hampson and three congeners, blood sucking. Many of the species also feed from lacrymal secretions of mammals in South-east Asia and may be of veterinary importance. The group as a whole is being studied by Banziger (1982, 1983), and *Gonodonta* has been reviewed by Todd (1959).

Anomis flava Fabricius, also illustrated above, belongs to another small group of genera; the forewings tend to have an angled margin and the male genitalia have coremata on the valve as in the *Ophiusa* group. The larvae are found mostly on the family Malvaceae so the genus contains cotton and jute pests (e.g. *Anomis sabulifera* Guenée) and defoliators of various ornamental and economic *Hibiscus* species. The *flava* group was revised by Tams (1935).

The last species illustrated, *Anticarsia gemmatalis* Hübner, is an important pest of soybean in the New World, though its close relative, *A. irrorata* Fabricius, has less impact in the Old World.

Another work on a major group in the subfamily deals with the tribe Armadini in Old World eremic regions (Wiltshire, 1979).

PLUSIINAE

The Plusiinae show relatively even but low diversity over tropical to temperate climatic belts and are usually associated with open habitats. Many of the species are migratory. The subfamily includes a high proportion of pest species. The majority of species have the following characters: slight crests of scales dorsally on the thorax and anterior of the abdomen; a pale or metallic streak or stigma at the centre of the forewing, or other iridescent marking; a falcate tornus to the forewing, an effect created mainly by an elongation of the fringe scales; a secondary counter-tympanal hood just ventral to the main, rather elongate one; rather slender male genitalia with a clavus on the sacculus; an aedeagus with the tube more heavily sclerotised in a ventral band, often basally bulbous and with an elongate, slender vesica. More definitive but less easily observed characters are listed by Kitching (1987: 139).

The larvae are mostly of the 'semi-looper' type with the anterior two pairs of prolegs reduced or absent. Pupation is in a loose cocoon. The pupa has the cremastral spines set at the apex of a rather elongate, sculptured apex to the abdomen.

The species illustrated below, *Chrysodeixis eriosoma* Doubleday, is the widespread Indo-Australian relative of the Afrotropical and European *C. chalcites* Esper. The relationship and status of these two taxa still requires clarification, but both have polyphagous larvae and are pests of a wide range of crops. The genus was reviewed by Dufay (1970); other members of it are also pests. Another polyphagous Old World pest is *Thysanoplusia orichalcea* Fabricius, and more specialised feeders of economic importance in the Old World are *Anadevidia peponis* Fabricius (Cucurbitaceae) and *Ctenoplusia albostriata* Bremer & Grey (Compositae).

Useful references on the taxonomy of the subfamily are, for the New World, Eichlin & Cunningham (1978) and, for the Old World, Dufay (1958, 1970, 1974) and Holloway (1985). The classification of the subfamily has been reviewed by Kitching (1987), who provides a comprehensive list of references to the subfamily.

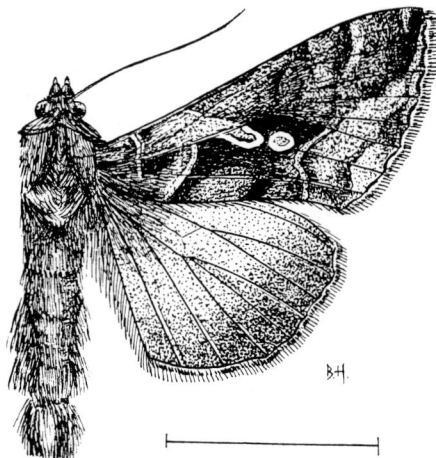

Chrysodeixis eriosoma

STICTOPTERINAE

This tropical subfamily is most clearly defined by the pupal cremaster which is a stout 'Y' shape, the arms of the 'Y' being conical. Other characteristics that hold in a majority of cases are: reticulate, often rather metallic forewing patterning; a translucent basal half to the hindwing; a slight secondary counter-tympanal hood, though not as well developed as in the Plusiinae and also seen in the Euteliinae; ovipositor lobes in the female genitalia with irregular, rather 'frayed' margins, semicircular and arranged together in a disc.

The larvae have a complete set of prolegs, are smooth, tend to feed on young foliage and pupate in the soil in a cocoon.

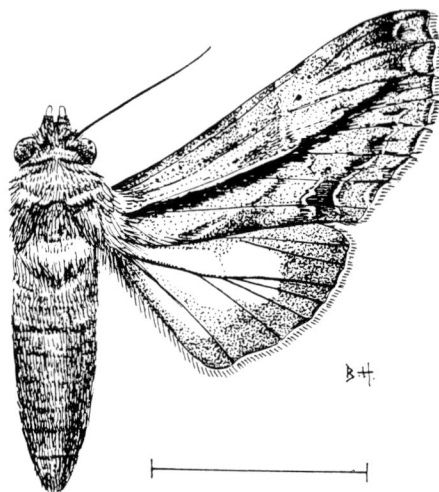

Aegilia sundascribens

160

The greatest diversity in the Oriental tropics, with extension into those of Australasia. Representation is very much lower in Africa and least in the New World. Two generic lineages can be recognised (Holloway, 1985), one of which appears to be restricted in larval host-plant to the family Guttiferae and includes minor pests of species of *Garcinia* (e.g. mangosteen) and *Calophyllum* that are utilised by man. The species illustrated above, *Aegilia sundascribens* Holloway, belongs to this lineage. The second lineage has a wider range of host-plants though is of interest in that it includes some of the few known defoliating specialists on Dipterocarpaceae, an important family of timber trees in the Oriental tropics. An extensive review of the subfamily was provided by Holloway (1985).

EUTELIINAE

The Euteliinae are perhaps most closely related to the Stictopterinae, sharing an unusual globular sclerotisation in the ductus ejaculatoris of the male aedeagus vesica (Holloway, 1985) and other features (Kitching, 1987). The subfamily is of approximately equal diversity to the Stictopterinae, again mainly tropical, but this diversity is more evenly distributed over the main tropical regions and is more characteristic of savanna and semi-arid habitats than the Stictopterinae which appear to be more or less restricted to tropical forests.

The subfamily is most clearly defined by the presence on the basal sternite of the abdomen of lenticular flanges running posteriorly from the apodemes and converging somewhat. Other noctuids merely have a groove or a slight depression in this postion. The euteliine structure is probably a modification to provide strengthening and additional muscle attachment for maintenance of the characteristic resting posture: the wings are scrolled and held against the substrate in a cruciform fashion, with the abdomen curving up above the rest of the body. Other distinctive features shown by many genera are: partially bipectinate male antennae; a laterally tufted apex to the abdomen, giving it a squarish appearance; and, more universally, lack of a pupal cremaster.

The larva is smooth, with four pairs of prolegs present. Pupation is in a cocoon in the soil or in rotten wood.

The larvae of many genera are recorded predominantly from Anacardiaceae, including those of *Penicillaria* Guenée and *Chlumetia* Walker, genera that include a number of mango pests. *Penicillaria jocosatrix* Guenée, illustrated below, is a mango pest throughout the Indo-Australian tropics. A section of the genus *Targalla* Walker appears to be restricted to Myrtaceae. The Oriental members of the subfamily were reviewed by Holloway (1985).

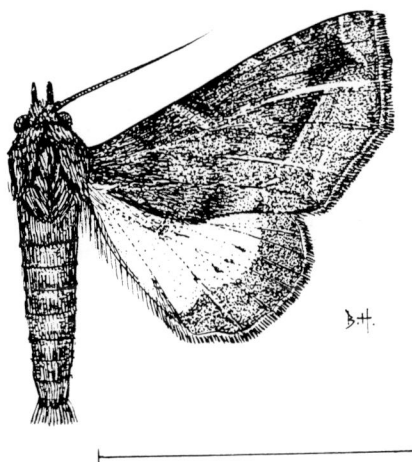

Penicillaria jocosatrix

161

SARROTHRIPINAE and CHLOEPHORINAE

These subfamilies together are characterised by a bar-shaped male retinaculum and a boat-shaped cocoon (Kitching, 1984). The pupa lacks a cremaster but often has a spined or corrugate transverse ridge at the anterior edge of abdominal segment 10 ventrally.

Distinction of the two subfamilies is at present based on presence (Sarrothripinae) versus absence (Chloephorinae) of raised scales on the forewing, an unsatisfactory state of affairs. Kitching reviewed the possibility that other subdivisions of the group may be likely, a task attempted by Mell (1943) for the Chinese fauna but which still requires a more broadly based study.

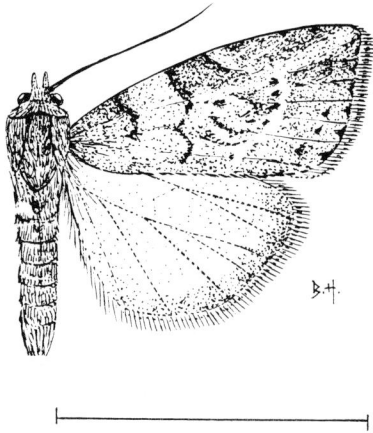

Characoma nilotica

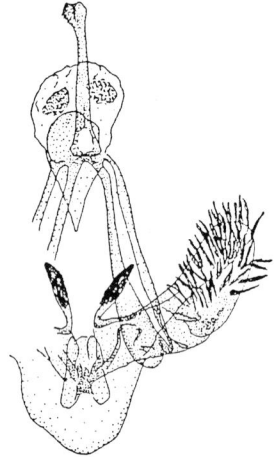

Characoma male genitalia

Characoma nilotica Rogenhofer (illustrated above) is a very widespread tropical species (or species complex), belonging to a genus of numerous species, the larvae of many of which have been recorded as boring in the pods of Leguminosae. The genus shares with others such as *Nanaguna* Walker the development of a darkly scaled process from the transtilla of the male genitalia.

A full survey of both subfamilies would probably reveal more groupings of interest. For example, both *Labanda* Walker and *Blenina* Walker species feed apparently exclusively on *Diospyros* in the Ebenaceae, but it is not yet known whether they are taxonomically related.

Selepa celtis Moore belongs to a genus where the valve has a distal harpe; *Selepa* probably belongs to another grouping. The species illustrated below is widespread in the Indo-Australian tropics and is a pest of a variety of crops in India.

Amongst the genera currently in the Chloephorinae there are further indications of host specialisation. The *Carea* Walker group of genera (e.g. also *Careades* Bethune-Baker and *Calymera* Moore) shows some association with the Myrtaceae. The genera *Aiteta* Walker and *Westermannia* Hübner are associated with *Terminalia* in the Combretaceae. The genera *Xanthodes* Guenee and *Earias* Hübner are strongly associated with the Malvaceae. The latter genus contains a number of serious cotton pests such as *Earias insulana* Boisduval, illustrated below.

162

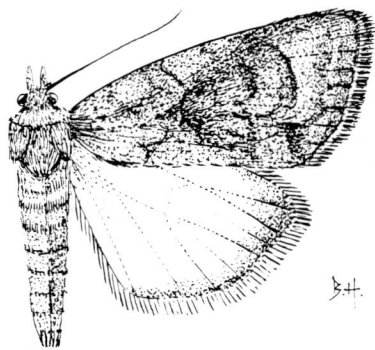

Selepa celtis

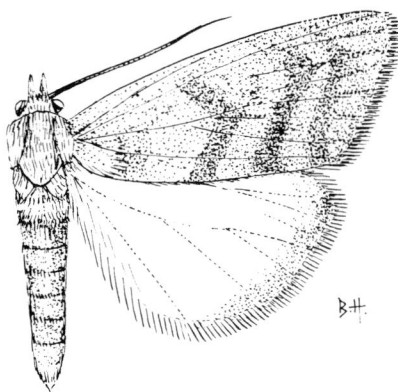

Earias insulana

Larvae of the *Carea* group of genera have a characteristically swollen thoracic region. *Xanthodes* and allies have 'semi-looper' larvae. *Earias* larvae have secondary setae that have led to them being classified with the Arctiidae in the past.

The two subfamilies as a whole are much more diverse in the Old World tropics than in the New, the Chloephorinae being absent from the latter. Many of the species in both groups are brightly coloured.

NOLINAE

The Nolinae have been accorded family status in the past but there are several indications of close relationship with the Sarrothripinae/Chloephorinae complex, such as a boat-shaped cocoon, raised scales on the forewing and tufted setae on the larvae (Kitching, 1984). The moths are generally small and grey in colour.

The larvae are often flower-feeders. The species illustrated below, *Celama analis* Wileman & West, has been recorded from the flowers of *Acacia*, *Nephelium* and *Lantana* amongst others, and is widely distributed in the Indo-Australian tropics.

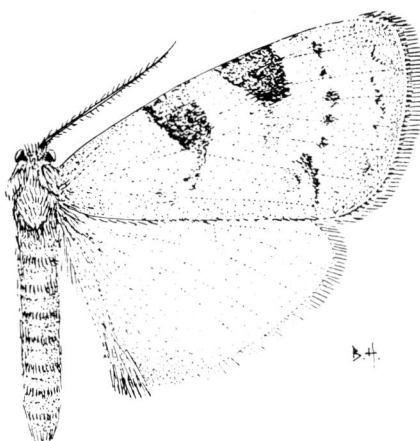

Celama analis

163

ACONTIINAE

This is a large group of small to very small, but often robust quadrifines. The grouping is taxonomically ill-defined (Kitching, 1984). Usually the larvae are of the 'semi-looper' type. Greatest diversity is in the tropics, particularly in open habitats and savanna. Important genera are *Eublemma* Hübner, *Amyna* Guenee, *Maliattha* Walker and *Acontia* Ochsenheimer.

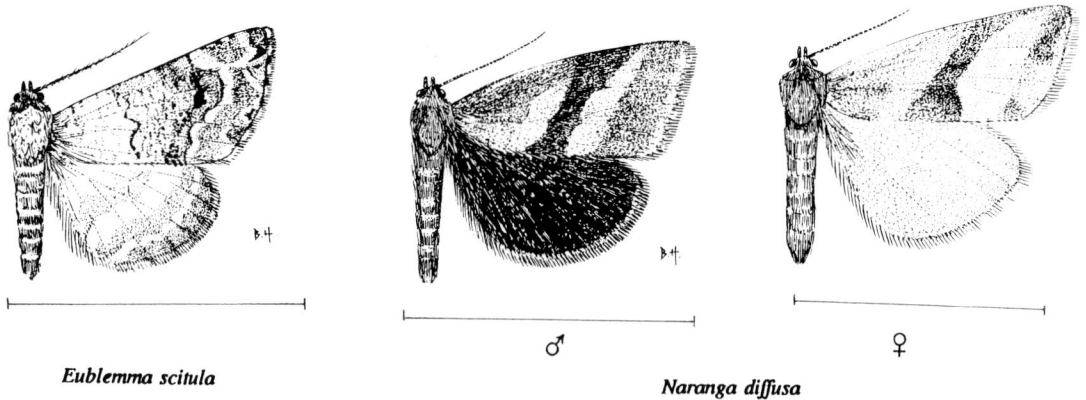

Eublemma scitula

♂   ♀

Naranga diffusa

Two species are illustrated above. *Eublemma scitula* Rambur is a Palaeotropical member of a taxonomically and biologically diverse genus. *E. scitula* is predatory on Coccidae and thus beneficial, but flower-feeding congeners are regarded as pests. *Eublemma radda* Swinhoe from Thailand has been reared from *Nepenthes* pitchers where it may feed on the insect prey of the plant. In Australasia larvae of members of the genus *Enispa* Walker live in spiders' webs where they feed on captured insects.

The second species is the sexually dimorphic *Naranga diffusa* Walker, a pest of rice in the Oriental tropics.

Two useful papers covering aspects of the subfamily are by Ueda (1984, 1987).

COCYTIINAE, RIVULINAE, HYPENODINAE

These three subfamilies are very small, based on a few quadrifine genera. The first consists of two genera of bee-mimicking species with clubbed antennae from New Guinea. The second is based on a moderately diverse genus of small species with unusually long mid-legs; it is not clearly defined (Kitching, 1984). The last is also weakly characterised, based on a small Holarctic genus.

PANTHEINAE

The Pantheinae are a small group of quadrifines with hairy eyes, some sections of which have characteristics in common with the Acronictinae (Kitching, 1984). They may well be polyphyletic and thus an unnatural group. South-east Asian taxa have recently been reviewed by Holloway (1985) and Kobes (1985).

TRIFINE SUBFAMILIES

NOCTUINAE

The subfamily is defined by the habit of holding the wings all in the same horizontal plane when at rest rather than at an angle over the back. Another character that has been used to distinguish the subfamily is the presence of rows of spines on the hind-tibia. Included are many species of economic importance, especially cutworms in the genera *Agrotis* Ochsenheimer and *Euxoa* Hübner.

The species illustrated below is *Agrotis ipsilon* Hufnagel, the greasy cutworm, a species that ranges throughout the tropics and subtropics to warmer temperate latitudes and attacks a wide range of crops (Carter, 1984). Works with general treatments of noctuine genera are Boursin (1954-1963), Common (1958), and a review of most N. American *Euxoa* by Hardwick (1970a), supplemented by a series of subsequent papers by J.D. Lafontaine in the *Canadian Entomologist*. The genus *Xestia* Hübner is under review by Lafontaine *et.al.* (e.g. 1983).

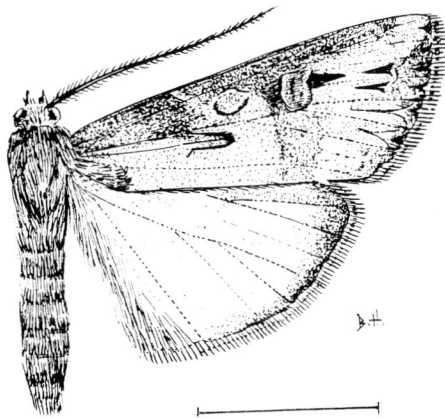

Agrotis ipsilon

HELIOTHINAE

Heliothines are defined by a multiple claw to the fore-tibia and, within the Trifinae, by biordinal crochets on the larval prolegs. The valve of the male genitalia is elongate, slender, coronate (a row of setae on the interior of the distal margin). The majority of species have larvae that feed on flowers or seeds and many are pests, e.g. in the genera *Helicoverpa* Hardwick, *Heliothis* Hübner and the *Heliocheilus* Grote-*Raghuva* Moore complex. The genus *Helicoverpa* embraces the 'corn earworm' complex revised by Hardwick (1965). The species illustrated below are *Heliothis virescens* Fabricius, a tobacco pest and member of a small complex of New World flower and fruit feeding species, and *Raghuva* albipunctella de Joannis, a serious pest of pearl millet in Sahelian Africa. North American species have been reviewed by Hardwick (1970b), and the whole subfamily is currently under study by M.J. Matthews.

165

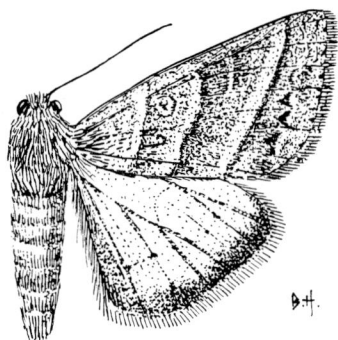

Heliothis virescens

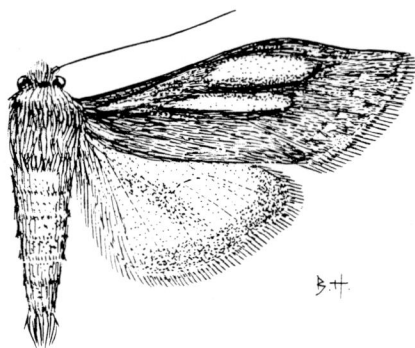

Raghuva albipunctella

Heliocheilus

HADENINAE

This subfamily embraces all trifines with hairy eyes; included species may comprise three or more natural groups (Kitching, 1984). The male genitalia usually have a marked corona to the valve, with the cucullus constricted subapically, and a double or triple harpe, though this configuration is also seen in the noctuine genus *Diarsia* Hübner.

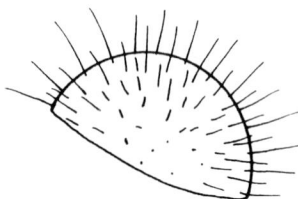

eyes hairy

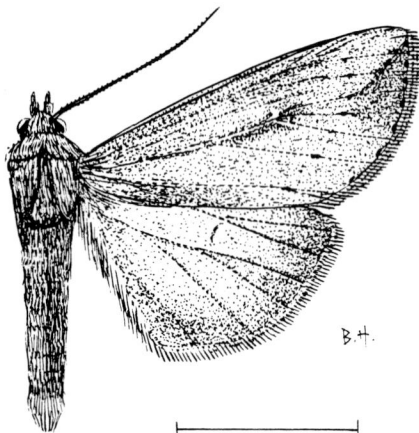

Mythimna unipuncta

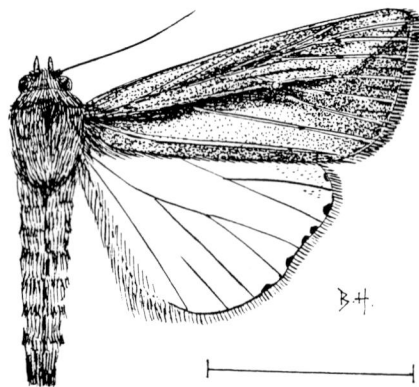

Mythimna loreyi

166

The most important within the subfamily is the *Mythimna* complex (embracing names such as *Acantholeucania* Rungs, *Aletia* Hübner, *Borolia* Moore, *Leucania* Ochsenheimer and *Pseudaletia* Franclemont). All the numerous species are grass feeders and many are pests of graminaceous crops or pasture, often being involved in 'armyworm' outbreaks. The species illustrated above are *Mythimna (Pseudaletia) unipuncta* Haworth and *Mythimna (Acantholeucania) loreyi* Duponchel, members of complexes of economic importance reviewed by Franclemont (1951) and Rungs (1953) respectively. Other important references are Berio (1985), treating the Italian fauna, and Calora (1966), describing the Philippines species.

CUCULLIINAE (including DILOBIDAE, after Minet (1982))

The Cuculliinae are probably also an unnatural group (Kitching, 1984). The species are predominantly temperate in distribution and include few serious pests. A large temperate section of the subfamily has an adult flight period in spring or autumn. The subfamily has been characterised by the possession of 'lashed' eyes. A broad selection of the Palaearctic fauna is covered by Berio's (1985) treatment of the Cuculliinae of Italy.

ACRONICTINAE

This subfamily name has been applied in a wide sense to embrace the next subfamily as well. In its more narrow application it includes a small number of genera such as *Acronicta* Ochsenheimer and *Craniophora* Snellen. The adults have grey, cryptic forewings. The larvae are usually hairy, having secondary setae on the larval trunk, sometimes dramatically so. They are mainly arboreal feeders and not of great economic significance.

AMPHIPYRINAE

This subfamily is most unsatisfactorily defined by loss and absence characters (Kitching, 1984), and may be considered to include the (large) residue of trifines that cannot be assigned clearly to other subfamilies.

However, it includes a number of genera of great economic importance such as: *Spodoptera* Guenée (Brown & Dewhurst, 1975; Todd & Poole, 1980), including armyworms and other pests like the species illustrated below, *S. littoralis* Boisduval; *Sesamia* Guenée, *Busseola* Thurau (Tams & Bowden. 1953; Bowden, 1956; Nye, 1960) and a large group of other genera with 'Mythimna-like' adults, and larvae that bore in the stems of Gramineae including cereal crops, e.g. *Sesamia penniseti* Tams & Bowden and *Busseola fusca* Fuller illustrated here; *Prospalta* Walker, feeders on Compositae. Other genera of importance in the Holarctic Region will be found described in Carter (1984: 258-274).

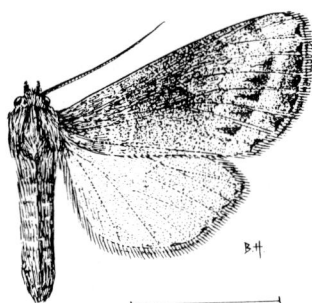

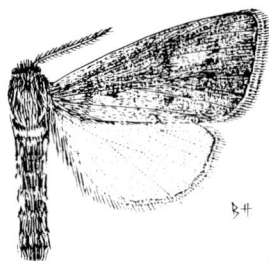

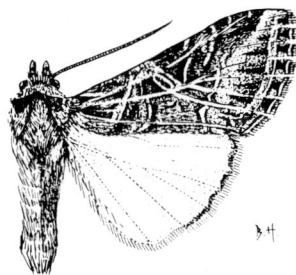

Busseola fusca   *Sesamia penniseti*   *Spodoptera littoralis*

The Amphipyrinae have greater diversity in the tropics than other trifine sub-families. The Trifinae generally are most diverse in temperate regions of the world and at high altitudes in the tropics. Those in the tropical lowlands include a high proportion of species of open habitats. The last subfamily is exceptional to this in that it is most diverse in the tropics and in forested habitats.

AGARISTINAE

The Agaristinae have been accorded full family status in the past. They are brightly coloured, though often predominantly black, and mostly day-flying. They may be most closely related to the previous two subfamilies and are probably monophyletic (Kitching, 1984). The male genitalia are uniform through the group, having a coronate, unconstricted valve with a simple harpe. The counter-tympanal hood is enlarged.

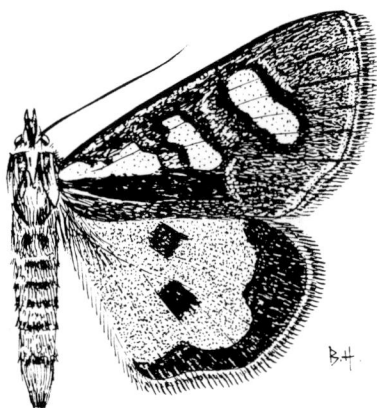

Aegocera bimacula

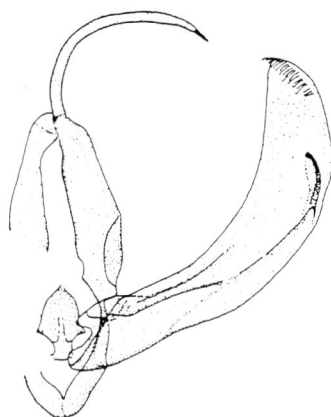

Agaristine male genitalia (*Sarbanissa*)

The group was catalogued with illustrations by Kiriakoff (1977). The species illustrated above is *Aegocera bimacula* Walker, a pest from the Indian subregion.

GEOMETROIDEA

This superfamily is here discussed in the old, broad sense, defined by the presence of abdominal tympanal organs. Nevertheless, a number of small families such as the Sematuridae and Apoprogonidae, lacking such organs, have traditionally been included and the Pyraloidea, having abdominal tympana, excluded.

This arrangement has been criticised by Minet (1983) who, from a detailed study of the morphology of such tympanal organs in traditional Geometroidea and Pyraloidea, has suggested that the abdominal tympanum has evolved several times in independent lineages. He therefore erected several more superfamilies and his reclassification of the geometroids

168

is as follows:

GEOMETROIDEA: Geometridae: Subfamilies as before

DREPANOIDEA: Drepanidae: Drepaninae
                               Oretinae
                               Thyatirinae
                               Cyclidiinae

URANIOIDEA: Uraniidae: Uraniinae
                               Microniinae
                               Epipleminae

AXIOIDEA: Axiidae

Unplaced: Pterothysanidae
               Hibrildidae
               Apoprogonidae
               Sematuridae
               Epicopeiidae

Minet suggested that the tympanum-like structures on the seventh abdominal segment of the small Meditarranean family Axiidae lack sensory organs and may therefore have no auditory function. The early stages he suggested were similar to those of Bombycoidea in several respects.

The Sematuridae are a small Neotropical group with high wing area to body size and strongly tailed hindwings. The apodemes of the basal abdominal segment are bifid and associated with a structure that could be a reduced tympanal organ (*Coronidia* Westwood); the male genitalia are reminiscent of those of some ennomine geometrids.

The Apoprogonidae contain a single South African species with clubbed antennae but a wing facies reminiscent of catocaline Noctuidae such as *Pericyma*, Herrich-Schäffer and *Zale* Hübner. The structure of the tegumen in the male genitalia is reminiscent of that in ennomine geometrids; the valves are simple with a saccular spur; there are lateral eversible scent pencils at the base of the abdomen.

The Epicopeiidae are Oriental, the genus *Epicopeia* Westwood consisting of large species that mimic troidine papilionids. Other genera mimic Pieridae and Chalcosiinae.

DREPANIDAE, THYATIRIDAE and CYCLIDIIDAE

These families were grouped together as one family by Minet (1983) on the basis of shared tympanal features. The hindwing vein Sc approximates to or joins Rs beyond the end of the cell in all three families. The forewing areole is elongate, markedly so and very narrow in the drepanid subfamily Oretinae.

The Drepanidae are often referred to as 'hook-tips' because of the falcate tips to the forewings, though a large number of Oriental species lack this character: a group of small, ligneous-patterned species with an obtuse angle to the forewing margin. Most drepanids are attractively and relatively brightly coloured. They rest with wings held flat against the substrate.

The subfamily Drepaninae has strikingly modified male genitalia, and usually the terminal abdominal sternites are also modified. The uncus is often bifid and accompanied, or replaced, by a pair of socii from the tegumen. The subfamily Oretinae contains generally more robust species than the Drepaninae. The male genitalia have the valves much simpler than in most Drepaninae, but usually with a strong saccular spur. The uncus is usually broad, setose, constricted subapically and often weakly bilobed; it is opposed by a single gnathal process similar to that seen in Limacodidae.

169

The drepanid larva appears to have two definitive features, though the early stages of many genera are unknown. Anal prolegs are absent, the suranal plate terminating in an acute process in both subfamilies. The mesoseries of crochets on the other prolegs is opposed by a rudimentary lateroseries (Fracker, 1916). The larva has no secondary setae. The family would appear to consist solely of arboreal defoliators.

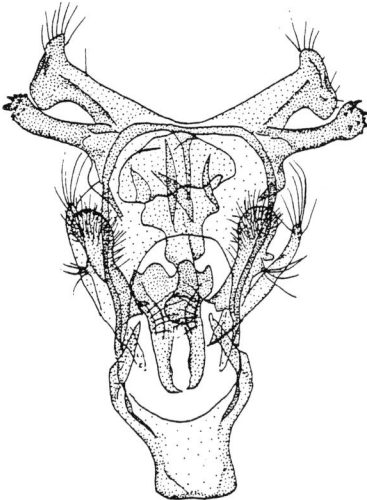

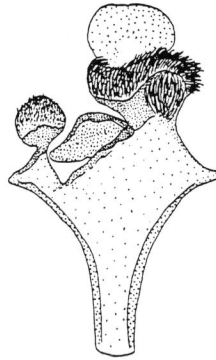

8th sternite

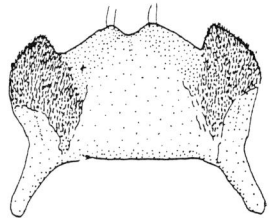

8th tergite

Drepanid male genitalia (*Paralbara*)

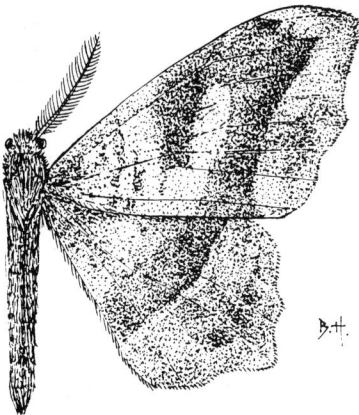

Epicampoptera tamsi

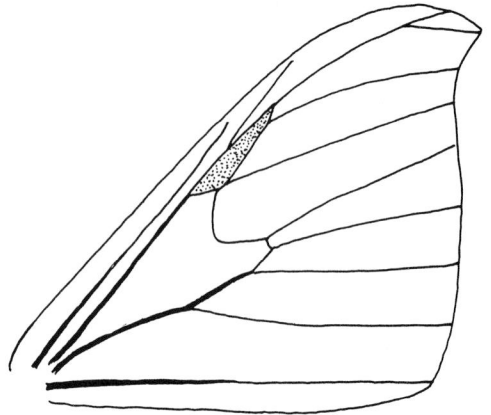

Drepanid (*Tridrepana*) forewing venation with areole stippled

The Drepaninae are most diverse in the Holarctic and Oriental Regions, and weakly represented in Africa, whereas the Oretinae are found throughout the Old World tropics and subtropics. The species illustrated above *Epicampoptera tamsi*, Watson, is an African oretine that is a pest of cocoa. The Drepanidae have been extensively revised by Watson (1957, 1961, 1965, 1967, 1968) and Wilkinson (1967).

The Thyatiridae are predominantly Holarctic, though are weakly represented in Africa and the montane forests of the Indo-Australian tropics. They are robust, noctuid-like moths, usually with striking, often colourful forewing facies. The male genitalia (illustrated below) are uniform through the family and characteristic with unornamented, oval valves and an uncus (or uncus and socii) consisting of three approximately equal, curved processes.

170

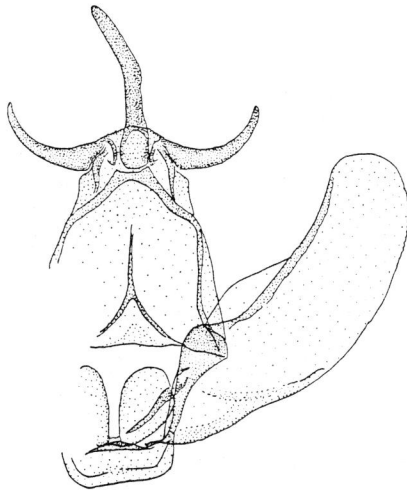

Thyatirid male genitalia (*Tethea*)

The thyatirid larva has no diagnostic features. The prolegs are all present and there are no secondary setae. The family defoliates trees and shrubs, the larvae loosely rolling or folding the leaves (Fracker, 1916).

The Thyatiridae were monographed by Werny (1966).

The Cyclidiidae are a small Oriental group of drepanid-like species. *Cyclidia* Guenée itself has simple valves and trifid uncus in the male genitalia as in the Thyatiridae. The only other genus, *Mimozethes* Warren, has a single uncus opposed by an ennomine-like gnathus, and a prominent process occurs on the valve sacculus. The two genera are probably unrelated. *Cyclidia* may be sister-group to the Thyatiridae, themselves defined by modification to the wing shape in conjunction with robustness. Chu & Wang (1985) have demonstrated that the Cyclidiidae have a sister-group relationship to Oretinae + Drepaninae but did not consider the Thyatiridae.

A further, small monophyletic group (*Hypsidia* Rothschild) referable to Minet's Drepanoidea has recently been recognised in Australia (Scoble & Edwards, in press). It does not fall within the concept of any of the families discussed above and hence would have also to be accorded family status unless all Drepanoidea were placed in the Drepanidae *sensu lato* of Minet.

URANIIDAE

This family and the Epiplemidae share a distinctive feature of the male tympanum: it is situated laterally between tergites 2 and 3 just posterior to the second abdominal spiracle. In the female the pair of tympana are at the base of the first abdominal sternite as in other geometroids. The two families also share a feature of forewing venation, with R5 having a common stalk with M1; the rest of the radial veins arise independently from the cell or Rs with at least R3 and R4 sharing a common stalk.

The morphology of both families is in need of review, and the results may well support Minet's (1983) contention that they should be combined.

The Uraniidae are usually divided into two subfamilies, the Uraniinae and the Microniinae. The first includes most of the large, tailed species such as the colourful, day-flying Neotropical *Urania* Fabricius (e.g. *Urania leilus* Linnaeus, illustrated below), Madagascan *Chrysiridia* Hübner and Papuan *Alcides* Hübner, and the drabber, mainly nocturnal Indo-Australian genus *Lyssa* Hübner (= *Nyctalemon* Dalman; Altena (1953)). The male genitalia of *Lyssa* have a distinctive upcurved uncus, simple, finely setose valves, and a sharply curved aedeagus that is fused at the base to the juxta. The female has a single scobinate signum.

171

Urania leilus

The larva of *Lyssa* was described by Corbett & Dover (1927), feeding on *Eugenia malaccense* (Myrtaceae) though this is undoubtedly an erroneous transcription of E[*ndospermum*] *malaccense*. It has no hairs or tubercles and is dark orange and black with longitudinal yellow stripes. There was no information on the prolegs.

The Microniinae are mainly small to medium sized, delicate species with white wings marked transversely with fine grey striae or grey bands. The hindwing is often tailed, the tail invested with black dots, presumably creating the effect of a 'false head' to deflect predators, an effect seen also in the ennomine geometrid genus *Ourapteryx* Leach. A survey of genitalic morphology has yet to be undertaken. The subfamily is restricted to the Old World tropics.

The larva of one genus (*Acropteris* Geyer) is somewhat spindle-shaped with no secondary setae and all abdominal prolegs present, the anal ones being somewhat cylindrical and directed backwards. There is some indication in the Indo-Australian tropics of a degree of specialisation on the family Euphorbiaceae, particularly *Endospermum*, though larvae have also been noted from Asclepiadaceae.

172

EPIPLEMIDAE

The Epiplemidae are moderately diverse in all tropical regions of the world but do not extend to temperate zones. The majority of species have the hindwing excavate between short tails sited where veins Rs and M3 meet the margin, perhaps also with another slight angle at M1. The postmedial fascia of the hindwing tends to run from the centre of the costa obliquely towards the posterior tail before angling sharply back towards the tornus.

The generic classification within the family is overdue for revision. Striking features of the male genitalia and female signum are likely to be of great value in this, the former varying from having a simple, slender uncus and ovate valves to having an almost tubular, multilobed uncus and/or a divided valve (see Holloway, 1976, 1979), and the latter being multiple, finely scobinate, or large but delicately stellate.

The cylindrical larvae have no secondary setae. The prolegs are unmodified, with the crochets arranged in a mesoseries approaching a penellipse (Fracker, 1916). They would appear to be tree and shrub defoliators, recorded from the families Bignoniaceae, Oleaceae, Rubiaceae and Verbenaceae in the Oriental tropics. On disturbance they tend to drop, suspended on a silken thread.

The species illustrated below is *Epiplema conflictaria* Walker, a species or species complex widespread in the Indo-Australian tropics (Holloway, 1979).

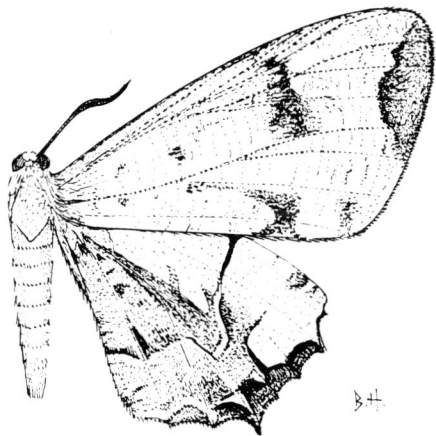

Epiplema conflictaria

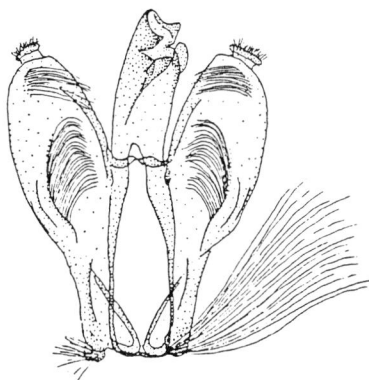

Epiplemid male genitalia (*Epiplema*)

GEOMETRIDAE

The Geometridae are, amongst the macrolepidoptera, second only to the Noctuidae in numbers of species. The tympanal structure is uniform within the family, possessing a sclerotised band termed by Minet (1983) the ansa.

The vast majority of species (exceptions being taxa considered primitive) have the familiar looper or inchworm larva, where the first three pairs of abdominal prolegs have been lost. The larva progresses by extending its anterior part, engaging the true legs, then bringing up its rear close to the thorax, looping most of the abdominal segments overhead. The impression given of a precise measuring motion has led both to the family name and to the American name for the larvae, inchworms.

The subfamilial classification of the Geometridae, as with that of the Noctuidae, is far from perfect. Again, some groups are more clearly defined than others and, within ill-defined groups, there are some clear-cut groupings of genera.

ARCHIEARINAE

This subfamily contains a small number of Holarctic and Chilean taxa with cryptic forewings and yellow, orange or red hindwings. The species are probably all day-flying and are not of economic importance.

OENOCHROMINAE

The Oenochrominae are based on the robust, Proteaceae feeding Australian genus *Oenochroma* Guenée. Most of the rest of the groups included are of very delicate, slender build, with long, slender legs, and the subfamily as a whole is probably polyphyletic. But one further feature linking various groups is behavioural and is unusual in Lepidoptera. Oriental species tend to be found in the rainforest understorey, partially day-flying or readily disturbed by day. This behaviour is seen in the group of small, fragile white species centred on the genus *Derambila* Walker, in the relatively robust yellow *Celerena* Walker, in the yellow, orange and red *Eumelea* Duncan and in the pantropical group of grey and brown genera (e.g. *Noreia* Walker) that was defined in terms of derived (apomorphic) characteristics of the male abdomen by Holloway (1984). There is a further large group of small grey species in Australasia such as *Dichromodes* Guenée.

The subfamily is currently under study by Dr M.J. Scoble.

GEOMETRINAE

The Geometrinae are the emerald moths and consist mainly of bright green or emerald green species. However, there is a group of more robust species in the Old World tropics with a more mottled, lichenous patterning, e.g. *Pingasa* Moore.

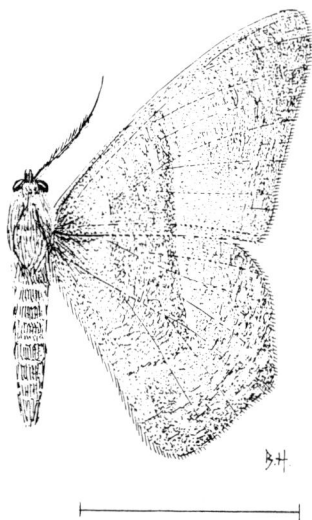

Thalassodes pilaria

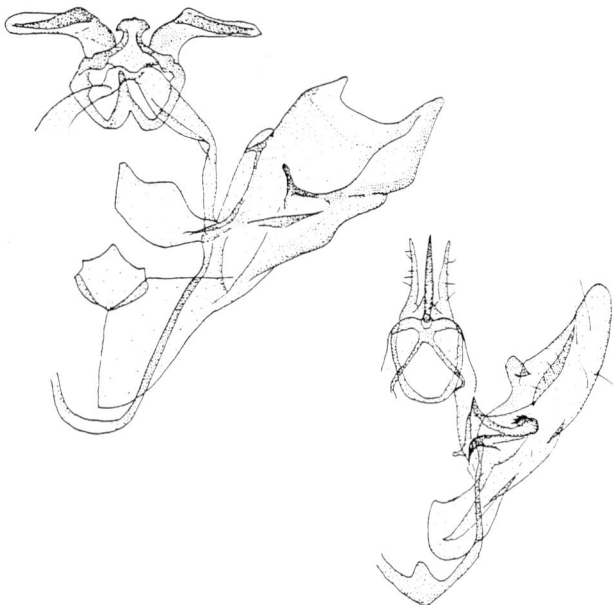

Geometrine male genitalia (*Archaeobalbis, Diplodesma*)

174

In the male abdomen there is a pair of fields of deciduous setae on sternite 3; these fields sometimes come together over the midline in the *Pingasa* group and are also seen in members of the *Noreia* group in the Oenochrominae. The uncus is accompanied by a pair of closely associated socii and, particularly in the *Pingasa* group, these may predominate over a reduced uncus.

The larvae are usually slender, green, with the head apically bifid into two acute processes.

The subfamily is much more diverse in the tropics and includes a number of minor pest species. *Thalassodes pilaria* Guenée, illustrated above, has been known to defoliate a wide range of hosts, including *Nephelium*, in tropical Australasia and the Pacific.

ENNOMINAE

This is by far the largest subfamily and is generally defined by a loss character, the absence of vein M2 in the hindwing. A number of genera widely scattered through the subfamily (e.g. *Semiothisa* Hübner, *Luxiaria* Walker, *Cleora* Curtis and *Abraxas* Leach) have a transverse band of posteriorly directed setae in the centre of sternite 3 of the male abdomen. In most genera this sternite is unornamented so the character is not of diagnostic value for the subfamily as a whole. The signum of the female bursa is usually a single, approximately circular sclerotisation, broadly spined, often set up on a short central stalk.

Several major tribal groupings are evident, at least in the Holarctic and Indo-Australian tropics. The *Hypochrosis* Guenée group has distinctive male genitalia with the vinculum narrow, but extensive, bulging ventrally each side of the mid-point; the valves give rise to a pronounced, asymmetric furca that usurps the function of the juxta. Other included genera are *Ourapteryx* Leach, *Sabaria* Walker, *Ctimene* Boisduval, *Garaeus* Moore and *Fascellina* Walker. The genitalia of a range of species from these genera have been illustrated by Holloway (1976).

The genera *Semiothisa* Hübner, *Tephrina* Guenee, *Luxiaria* Walker, *Probithia* Warren and allies have a divided valve in the male genitalia and probably form a natural group. Several *Semiothisa* and *Tephrina* have been recorded as minor pests, and a significant proportion of host records for these genera have been from the family Leguminosae. Members of this group have been recorded feeding as adults from mammal lachrymal secretions in South-east Asia (Banziger, 1972).

The *Hypomecis* Hübner group (Boarmiini of other authors) has been investigated in detail in Japan by Sato (1984). He defined the group by: the presence of a fovea (blister) at the base of the male forewing just posterior to the cell; four external setae on the ventral proleg of the 6th larval abdominal segment; a pupal cremaster ending in two spines. A number of genera have the band of setae on abdominal sternite 3 mentioned above. The fovea also occurs in the *Noreia* group of Oenochrominae (Holloway, 1984: 140), and in a widely distributed array of Australasian ennomines (Holloway, 1986b), and hence either this character is suspect as an apomorphy or the classification needs re-examination. The male genitalia perhaps most typically have the valve with a strongly sclerotised costa, overlapped over the apical portion by inwardly and dorsally directed setae arising from a somewhat arcuate zone over the dorsal half of the apex of the valve.

Sato presented a number of generic groups within the complex, each defined by character states presumed to be apomorphic. Important genera he included are *Hypomecis* Hübner (= *Boarmia* Treitschke), *Alcis* Curtis, *Cleora* Curtis, *Ascotis* Hübner, *Diplurodes* Warren, and *Ectropis* Hübner. The species tend to be polyphagous defoliators as larvae, many feeding on both angiosperms and gymnosperms. This is seen in the species illustrated below, *Ectropis bhurmitra* Walker, a South-east Asian member of an Old World complex that includes a number of recognised pest species, often of conifers.

Some ennomine larvae have only slightly reduced abdominal prolegs. McQuillan (1985) revised the small Australian genus *Mnesampela* Guest where only the first pair is slightly reduced. The adults lack hindwing vein M2 and have a typical bar of setae on the third sternite of the male abdomen.

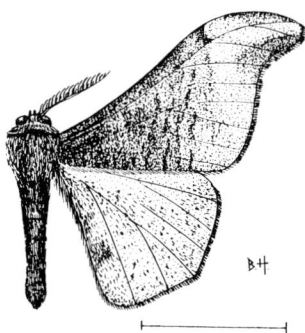

Hyposidra talaca

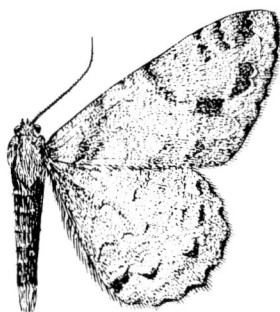

Ectropis bhurmitra

Milionia basalis

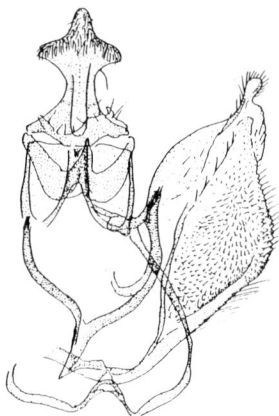

Hypochrosis

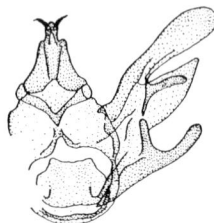

Visitara

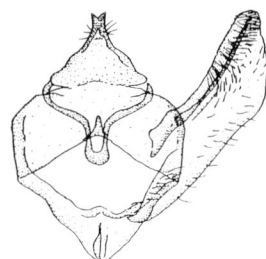

Xandrames

Ennomine male genitalia

The other species illustrated here are *Hyposidra talaca* Walker, a polyphagous defoliator found throughout the Indo-Australian tropics and often recorded as a minor pest, and *Milionia basalis* Walker, a widespread Oriental member of a diverse and predominantly Papuan genus that appears to be restricted to conifers in larval diet and thus includes a number of forestry pests.

The Ennominae are diverse throughout the world, perhaps with greatest richness in the subtropics, and giving way to the Larentiinae in diversity at high altitudes and latitudes (Holloway, 1986b). Many publications cover large and important groups of Ennominae, such as those by Prout (1929, 1937), Fletcher (1953, 1967, 1974), Inoue (1970, 1972), Rindge (1961-1978 and many smaller papers). McGuffin (1972-1981) described the Canadian fauna in detail.

176

STERRHINAE

The Sterrhinae are defined in the key by a minor character of hindwing venation. They are usually pale, delicate species, yellow, white, straw or pinkish in, colour, with wings multifasciate, usually obliquely so. Discal spots are prominent on both wings; in *Anisodes* Guenée and allies the discal spot of the hindwing is usually larger, often a pale disc ringed finely with dark colour. In the other large genera, *Scopula* Schrank and *Idaea* Treitschke, both discal spots are punctate. The *Anisodes* group also has a hammer-headed ansa to the tympanum, a character used to define the Larentiinae by Minet (1983).

Scopula rubaria with key to elements of forewing pattern

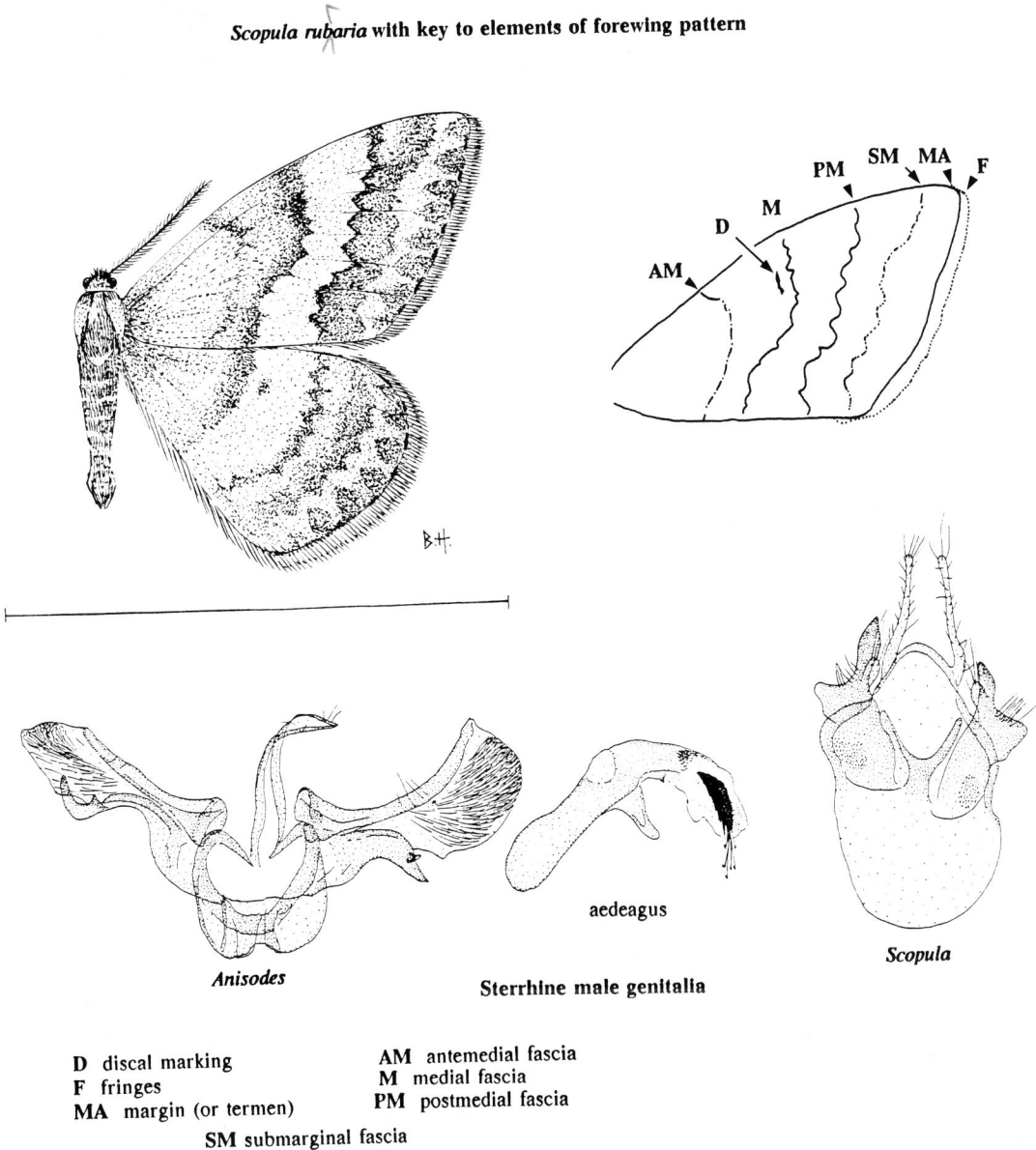

Anisodes

aedeagus

Scopula

Sterrhine male genitalia

D discal marking
F fringes
MA margin (or termen)
SM submarginal fascia

AM antemedial fascia
M medial fascia
PM postmedial fascia

177

The male genitalia of the *Anisodes* group usually have the uncus reduced, the vinculum very much broader and elaborate than the tegumen, and the valves usually highly modified. The gnathos is absent, also in *Scopula* and allies, but is present in *Idaea*. In *Scopula*, *Problepsis* Lederer and allied genera the male genitalia are compact, the valves small, bifid, and the uncus reduced, replaced in *Scopula* by a pair of slender socii. In *Idaea* the valves are entire, slender, variable in structure and modification.

The signum of the female genitalia may be helpful in grouping genera within the subfamily. In *Idaea* and some *Anisodes* there is general scobination, in *Scopula* a small ovate field of spines and in *Cyclophora* Hübner a single signum.

Pupation in the *Anisodes* group is unusual in that the pupa is supported by a pad at the cremaster and a silken girdle as in Papilionoidea, though the girdle is located abdominally rather than at the thorax (data located for *Cyclophora*, *Anisodes*, *Ptochophyle* Warren and *Chrysocraspeda* Swinhoe). Pupation in other sterrhines is in a silken cocoon, often web-like.

In general in the Old World the *Anisodes* group consists of arboreal feeding species, found mainly in forest, and the *Scopula* group and *Idaea* are more diverse in open habitats and include mainly herbaceous feeders. Diversity is greatest in the tropics though the subfamily is much more strongly represented in temperate latitudes than the Geometrinae. The species illustrated above is *Scopula rubraria* Doubleday, a widespread species of grasslands and herbaceous vegetation in Australasia. Members of the *Scopula* group have been recorded feeding from lacrymal secretions of mammals in South-east Asia (Banziger, 1972). The Canadian fauna has been described by McGuffin (1967).

LARENTIINAE

The Larentiinae were suggested by Minet (1983) to be defined by the shape of the apex of the tympanal ansa, expanded into a broad hammer-head. However, this feature is seen in the *Anisodes* group of the Sterrhinae as mentioned above.

The forewings are strongly marked with fine, transverse fasciae that tend to meet the dorsum at right angles rather than obliquely (Sterrhinae). A broad medial area is often darker than the rest of the wing. This medial area includes, and is bounded by, multiple postmedial and antemedial fasciae between which is a band unmarked except for the dark discal dash. Exterior and interior to this medial zone are moderate pale zones, then further fine fasciation. The hindwings are generally much more uniform, though reflect the forewing pattern in a minority of genera such as *Asthena* Hübner, *Acolutha* Warren, *Collix* Guenee, *Eois* Hübner and *Poecilasthena* Warren. The hindwing discal spot is weak or absent in most larentiines.

In the male genitalia there is a great variety of structure, but frequent features are: eversible coremata in the abdomen, particularly between segment 8 and the genitalia; a pair of processes on the eighth sternite (octavals) that are relatively central within it (asymmetric processes in *Scopula* are lateral); elaboration of juxta and transtilla in association with each other, particularly in the tribe Eupitheciini (*Eupithecia* Curtis, *Chloroclystis* Hübner, *Gymnoscelis* Mabille), the transtilla being divided into dorsal and ventral structures termed labides.

The bursa of the female genitalia exhibits a diversity of ornamentation from general scobination to most elaborate signa; it will thus be of greater use for characterising and grouping genera than for defining the subfamily.

The subfamily is most diverse in temperate regions and at high altitudes in the tropics (Holloway, 1986b). Many of the species are associated with open habitats. The species illustrated below is *Xanthorhoe sodaliata* Walker, a widespread species of pastures and herbaceous vegetation in temperate Australasia. Treatments covering a good selection of the subfamily may be found for the New Zealand fauna (Dugdale, 1971), the Indo-Australian tropics (Holloway, 1976; Dugdale, 1980), Japan (Inoue, 1979-80; Inoue *et al.*, 1982), Scandinavia (Skou, 1984) and Canada (McGuffin, 1958).

D discal marking **AM** antemedial fascia
F fringes **M** medial fascia
MA margin (or termen) **PM** postmedial fascia
SM submarginal fascia **SB** subbasal fascia

Note that most fascia are multiple and meet the dorsum approximately at right angles

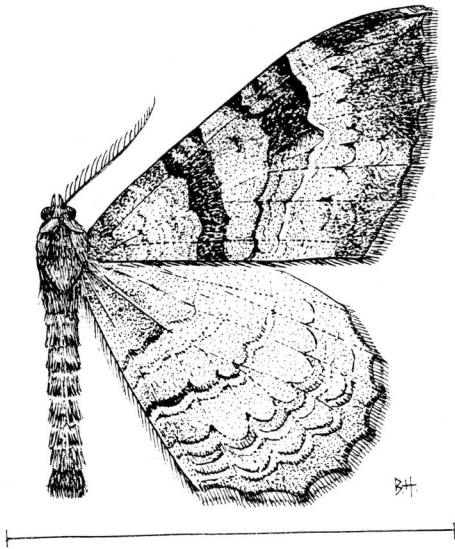

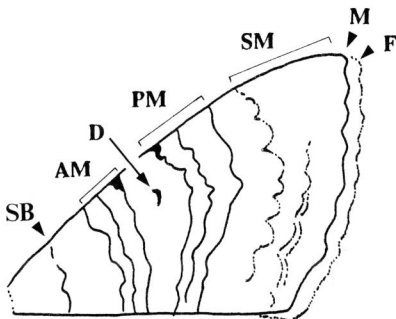

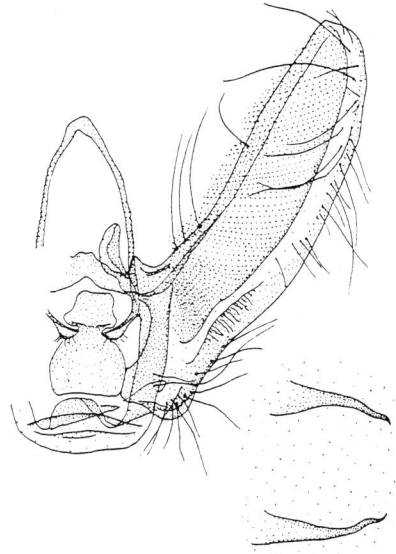

Chloroclystis male genitalia and octavals

Xanthorhoe sodaliata with diagram of elements of forewing pattern

CALLIDULOIDEA

This superfamily is based on the CALLIDULIDAE, a small group of butterfly-like moths with orange, yellow and brown forewings. The family is restricted to the Indo-Australian tropics. The species fly by day in the understorey of rainforest, though a few species come to light at dusk. They rest on low foliage with the wings held vertically like a butterfly. Some have swollen, almost clubbed antennae. The cells of both wings have the cross-veins vestigial or absent.

In the male genitalia (*Petavia* Horsfield) the uncus is slender, terminating in a slight hook, and the valves are very large, somewhat oval, without ornament.

The eggs of *Tetragonus* Geyer (= *Cleosiris*) are flattened, scale-like, laid in loose clusters. The larva has the vertex of the head covered by a thin but complete, chitinised, semicircular shield on the first thoracic segment. The body of the larva is cylindrical, the segments well defined, bearing erect, dark hairs from large, broadly conical tubercles; all prolegs are present (T.R.D. Bell, unpublished). In the pupa the head is prominent. The behaviour of the larva is similar to that of many pyralids: it rolls a pinna of the host-plant (a fern) to make a cell, scaled with silk (Bell). Fern feeding may be general to the family.

The family Pterothysanidae has been associated with the Callidulidae (e.g. Janse, 1932-1964, Vol. I), both being included in the Geometroidea by Fletcher (1979). The family was unplaced by Minet (1983) who supported the separation of the African genus *Hibrildes* Druce from it in the Hibrildidae. In the Himalayan *Pterothysanus* Walker the cell cross-veins are weak, the facies of the wings mottled black and white, the build geometrid-like. The hindwing dorsum bears a prominent fringe of fine, long, hairlike scales. This fringe is absent in the grey-spotted, white, monotypic Madagascan genus *Caloschemia* Mabille, a taxon otherwise similar in size and build. In both genera the abdomen is yellow with lateral and dorsal rows of black bars.

The abdominal patterning of the African *Hibrildes* Druce is not compatible with that in the two genera above. The wings are more elongate, apically more rounded, the veins picked out prominently darker. One species, *norax* Druce, is whitish, and the other, *venosa* Kirby, is sexually dimorphic and mimics acraeine and danaine butterflies (Janse, 1932-1964, Vol. 3). In the male genitalia the uncus is absent, but a pair of strong socii arises from the tegumen; the valves have two prominent processes from the costa (Janse).

Minet (1986) has placed the Hibrildidae in the Bombycoidea, mainly on absence of features.

CASTNIOIDEA

One family, the CASTNIIDAE, is included. It has its greatest diversity and its largest species in the Neotropics, but there are a few rare species in the Oriental tropics (*Tascina* Boisduval) and a rather distinct group of small species in Australia (*Synemon* Doubleday). The species are day-flying.

The species have clubbed antennae in conjunction with a strong frenulum. On both wings the stem of vein M is strong within the cell and lies very close to the cubital stem, forming a narrow cell with it from which the cubital and posterior two medial veins arise in a quadrifid array. The forewing radial system includes an areole.

The most recent publication on the family is by Miller (1972) who is currently reviewing it in toto. She was critical of the previous monograph by Houlbert (1918) and suggested that all Neotropical species be retained in *Castnia* Fabricius *sensu lato* rather than Houlbert's genera until the latter can be adequately defined. Minet (1986) has placed the Castniidae in the Sesioidea on the basis of shared characteristics of the eye and metafurca. Common & Edwards (1981) compared them with the Cossoidea and Sesioidea.

The male genitalia illustrated by Miller show a robust uncus, gnathal processes, a deep, rather square valve, a moderate saccus from the vinculum and a large aedeagus that is sharply curved over the basal half such that both ends are directed posteriorly. The female genitalia have very long, slender ovipositor lobes and apophyses. The Australian *Synemon* has a comparable aedeagus that is sharply flexed subbasally; the valves are shallower; the saccus is short, bifid, the two processes well separate from each other.

The egg of *Castnia* is slender, spindle-shaped, with five prominent longitudinal ridges or carinae, giving it a stellate cross-section. Those of Synemon have from four to numerous ridges (Common & Edwards, 1981). The larva is smooth, almost hairless, whitish or

Due to a recurring error in production some of the page references are incorrect (mostly in the keys and index). Most other errors listed are editorial in origin.

Page 7   Line 23 "catocaline noctuids"
   21   Line 8: "Himantopteridae"
   28   Line 6 from bottom:"without mandibles"
   31   Line 3:"families (e.g. Papilionoidea) males..."
   38   Line 4:"openings"
   41   NYMPHALIDAE (p.184); LIBYTHEIDAE (p.185)
   42   PAPILIONIDAE (p.186)
   43   CASTNIIDAE (p.180)
   45   DUGEONEIDAE (p.138)
   47   EPIPYROPIDAE (p.138)
   51   NOCTUIDAE (p.155)
   55   URANIIDAE (p.171); EPIPLEMIDAE (p.173); ENNOMINAE (p.175)
   56   LARENTIINAE (p.178)
   57   STERRHINAE (p.177); HYBLAEIDAE (p.125); THYRIDIDAE (p.125)
   58   CALLIDULIDAE (p.179)
   59   Couplet 49, second half, line 2: "Rs" (not "R5")
   60   ENDROMIDAE (P.142); EUPTEROTIDAE (p.144); OXYTENIDAE (p.142)
   61   CERCOPHANIDAE (p.142); BRAHMAEIDAE (p.142); LEMONIIDAE (p.142);
         CARTHAEIDAE (p. 142)
   103   Line 6 from bottom: "Neopastis"
   106   Line 8 from bottom: remove "which is illustrated"
   112   Line 2 from bottom: for "illustrated above" read "illustrated below"
   119   Line 5 "Conifrons"; line 9 from bottom: "Pycnarmon"
   135   Line 10: "Minet"
   139   "Brachycodella"; line 2 from bottom: "Megalopygidae"
   150   Line 14: "Introduction"; line 23: "(see p.153)"
   151   Line 3: "(see p.52)"
   152   For "Syntomis" read "Amata"; line 13: "(p.52)"
   153   Line 2: "(see p.150)"
   154   Line 14: (p.150); line 21: "position"
   157   Line 9 from bottom: "(p.148)"; line 14: "...of the abdomen in the males."
   165/166 "Heliocheilus albipunctella"
   177   Figure caption: "rubraria"
   179   Figure key: "M margin (or termen)"
   180   Line 2 from bottom: "Synemon"
   192   Figure caption: swap "biordinal" with "triordinal"
   205   Couplet 50: delete "(figure below)"
   208   "homoideous/heteroideous" (not defined in glossary)= of even/varied length
   225   Line 16: "members"
   255   Acanthopsyche 80; Acleris 130; Acrocercops 89,66; Agonoxena 72; Agrotis 156;
         Anadevidia peponis 159; Anthophila 97; Azanus ubaldus 183, 184
   256   Brachmia convolvuli 112, 113; Celerena 174; Ceratomia 142; Crocidolomia binotalis 4,
         119
   257   Dasychira mendosa 151, 152; Dicymolomia julianalis 114; Earias insulana 162, 163;
         Eios 54, 56, 178; Epicephala 89; Erionota thrax 5, 182, 183; Erythrina 4; Euschemon
         39; Euzophora coccidophaga 114; Gonimbrasia belina 146
         Column 1: "Epicerura"; column 2: "Erythrina"; "Fascellina"
   258   Lophocorona 27; Lophomachia 56; Luffia 76; Lyonetia 65
   259   Maliattha 164; Mythimna loreyi, unipunctata 167; Nacoleia 119; Ocinara 60, 145;
         Operophthera 76; Papilio 186; Parasa lepida 140, 141; Parenthrene 97; Pectinophora 112
   260   Pericallia ricini 154; Phthorimaea 75; Polygrammodes 119; Praesetora 140; Prolimacodes
         49; Scopula 57, 177,178
         Column 1: "Pelopidas"; "Psalis" (not "Psara")
   261   Spilosoma 154; Squamura disciplaga 137, 138; Tiquadra 84; Trichophaga 82

flesh-coloured as is usual in stem-boring species: the prolegs bear two transverse rows of stout, curved crochets (Marlatt, 1905). in later instars crochets and prolegs are lost entirely (Common & Edwards, 1981).

The South American species would appear to be stem-borers, such as the illustrated (below) *Castnia licoides* Boisduval, a pest of sugar. Australian species have larvae that live in the soil, feeding on roots or (early instars) within tillers of sedges (Common, 1970; Common & Edwards, 1981). Adults of most S. American species are robust, similar in general appearance to *licoides*, but others have more elongate wings, often with transparent patches, and probably participate in some of the characteristic mimicry rings of the area.

Castnia licoides

PAPILIONOIDEA *sensu lato* (Brock, 1971)

The definition of this superfamily (see Ehrlich, 1958; Kristensen, 1976; Scott, 1984) has recently been modified by the investigation by Scoble (1986) of the Neotropical family Hedylidae that had, until recently, been assigned to the oenochromine Geometridae (e.g. Fletcher, 1979).

The alternative to defining an enlarged superfamily Papilionoidea in terms of a pouch over the first abdominal tergite (doubtfully in Hesperiidae), a post-spiracular bar on the first abdominal segment, a tapering of the abdomen from the centre towards the base, a thoracically girdled pupa and an anal comb in the larva, is to recognise three superfamilies: Hedyloidea (Hedylidae); Hesperioidea (Hesperiidae and Megathymidae - skippers); Papilionoidea (Lycaenidae, Nymphalidae, Libytheidae, Pieridae, Papilionidae - 'true' butterflies). The last two groups have been termed Rhopalocera (knob-horned) as

181

distinct from the moths, Heterocera (other-horned), a reflection of their clubbed antennae. In all three groups the wings tend to be held vertically at rest rather than flat over the body or against the substrate, though more information is needed on the Hedylidae. The Callidulidae also show this behaviour.

HEDYLIDAE

This family is primarily Neotropical, consisting of fragile greyish species mostly with relatively long, triangular forewings. Paler markings on the wings tend to be longitudinal rather than transverse. The forewing has veins R2 and R3 diagnostically sinuous. In the hindwing, vein Sc is well separated from Rs. A frenulum is present. The antennae are sometimes bipectinate, otherwise filiform.

The head capsule of the larva is extended dorsally into two long 'horns'. The pupa is formed in or on a folded leaf, held to it by a silken pad at the cremaster and a thoracic girdle.

The family has recently been reviewed by Scoble (1986).

MEGATHYMIDAE and HESPERIIDAE (skipper butterflies)

The Megathymidae are a small group of large, robust skippers found in the United States, Mexico and Central America. The larvae bore in the succulent leaves of agave plants. The family has been reviewed by Freeman (1969).

The Hesperiidae are found worldwide, though most diversely in the tropics. In both families the antennal bases are well separated and the forewing veins arising from the cell do so independently. The basic wing colour is dark brown in most cases, broken by rows of spots in white, yellow or orange, often translucent; banding is rarer. The antennal club is usually slightly hooked at the tip.

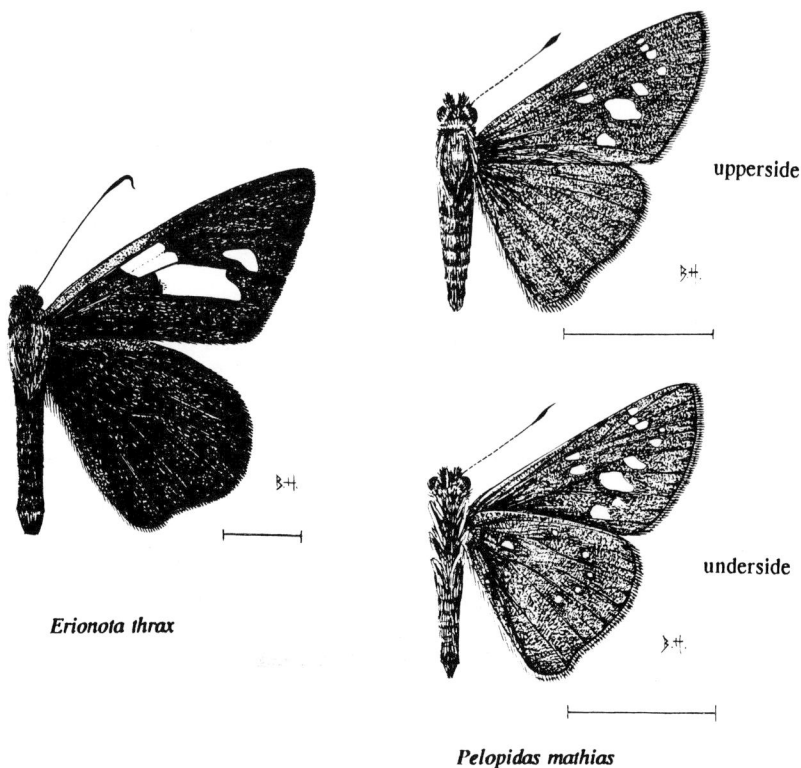

upperside

underside

Erionota thrax

Pelopidas mathias

Skippers are generally robust, and are powerful, highly agile fliers (see Betts & Wootton, in prep.). The majority are day-flying though there are crepuscular tropical groups such as *Erionota* Mabille and *Hasora* Moore.

The larva has a full set of abdominal prolegs. The thoracic zone tapers towards the anterior and is extensile; the abdomen also tapers. The body is often covered with a waxy or filamentous secretion. The pupa is somewhat mothlike, lacking angular projections that are seen in other butterfly groups and Hedylidae, and, though girdled, is often in a cocoon or rolled in leaves.

A number of subfamilies have been recognised, and the reader is referred to Evans' (1937, 1941, 1951-1955) series of monographs on the family for definitions of these. Many groups are restricted to monocotyledonous plants, particularly Gramineae. The species illustrated above are: *Erionota thrax* Linnaeus, a widespread Oriental pest of banana where the larva, as well as defoliating the plant, causes further damage by rolling a section of the leaf into a shelter; *Pelopidas mathias* Fabricius, a widespread Oriental rice pest, one of a large group of taxonomically difficult rice skippers for which Corbet & Pendlebury (1978) provide a partial diagnosis.

LYCAENIDAE (including RIODINIDAE)

This family includes the 'blue' butterflies, the 'hairstreaks' and the 'coppers', and is most diverse both in terms of subfamilies and species in the Old World; however, the Riodininae are very much more diverse in the Neotropics, where they are accompanied mainly by numerous Theclinae (hairstreaks).

The upper surface of the wings is very often brightly iridescent in blue, green, purple or copper, particularly in the male. The underside pattern is usually finely lineate or maculate on brown or grey, often with a row of brighter (e.g. orange) lunules submarginally, or just a few in association with a tail and black dots, giving the impression of a false head when the butterfly is at rest. The tail, when present, is associated with vein CuA2.

The eyes are emarginate to accommodate the antennal base, or the antennal base abuts the eye. The forelegs are somewhat reduced relative to the other two pairs, the foretarsi of the male being fused into a single segment. In the Riodininae the male foreleg is relatively much more reduced than that of the female. The male genitalia have the tegumen very broad and deep, developed into two lobes between which is the reduced uncus; the gnathos is present as pair of distinctive curved hooks, termed brachia or falces.

The larvae are rather stout, flattened, the head somewhat obscured by the thorax. Most species have a honey gland on the seventh abdominal segment and a pair of eversible tubercles on the eighth. These structures are connected with the very frequent association between the larvae of Lycaenidae and ants (Cottrell, 1983). This association appears to function partly as a means of defence against predators and partly as a means whereby the larvae, with carnivory frequently found through the family, bribe their way to the ant grubs by presentation of honeydew from the abdominal structures. The exceptionally thick cuticle of the larvae is a modification for protection against ant jaws.

The higher classification of the family has been investigated by Eliot (1973), who recognised a number of subfamilies and numerous tribes. The Miletinae, fragile, long-winged, brownish species with white patches and an irregularly striate underside, have larvae that are predatory on Homoptera and thus beneficial. The larvae of many species in other subfamilies, especially the Polyommatinae, defoliate Leguminosae or feed in the developing pods and thus may be detrimental to legume crops. *Azanus ubaldus* illustrated below, is a legume feeder found in the semi-arid tropics of the Old World. Some Theclinae are fruit or seed feeders and thus may be of economic importance.

Major works on sections of the family are by Stempffer (1967), Eliot (1973, 1986), Eliot & Kawazoe (1983), Evans (1957), Tite (1963).

Azanus ubaldus

upperside   underside

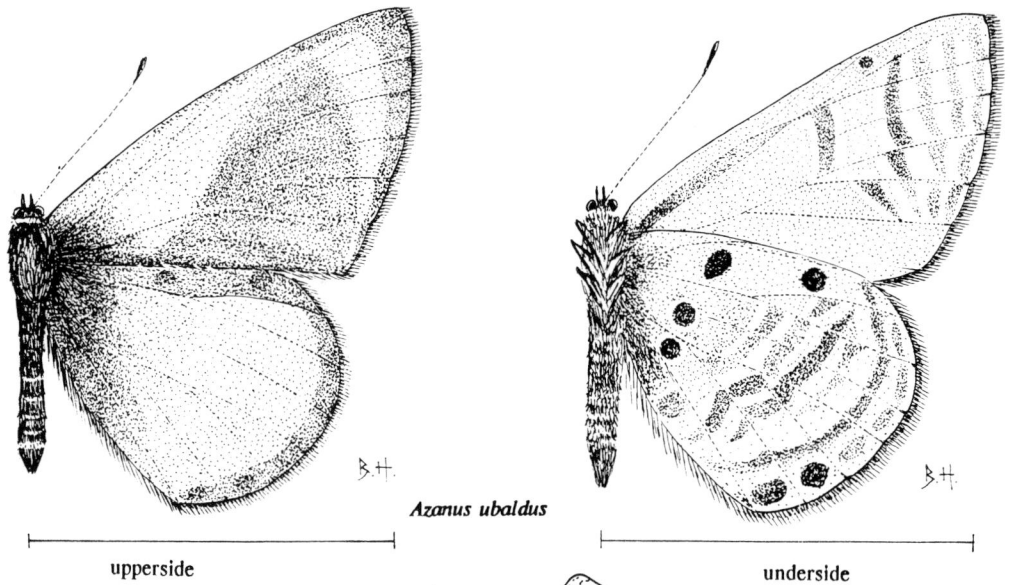

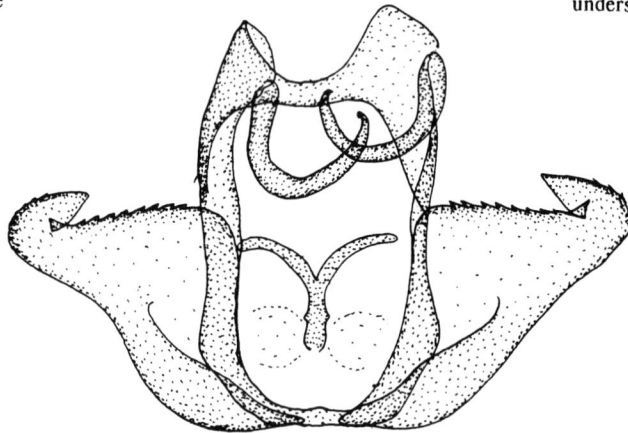

Lycaenid male genitalia

NYMPHALIDAE *sensu lato*

The treatment of this family here follows current practice (e.g. Ehrlich, 1958) in including many groups that have been regarded as families in the past, such as Satyridae, Danaidae, Brassolidae, Amathusiidae, Morphidae, Acraeidae, Heliconiidae and Ithomiidae. The process of reappraisal of the higher classification of the Nymphalidae is continuing as exemplified in De Vries *et al.* (1985).

The majority of species are medium to large, brownish or reddish, many with a row, often reduced in number, of ocelli submarginally, particularly on the underside of the wings, more so the hindwing. Many have a fast, direct flight involving periods of gliding, but the more slightly built Satyrinae, for example, have a more untidy, flipping flight, frequently changing direction (see Betts & Wootton, in prep.).

The forelegs of both sexes are considerably reduced, more so in the male where, as in the Lycaenidae, the tarsal segments are fused.

The larvae are of diverse form, generally more strikingly ornamented than in other butterfly families, e.g. with head horns (Apaturinae, Morphinae, Amathusiinae, Satyrinae, Charaxinae, Brassolinae), filaments (Danainae), a posterior bifurcation (Satyrinae, Apaturinae, Morphinae) and bristled spines (Nymphalinae, Argynninae, Limenitinae,

184

Heliconiinae).

The pupae are similarly diverse, but in all groups the girdle has been lost and the pupa is suspended by the cremaster from a silken pad.

Several subfamilies show a marked degree of host-plant specialisation, such as the Satyrinae, Brassolinae and Amathusiinae, all virtually restricted to monocotyledons (some Satyrinae are on club-mosses), thus including pests of palm and grain crops. The Heliconiinae are almost entirely restricted to *Passiflora* as are about half the Acraeinae. The Danainae feed mainly on Asclepiadaceae, Moraceae and Apocynaceae (Ackery & Vane-Wright, 1984).

Important reviews of major sections of the Nymphalidae or economic groups are by Eltringham (1912), Ackery & Smiles (1976), Ackery & Vane-Wright (1984), Eliot (1969), Higgins (1981), Miller (1968), Van Someren (1963-1975), Smiles (1982), and Pierre (1983). Tsukada (1985) has monographed some of the South-east Asian fauna.

LIBYTHEIDAE

This small family is found worldwide in tropical to temperate latitudes. The medium-sized butterflies have rather emarginate forewings, usually dark brown with orange patches, and prominent, forwardly directed (porrect) palps, leading to their name of 'beaks'. There is a strong association with *Celtis* (Urticaceae) as a larval host-plant. Recent reviews of the family are by Shields (1984, 1985).

The forelegs are imperfect in the male but normal and functional in the female. The larva is cylindrical, with a small head. The pupa is suspended as in the Nymphalidae.

The family has beeen placed as subfamily of the Nymphalidae by Kristensen (1976), but Scott (1984) retained it as a family and presented evidence for a sister-relationship with the Nymphalidae, reverting to the system of Ehrlich (1958).

PIERIDAE

The Pieridae include the 'whites' and the 'yellows'. The family is characterised by absence of a cleaning spur on the fore-tibia and the loss of a forewing radial vein (with a few exceptions). All pairs of legs are fully developed as in the Papilionidae, but in the latter family there has been loss of one of the hindwing anal veins, and retention of a separate, short, second anal vein in the forewing; the Pieridae resemble other papilionoids in having one forewing anal vein and two on the hindwing. Flavone pigments are frequent through the family, and urea is also incorporated in the pigments. The tarsal claws are strongly bifid.

The larvae are elongate, smooth or with very short hairs, usually green and cryptic. The pupae are attached, head uppermost, to the substrate by an anal pad and a thoracic girdle; the head capsule is often pointed.

The two major subfamilies are the Pierinae and the Coliadinae. The former includes many species that feed on one or more of the related plant families Cruciferae and Capparaceae, and also on the unrelated but chemically convergent Tropaeolaceae, favouring the Capparaceae in the tropics and the Cruciferae in temperate latitudes. Many pest species are included, particularly the 'cabbage whites' in the genus *Pieris* Schrank. Another large pierine group includes the Indo-Australian genus *Delias* Hübner (Talbot, 1928-1937) and the Afrotropical *Mylothris* Hübner, and feeds mainly on parasitic plants in the family Loranthaceae. The species may thus be beneficial when mistletoes attack shrub and tree crops.

The Coliadinae, the 'yellows', have a high proportion of species defoliating Leguminosae, such as in the genera *Eurema* Hübner, and *Catopsilia* Hübner and *Colias* Fabricius; *Eurema hecabe* Linnaeus, a widespread Old World tropical species, is illustrated below. Many of these can be pests of crops.

Like the Papilionidae, Pieridae are frequent members of large 'mud-puddle clubs' at seepages and on river banks in the tropics. Very many species are migratory.

185

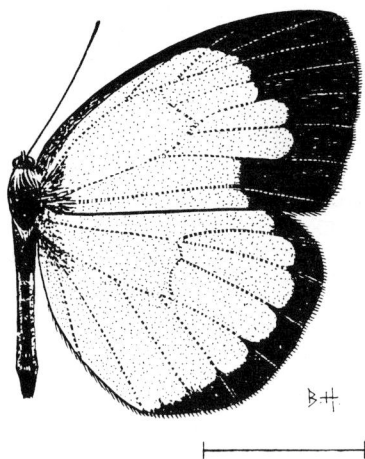

Eurema hecabe

PAPILIONIDAE

The Papilionidae, or swallowtails, are generally regarded as being the most spectacular of the Lepidoptera, rivalled perhaps only by the day-flying Uraniinae. They are much prized by collectors, being strikingly marked in a variety of colours, usually on a black ground. In various parts of the world, such as Papua New Guinea, it has been found profitable to farm them and supply this demand. The Papilionidae are among the few insect families so far to be the subject of an IUCN Red Data Book (Collins & Morris, 1985).

Venation features peculiar to the family are the presence of a separate second anal vein on the forewing that meets the dorsum subbasally, and the loss of one anal vein in the hindwing. Not all species have a 'swallowtail', but, when present, it occurs as an extension based on vein M3. In the male genitalia the eighth tergite is produced into a superuncus, partially usurping the role of the reduced uncus and tegumen.

The larvae are smooth or invested with simple spines. A family characteristic is the presence of the osmeterium, an extrusible process on the prothorax. The pupa is upright and girdled as in the Pieridae.

The adults are powerful fliers and often come to mud, puddles, river banks or seepages in the company of Pieridae (mud-puddle clubs) to obtain essential salts.

There are several plant families utilised as hosts, often with major lineages restricted to them, such as the Troidini and Zerynthiini on the Aristolochiaceae. Many species of the genus *Papilio* Linnaeus, such as the Afrotropical *P. demodocus* Esper, feed on Rutaceae so may be pests of *Citrus*. Others feed on Umbelliferae and may be pests of crops in this family. The families Lauraceae and Annonaceae are also frequently utilised.

A comprehensive review and bibliography for the family may be found in Collins & Morris (1985), but other useful references are Igarashi (1979) for illustrations of early stages, and D'Abrera (1975) and Haugum & Low (1978-1983) for accounts of the spectacular Birdwing butterflies. A recent publication on the higher classification is by Hancock (1983). Ackery (1975) has reviewed the Parnassiinae.

Papilio demodocus

BUTTERFLY LITERATURE

There is copious literature on the butterflies, the most popular of insect groups, so workers in most parts of the world are likely to be able to identify perhaps the majority of species that they encounter, provided they can gain access to the works concerned. The following account gives reference to a selection of the more important faunistic works published since the Second World War (see also Ackery, 1984).

General 'dictionary' style works have been published by Lewis (1973) and Smart (1975). The most comprehensive treatment of the world fauna is currently being produced by D'Abrera (1975-1986), with only some of the Neotropical fauna and all that of the Holarctic still to be covered; skippers are excluded, but the works of Evans on that family have been mentioned.

The North American fauna is treated by Howe (1975) and Pyle (1981). DeVries (in press) is likely to be most helpful for Central America, whilst Riley (1975), Barcant (1970) and Brown & Heinemann (1972) deal with the Caribbean.

For Africa there are general works by Carcasson (1981) and Williams (1969), and more regional treatments for West (Fox *et al.*, 1965) and Southern Africa (Cooper, 1973; Pennington, 1978; Pinhey, 1965).

European butterflies are covered by Higgins & Riley (1980), and the Middle East fauna is partly covered by Larsen (1974, 1984), though the region from Turkey to Afghanistan is poorly served apart from the latter (Sakai, 1981).

The Indian Subregion is covered by Talbot (1939), Woodhouse (1950) and Wynter-Blyth (1957), and the Far East by Johnston & Johnston (1980), Kawazoe & Wakabayashi (1976), Shirozu (1960) and Shirozu & Hara (1960-1962). References to local Indian check-lists have been collated by Varshney (1977).

South-east Asia and Oriental Indonesia are served by a good range of regional and local faunas such as Corbet & Pendlebury (1978), Fleming (1975), Lekagul et al. (1977), Morishita (1981), Pinratana (1981-1983), Tsukada & Nishiyama (1982), and Yata (1981).

187

In Australasia, Australia is covered by Common & Waterhouse (1972) and New Zealand by Gibbs (1980). A work on the New Guinea butterflies is in preparation by M.J. Parsons (see also Parsons, 1986a, b). Some of the larger Pacific island faunas have also been studied, such as Vanuatu (Samson, 1983), New Caledonia (Holloway & Peters, 1976), Fiji (Robinson, 1975) and Samoa (Hopkins, 1927).

The generic names have been catalogued by Hemming (1969).

IMMATURE STAGES
(DJC & JDH)

189

INTRODUCTION TO LARVAE
(D.J.CARTER)

Most pest Lepidoptera are initially encountered as larvae as this is the stage that damages various crops. Unfortunately, the greater part of Lepidoptera taxonomy is based on characters of the adults and consequently the literature relating to identification of larvae is very restricted and largely confined to well known species of the Holarctic fauna.

Some of the most important characters for identifying larvae are relative setal positions on the head and body. The study of setal arrangement has been complicated by the proliferation of systems for naming and numbering body setae and setal groups. The most widely adopted modern system of setal nomenclature is that of Hinton (1946), which is used here. Setal arrangement on the head is easier to understand as there are fewer systems of nomenclature and that of Heinrich (1916), modified by Gerasimov, (1935) has been generally adopted.

The study of setae is further complicated in many groups by the presence of secondary setae. Secondary setae usually appear in the second instar although some larvae possess secondary setae in the first instar. These tend to mask the primary setae so important for recognising many groups. However, secondary setae are often grouped on plates or verrucae corresponding in position to primary setae. The presence of secondary setae may in itself be an important character and the distribution and postion of secondary setae on verrucae or scoli and the structure of the setae themselves can all provide useful recognition characters. Care should be taken when handling hairy larvae as many possess poisonous hairs that can cause a painful rash.

Other useful characters on the head are the structure of the mandibles, antennae, labrum and spinneret, the arrangement of the ocelli and the size of the frons relative to the rest of the head. The number and arrangement of prolegs on the abdomen is a useful guide to many families and smaller groupings and the form and arrangement of hooks (crochets) on the prolegs provide further useful characters that are relatively easy to study.

One of the most obvious identification characters, colour pattern, is extremely unreliable for lepidopterous larvae. Many species have a wide range of colour forms while most species that are internal feeders are without colour or distinctive markings. The colours of larvae are very difficult to preserve and most patterns soon fade when specimens are stored in alcohol or similar preservatives. Most larvae discolour very rapidly after death and larvae for identification have often dried, shrivelled and blackened.

The fact that lepidopterous larvae pass through several instars (usually 5 or 6) during which they may undergo considerable changes in colour and form is an added complication to larva identification. For the most part, the key to families provided applies only to fully grown (final Instar) larvae. Where possible, young larvae should be reared until they are fully grown.

The key provided is largely biased towards groups of economic importance and a number of families are omitted. Due to the lack of larval material of some groups, this key is often based on the study of only a few examples from each family and so should be regarded as provisional. Many years of further study of larvae will be needed before a more reliable and comprehensive key can be provided. An illustrated resume of information on the larvae of the more important families of Lepidoptera of economic importance follows the key.

In order to identify larvae, it is important that material is carefully collected and preserved. Although the shortcomings of colour patterns as identification characters have already been mentioned, they may be useful at species level and so should be recorded before preservation. Colour photography is extremely useful for this purpose but, failing this, brief colour descriptions or sketches are worth making.

A simple method of killing and preserving larvae is to drop them into boiling water for a few seconds until they are fully extended and to blot them dry on absorbent paper

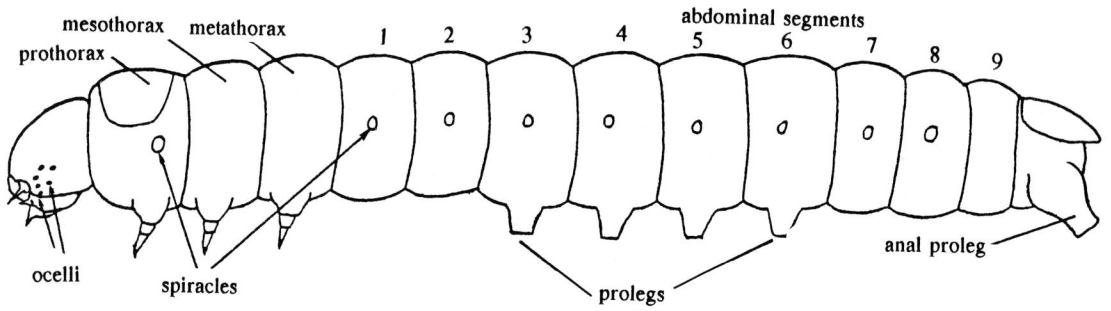

General features

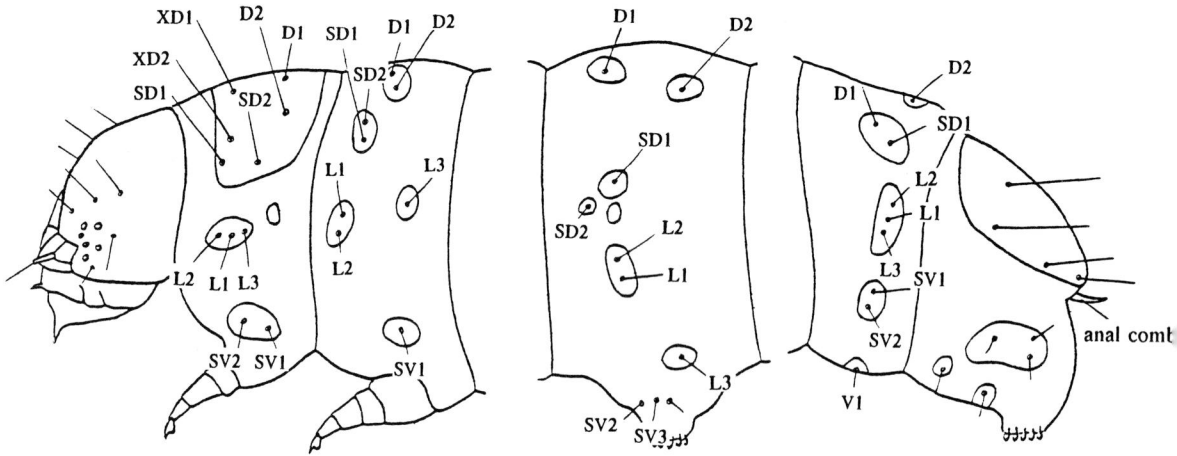

Setal nomenclature of body segments

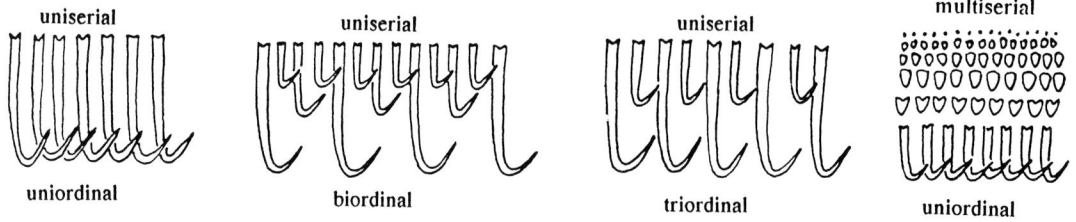

prothorax   mesothorax   4th abdominal segment   9th & 10th abdominal segments

Arrangements of crochets on prolegs

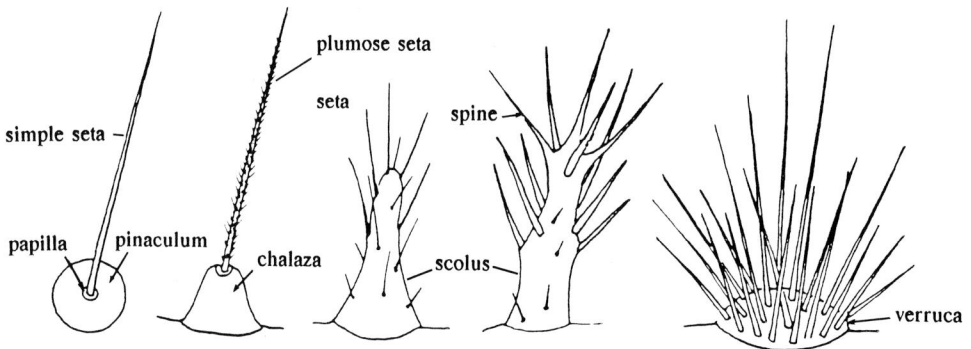

uniserial   uniserial   uniserial   multiserial

uniordinal   biordinal   triordinal   uniordinal

Types of setae

192

before transferring them to 80% alcohol to which a trace (about 2%) of glycerine has been added. Pale coloured larvae usually blacken if dropped direct into alcohol without the boiling water treatment. An alternative method, and one that is useful in the field, is to kill and preserve larvae in a fixing fluid composed of nine parts 80% alcohol and one part glacial acetic acid. Larvae should remain in this fluid for at least 24 hours but can safely be stored in it for several weeks. They should be finally transferred to 80% alcohol with glycerine. If larvae preserved in this way tend to retract their prolegs, they can be slightly inflated by injecting alcohol through the anus with a hypodermic syringe after killing. Other methods of preserving larvae are described by Peterson (1962).

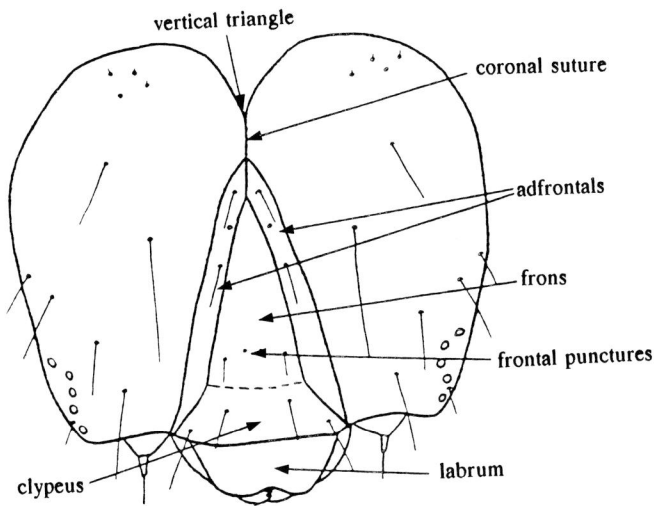

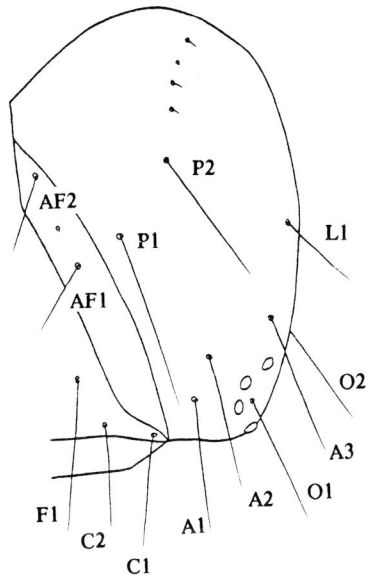

Anterodorsal view

General features of head

Setal nomenclature of head

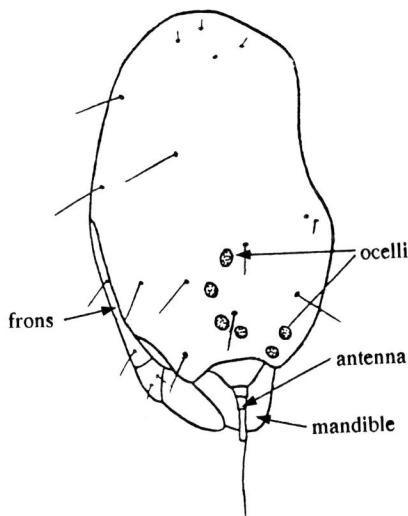

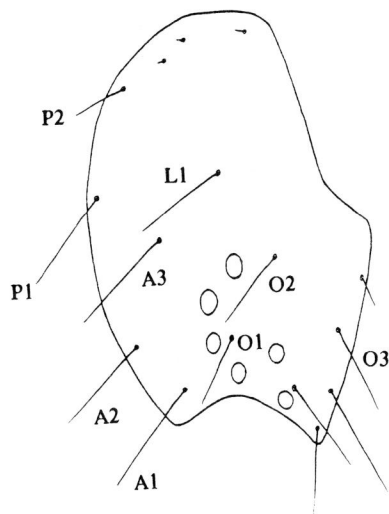

Lateral view

193

Larvae that have dried and shrivelled can usually be reconstituted to some degree by soaking in a warm 10% solution of potassium hydroxide. Shed larval skins may be treated in the same way, although these are extremely delicate and difficult to handle. A useful method of dealing with shed skins is described by Hinton (1956). In order to study the setae of small larvae, it may be necessary to mount the skin on a microscope slide. The larva is first treated in hot potassium hydroxide solution until the body contents are dissolved and the skin is slit along one side with a fine blade. After removal of the head, the skin can be spread out for mounting. Permanent mounts of larvae are seldom satisfactory as it is necessary to rotate the head and manoeuvre the skin for examination. Probably the best method is to preserve dissected larvae in tubes of alcohol and to examine them in temporary glycerine mounts.

KEY TO FAMILIES OF LARVAE
(DJC)

1.   Case-bearing 2

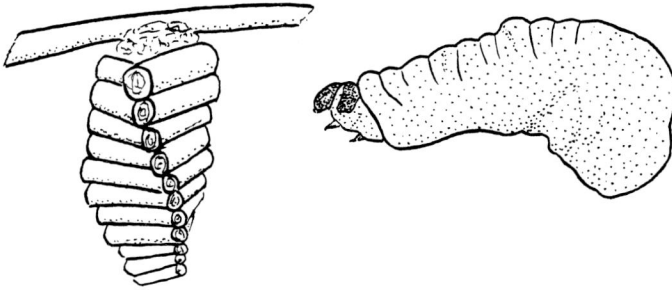

--   Not case-bearing 5

2.   Prothorax with prespiracular (L) pinaculum fused with thoracic plate 3

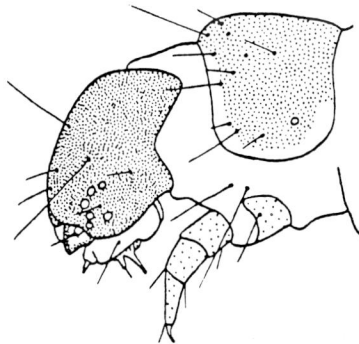

--   Prothorax with prespiracular (L) pinaculum separate from thoracic plate or not developed 4

3.   Meso- and metathorax with SV group unisetose. Small larvae in flat, oval cases on the ground ....INCURVARIIDAE (ADELINAE)

--   Meso- and metathorax with SV group bisetose (see figure below) ....PSYCHIDAE

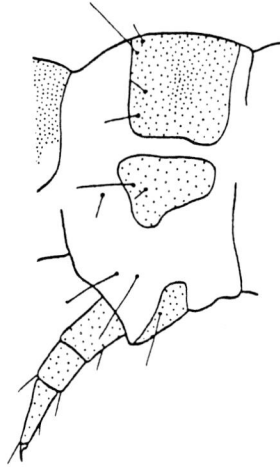

4.   Abdominal prolegs with crochets arranged in uniserial, transverse bands; case formed of plant material ....COLEOPHORIDAE

Coleophoridae

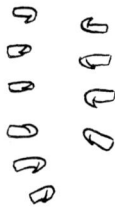

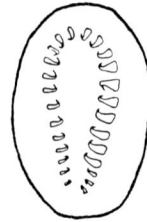

Tineidae (part)

--   Abdominal prolegs with crochets arranged in circle or penellipse; case formed of fragments of animal origin ....TINEIDAE (part)

5.   Thoracic legs absent or very weakly developed; prolegs with or without crochets ....
....6

--   Thoracic legs present 12

6.   Adfrontal suture situated behind the antenna ...........................ERIOCRANIIDAE

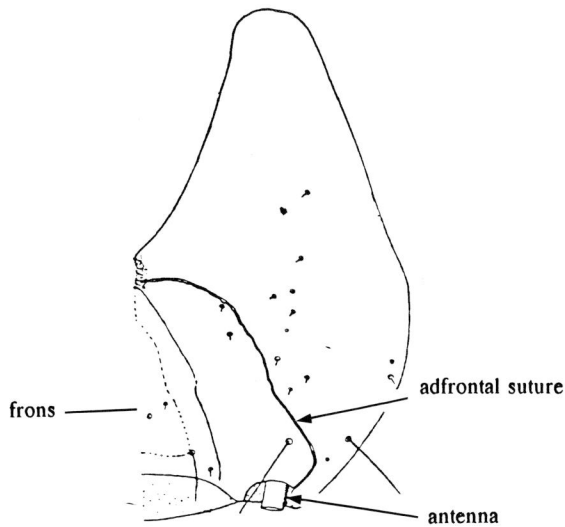

frons

adfrontal suture

antenna

-- Adfrontal suture situated in front of the antenna or not visible 7

7. Head with frons parallel-sided or shaped as an inverted trapezium 8

-- Head with frons triangular, its upper edge not or hardly included in the formation of the hind edge of the head capsule ....HELIOZELIDAE

8. Prolegs with crochets, at least on anal prolegs .............................TISCHERIIDAE

-- Prolegs without crochets 9

9. Head with frons rectangular; weakly developed, stump-like, unsegmented legs present on meso- and metathorax ....NEPTICULIDAE

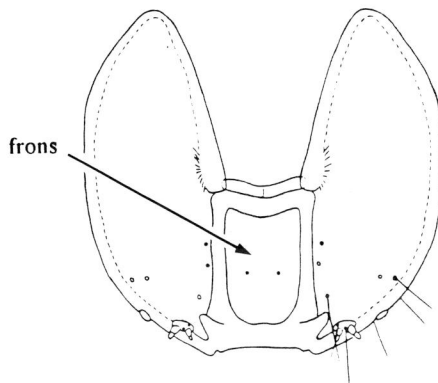

frons

-- Head with frons shaped as an inverted trapezium; legs absent 10

10. Frontal bridge connecting hind margins of adfrontal ridges present 11

adfrontal ridges present

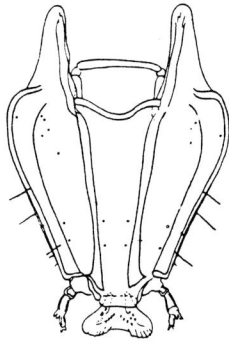

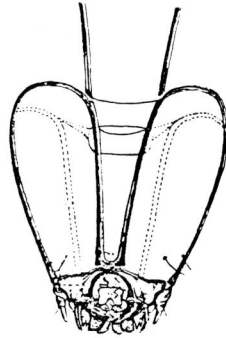

adfrontal ridges absent

--   Frontal bridge connecting hind margins of adfrontal ridges absent; mandible some-times with a large, dorsal conical projection; long very slender larvae ....OPOSTEGIDAE

11.   Antennae with several projections in addition to the sensory cones ....GRACILLARIIDAE (early instars)

antennal sensory projections

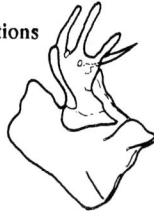

--   Antennae without projections other than the sensory cones ....GRACILLARIIDAE (PHYLLOCNISTINAE)

12.   Larvae with thoracic legs but without crochet-bearing prolegs or with only a few crochets (less than 5 on each proleg) 13

--   Larvae with thoracic legs and prolegs, the latter usually with crochets 16

13.   Larvae slug-like, often with scoli; head strongly retracted into prothorax ....LIMACODIDAE

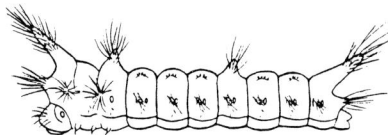

head retracted

--   Larvae subcylindrical; without secondary setae; head not retracted into prothorax 14

14.   Small larvae, less than 3cm long, without patches of spicules on the dorsal surface 15

--   Large larvae, more than 5cm long, with prominent patches of spicules on the dorsal surface ....CASTNIIDAE (part)

15.   Prothorax with two prespiracular (L) setae and with spiracle on a sclerotised plate ....GLYPHIPTERIGIDAE

--   Prothorax with three prespiracular (L) setae and spiracle not on a sclerotised plate ....GELECHIIDAE (part)

16.   Crochets of abdominal prolegs arranged in medial longitudinal bands (mesoseries), sometimes with a rudimentary lateroseries forming a pseudocircle (see figure below) 64

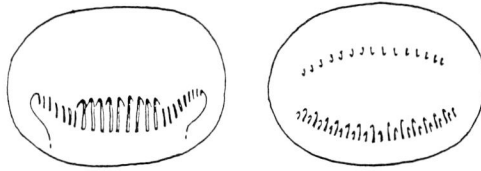

--   Crochets of abdominal prolegs not arranged in mesoseries (but see 64) 17

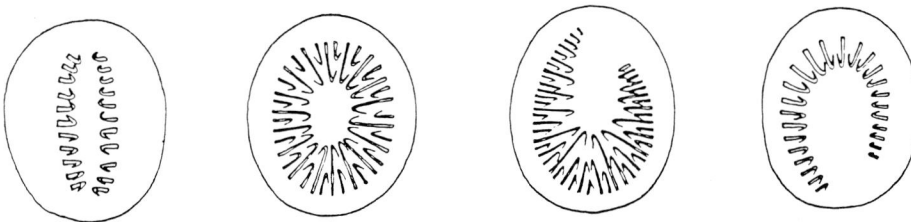

17.   Body with secondary setae, sometimes only in SV region 18

--   Body without secondary setae 23

18.   Head large and rounded with many secondary setae, sometimes on chalazae; prothorax narrowed to form a distinct "neck"; anal comb sometimes present ....HESPERIIDAE

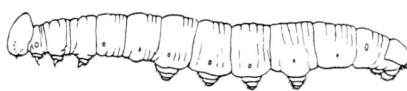

199

--   Head not large and rounded, without secondary setae; anal comb absent, small larvae 19

19.   Secondary setae generally distributed over the body 20

--   Secondary setae only in small groups, often confined to the SV region 21

20.   Tarsi of thoracic legs with specialised flattened setae; abdominal segment 9 without a dorsal plate ....AGONOXENIDAE (*A. argaula*)

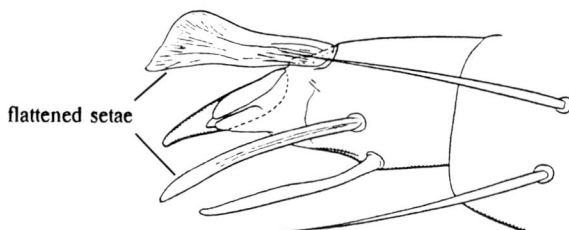

flattened setae

--   Tarsi of thoracic legs without specialised setae, abdominal segment 9 with a dorsal sclerotised plate ....COSMOPTERIGIDAE (*Blastodacna*)

21.   Abdominal prolegs with crochets in a uniordinal circle; tarsi of thoracic legs with specialised flattened setae ....AGONOXENIDAE

--   Abdominal prolegs with crochets biordinal or triordinal (possibly multiordinal); tarsi of thoracic legs without specialised setae 22

22.   Abdominal prolegs with crochets biordinal; secondary setae not confined to SV group; thoracic segments with SV group trisetose ................................SCYTHRIDIDAE

--   Abdominal prolegs with crotchets triordinal to multiordinal; secondary setae confined to SV group; meso- and metathorax with SV group unisetose ....
....OECOPHORIDAE (XYLORYCTINAE)

23.   Abdominal segment 6 without prolegs ...................................GRACILLARIIDAE

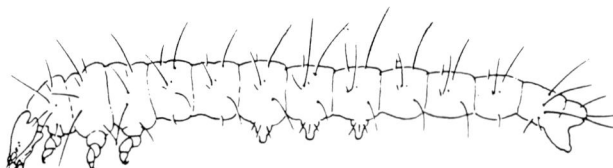

--   Abdominal segment 6 with prolegs 24

24.   Body globular; head reduced to a small triangular capsule; ocelli grouped on a raised tubercle. Larvae parasitic on Hemiptera ...........................EPIPYROPIDAE

--   Body not globular; head normal; ocelli not on a raised tubercle 25

25.   Abdominal spiracles on pinacula 26

--   Abdominal spiracles not on pinacula 27

26.   Pinacula surrounding spiracles extended dorsocranially; prolegs with few crochets or without crochets ....OCHSENHEIMERIIDAE

--   Pinacula surrounding spiracles not extended dorsocranially; prolegs with crochets arranged in a circle .............................YPONOMEUTIDAE (ACROLEPIINAE)

27.   Prothorax with prespiracular (L) group trisetose 28

--   Prothorax with prespiracular (L) group bisetose 58

28.   Prothorax with prespiracular (L) pinaculum fused with thoracic plate 29

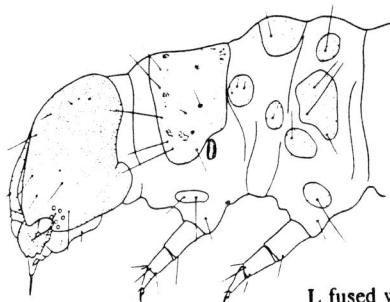

L fused with thoracic plate

--   Prothorax with prespiracular (L) pinaculum separate or not developed 30

29.   Prolegs with crochets in a uniserial circle; small larvae ....TINEIDAE (part)

--   Prolegs with crochets in a biserial or multiserial circle; larvae medium size to large ....HEPIALIDAE

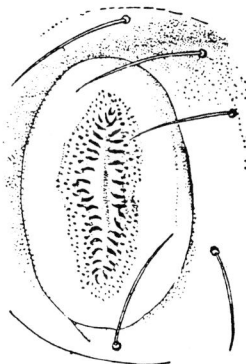

Hepialidae proleg

30. Abdominal prolegs with crochets arranged in uniserial transverse bands 31

-- Abdominal prolegs with crochets arranged in circles or penellipses 35

31. Abdominal prolegs usually with one transverse band of crochets, occasionally with two bands; anal plate with six setae; small larvae in fruits or buds ....
....INCURVARIIDAE (INCURVARIINAE)

-- Abdominal prolegs each with two transverse bands of crochets; anal plate with eight setae 32

32. Abdominal segments with setae L1 and L2 always remote from each other ....
....LYONETIIDAE (BUCCULATRICINAE)

-- Abdominal segments with setae L1 and L2 close 33

33. Each anal proleg with crochets divided into two groups ....GELECHIIDAE (part)

-- Each anal proleg with crochets in an undivided band 34

34. Dorsal surface of body with raised patches of spicules; large larvae c7cm ....
....CASTNIIDAE (part)

-- Dorsal surface of body without raised patches of spicules, small to medium sized larvae ....SESIIDAE

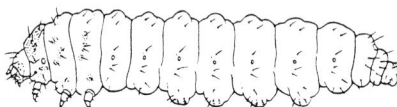

35. Small larvae mining in grasses ....ELACHISTIDAE

-- Not mining in grasses 36

36. Abdominal segment 3 with L1 and L2 remote, or in very small larvae L2 missing 37

-- Abdominal segment 3 with L1 and L2 close, often on the same pinaculum 42

37. Crochets arranged in a uniserial circle or penellipse 38

-- Crochets arranged in a multiserial circle or a penellipse surrounded by a circle of small crochets 41

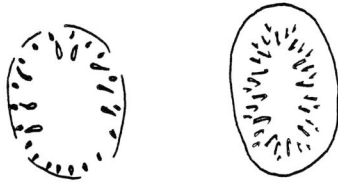

38.   Prolegs long, almost as long as thoracic legs ............YPONOMEUTIDAE
                                                            (PLUTELLINAE part)

--   Prolegs considerably shorter than thoracic legs 39

39.   Prothorax with prespiracular (L) setae about as far from the spiracle as the setae
       are from each other 40

--   Prothorax with prespiracular (L) setae about twice as far from the spiracle as the
       setae are from each other 41

40.   Abdominal segments with SD2 absent ......................YPONOMEUTIDAE
                                                    (ARGYRESTHIINAE)

--   Abdominal segments with SD2 present ....TINEIDAE (part)

41.   Abdominal prolegs with crochets arranged in a multiserial circle...................
       ....YPONOMEUTIDAE (YPONOMEUTINAE)

Yponomeutinae

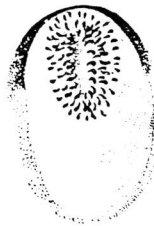

Plutellinae (part)

--   Abdominal prolegs with crochets arranged in a penellipse surrounded by a circle of
       small crochets ...............................YPONOMEUTIDAE (PLUTELLINAE part)

42.   Abdominal segment 9 with setae D2 closer together than setae D1 of abdominal segment
       8; often on a common pinaculum 43

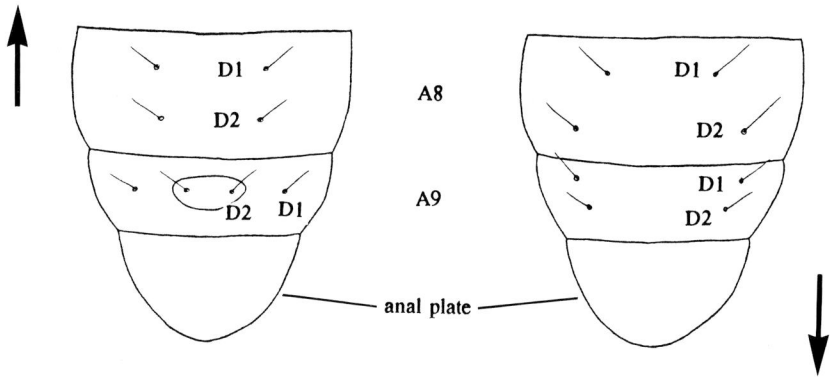

— Abdominal segment 9 with setae D2 at least the same distance apart as setae D1 of abdominal segment 8 51

43. Prolegs pencil-like, almost as long as thoracic legs 44

— Prolegs short, essentially smaller than thoracic legs 45

44. Abdominal prolegs each with a sclerotised collar ....GELECHIIDAE (part)

— Abdominal prolegs without sclerotised collars ....CHOREUTIDAE

45. Prothorax with spiracle situated dorsad of L pinaculum................BLASTOBASIDAE

— Prothorax with spiracle situated posterior to L pinaculum 46

46. Metathoracic legs swollen and club-shaped .............................OECOPHORIDAE
(CHIMABACHINAE)

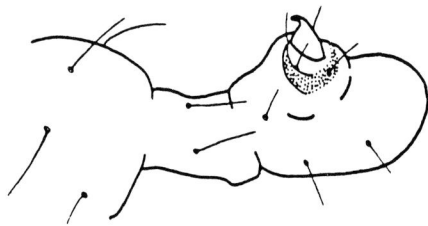

— Metathoracic legs not swollen 47

47. Head rugulose; each abdominal segment with all four D setae on a large sclerotised plate ....METARBELIDAE (*Indarbela*)

— Head smooth; D setae of abdominal segments not on large sclerotised plates 48

48.   Head wedge-shaped; large larvae ....COSSIDAE (ZEUZERINAE)

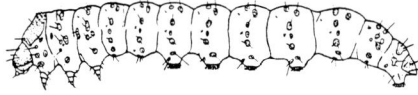

--   Head normal, round or oval shaped 49

49.   Prolegs each with a surrounding collar or with a sclerotised, central foot-patch ...
    ....GELECHIIDAE (part)

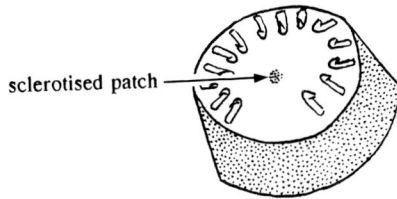

sclerotised patch

--   Prolegs without sclerotised collars or foot-patches 50

50.   Abdominal segment 8 with SD1 usually distinctly anterior to spiracle; anal comb
    sometimes present (figure below)..TORTRICIDAE

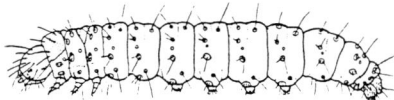

--   Abdominal segment 8 with SD1 anterodorsad of spiracle; anal comb absent ....
    ....OECOPHORIDAE (part)

51.   Head with L1 further from A3 than A3 is from A2 ....GELECHIIDAE (part)
    OECOPHORIDAE (part)

--   Head with L1 nearer to A3 than A3 is to A2 or distance is the same 53

52.   Metathorax with distance between coxae twice as great as their breadth ....
    ....COSMOPTERIGIDAE (part)

--   Metathorax with distance between coxae less than twice their breadth 53

206

Hyblaeidae

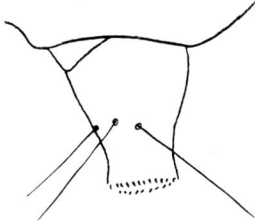

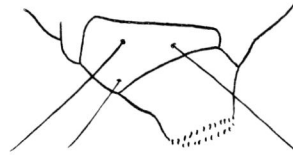

SV setae not on proleg

--   SV setae situated at base of abdominal prolegs; metathorax with SD1 considerably longer than SD2 63

63.   Crochets of prolegs biordinal or triordinal (only exceptionally uniordinal) ....
        ....PYRALIDAE (part)

--   Crochets uniordinal ....ALUCITIDAE

64.   Mesoseries strongly curved, closely resembling a penellipse, prolegs as long as thoracic legs, with strongly sclerotised lateral plates ....EPIPLEMIDAE

--   Mesoseries not curved, not resembling a penellipse 65

65.   Well developed prolegs present on abdominal segments 6 and 10 only, sometimes with rudimentary prolegs on abdominal segments 4 and 5 ....GEOMETRIDAE

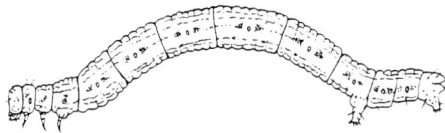

--   Well developed prolegs present at least on abdominal segments 5 and 6 66

66.   Larvae without secondary setae, except in some cases on the prolegs 67

--   Larvae with many secondary setae (some Sphingidae have microscopic secondary setae but larvae of this family are recognisable by their large size and the presence of a dorsal horn on abdominal segment 8) 72

67. Prolegs with secondary setae 68

-- Prolegs without secondary setae 69

68. Setae on raised pinacula; epicranial index (length of frons/length of coronal suture) more than 0.8 ....DILOBIDAE

-- Setae not on raised pinacula; epicranial index less than 0.6 ....NOTODONTIDAE (part)

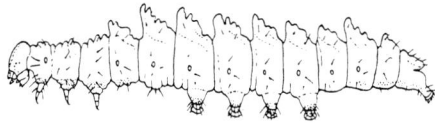

69. Abdominal prolegs with crochets arranged in a pseudocircle ....DREPANIDAE

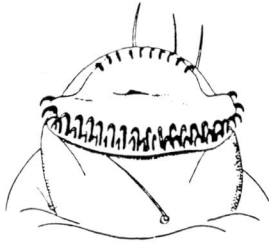

-- Abdominal prolegs with crochets in simple mesoseries 70

70. Meso- and metathorax with SV group bisetose 71

-- Meso- and metathorax with SV group unisetose ....NOCTUIDAE (part)

71. Body covered with minute spinules; D2 and L1 arising from elongate tubercles or conical projections, which are especially conspicuous on the meso- and metathorax; crochets homoideous ....NOCTUIDAE (*Earias*)

-- D2 and L1 not arising from elongate tubercles; prolegs with ends of plantae drawn out, crochets usually heteroideous ....ARCTIIDAE (part)

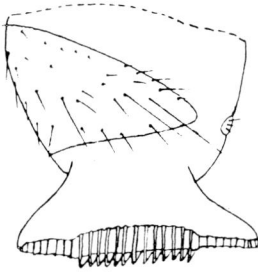

72. Additional rudimentary prolegs, without crochets, present on abdominal segments 2 and 7 ....MEGALOPYGIDAE

-- Abdominal segments 2 and 7 without prolegs 73

73. Abdominal segments 6 and 7 with mid-dorsal, eversible glands or if abdominal segement 6 is without a dorsal gland, then abdominal segment 8 with a large mid-dorsal hair pencil; body usually with a series of dorsal hair tufts or pencils ....LYMANTRIIDAE

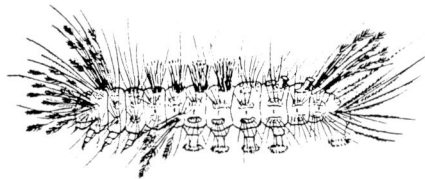

-- Abdominal segments 6 and 7 without mid-dorsal glands 74

74. Prolegs with crochets heteroideous; ends of plantae drawn out 75

-- Prolegs with crochets homoideous 76

75. Mesothorax with only one verruca above the spiracle ....ARCTIIDAE
(CTENUCHINAE)

-- Mesothorax with more than one verruca above the spiracle ....ARCTIIDAE (part)

76. Prolegs with a row of soft basal plates below the crochets; head retracted into the thorax ....ZYGAENIDAE

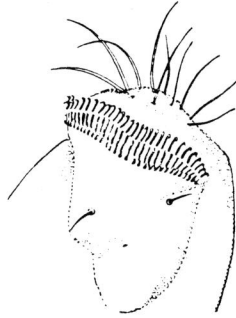

--   Prolegs without soft basal plates below the crochets 77

77.   Crochets uniordinal 78

--   Crochets biordinal or triordinal 83

78.   Anal plate produced to form a pair of anal projections; head with ocellus 3 enlarged
....NYMPHALIDAE (SATYRINAE)

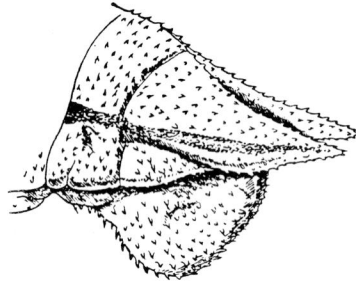

--   Anal plate not forked 79

79.   Spiracles small and circular; prolegs long and thin, with enlarged plantae ....
....PTEROPHORIDAE

--   Spriracles large and usually elliptical; prolegs short 80

80.   Some setae arranged on verrucae or flat, verruca-like plates; mandibles with secondary setae 81

--   Setae not grouped on verrucae or similar plates; mandibles without secondary setae
....NOTODONTIDAE (PHALERINAE)

81. Abdominal segment 8 with spiracle at least 1.5x the height of that on abdominal segment 7..82

-- Abdominal segment 8 with spiracle less than 1.5x (usually about 1.25x) the height of that on abdominal segment 7 ....THAUMETOPOEIDAE

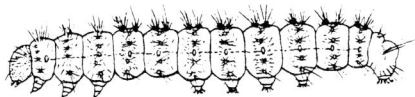

82. Head with secondary setae ....EUPTEROTIDAE

-- Head without secondary setae ....NOCTUIDAE (part)

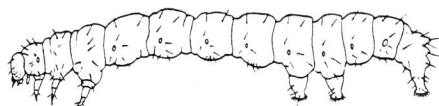

83. Secondary setae very variable in length (some at least 5x the length of others), not arising from verrucae or scoli ....LASIOCAMPIDAE

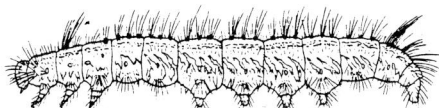

-- Secondary setae generally short and fairly equal in length or larvae with distinct warts or spines 84

84. Abdominal segment 8 with a mid-dorsal horn, tubercle or spine 85

-- Abdominal segment 8 without a mid-dorsal horn or spine 88

85. Body with spines or large protuberances on thorax and elsewhere 86

--   Body without spines or prominent projections on thorax, rarely elsewhere, except on abdominal segment 8 87

86.   Head angular or covered with spines; crochets usually triordinal ....NYMPHALIDAE (part)

--   Head rounded and smooth; crochets biordinal ....SATURNIIDAE (part)

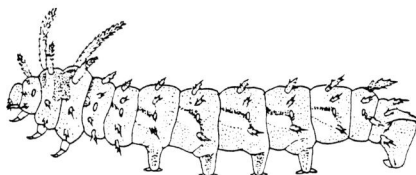

87.   Abdominal segments each divided into six to eight annuli; prolegs not widely separated ....SPHINGIDAE (part)

--   Abdominal segments indistinctly divided into two or three annulets; prolegs very widely separated ....BOMBYCIDAE

88.   Crochets arranged in a mesoseries divided by a central gap or reduced centrally, with a pad-like protuberance in the middle ....LYCAENIDAE

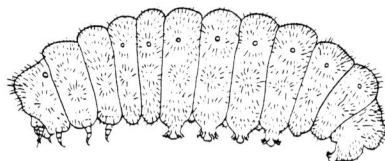

--   Crochets arranged in an undivided mesoseries 89

89.   Prothorax with a dorsal eversible gland situated in a groove or pit ....PAPILIONIDAE

212

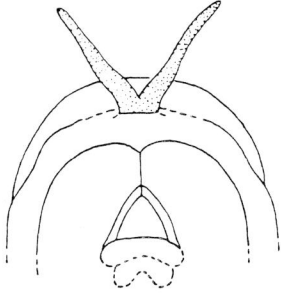

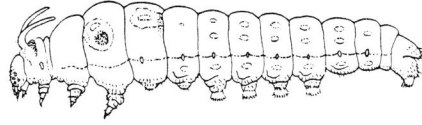

--   Prothorax without a dorsal eversible gland 90

90.   Head and body without protuberances 91

--   Head or body with spines or other protuberances 92

91.   Body and head thickly covered with short hairs, usually on small warts ....PIERIDAE

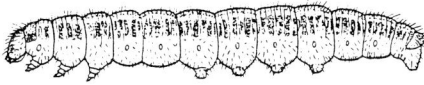

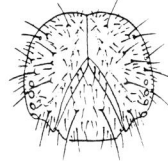

--   Body with minute secondary setae, appearing completely smooth; abdominal segment 8 with a sclerotised dorsal patch ....SPHINGIDAE (part)

92.   Head with protuberances ....NYMPHALIDAE (part)

--   Head without protuberances ....SATURNIIDAE (part)

FAMILY ACCOUNTS OF LARVAE
(DJC)

ARCTIIDAE

Medium sized to large larvae feeding on foliage, some constructing silken webs. Most have well developed verrucae bearing plumose setae. Head without secondary setae. Prolegs with expanded plantae; crochets uniordinal, arranged in heteroideous mesoseries (except in Lithosiinae where they are homoideous).

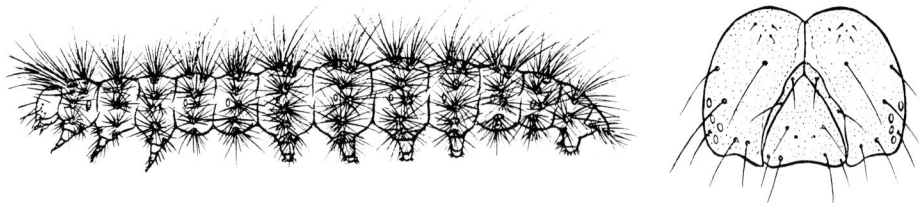

COLEOPHORIDAE

Small case-bearing larvae mine leaves or feed externally on a wide range of plants. Head with adfrontals reaching the vertical triangle. Prothorax with a strongly developed prothoracic shield. Meso and metathorax sometimes also with dorsal shields or plates. Primary setae only present and those sometimes inconspicuous or absent. Prolegs with uniordinal crochets in two transverse bands or without crochets.

COSSIDAE

Medium sized to large larvae mostly boring in wood, roots or bulbs. Primary setae only present (except for a few subdorsal secondary setae in some species). Head with frons 1/3 to 1/2 length of head, sometimes wedge-shaped. Prothoracic shield sometimes with a posterior band of spicules or platelets. Prolegs with biordinal or triordinal crochets in a complete circle.

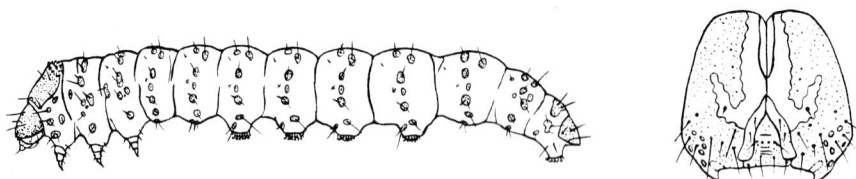

DREPANIDAE

Small larvae feeding on the foliage of trees and shrubs. Body without secondary setae but integument often densely spiculate. Abdominal prolegs with uniordinal or biordinal crochets in a mesoseries and a rudimentary lateroseries, forming a pseudocircle. Anal prolegs absent. Anal plate terminating in a spine or pointed process. Head sometimes with a pair of dorsal tubercles. Illustrated below.

214

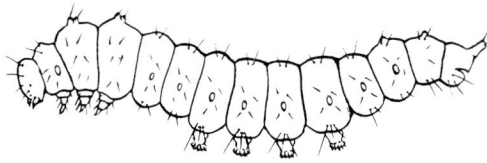

GELECHIIDAE

Small larvae mining or boring in fruits, seeds, tubers, stems, leaves and twigs of various plants or rolling or webbing foliage. Secondary setae absent. Crochets of prolegs sometimes greatly reduced in number. Prolegs often with strongly sclerotised collars. Anal prolegs with crochets sometimes divided into two groups. An anal comb sometimes present. Distinguished from similar families by setal arrangement.

GEOMETRIDAE

Small to medium sized larvae feeding openly on foliage and flowers. Body slender, sometimes with numerous secondary setae below the level of the spiracles. Only two pairs of fully developed prolegs present on abdominal segments 6 and 10. Anal prolegs and paraprocts often strongly developed. Crochets usually biordinal, arranged in mesoseries.

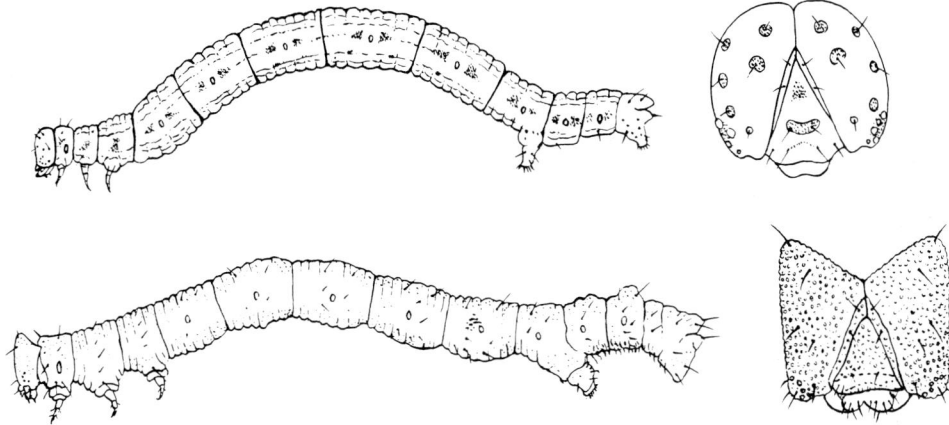

GRACILLARIIDAE

Small or very small larvae mostly mining leaves, bark or fruit, although many species feed openly in later instars. Mining larvae are strongly flattened with specialised, wheel-like mandibles. Later instars are cylindrical but can be distinguished

215

from other small larvae by the absence of prolegs on abdominal segment 6. Primary setae only present. Ocelli often reduced to 1 pair.

HEPIALIDAE

Medium sized to large larvae boring in roots, stems and tubers of various plants and sometimes behaving as cutworms. Primary setae only present, often on large pinacula. Prespiracular (L) pinaculum of prothorax fused with prothoracic shield. Head with ocelli tending to be grouped in two vertical rows. Prolegs with multiordinal crochets arranged in circles.

HESPERIIDAE

Medium sized to large larvae chiefly in cereals, grasses, palms and legumes, webbing and rolling the foliage. Head very large and conspicuous as prothorax is distinctly narrower than rest of body. Segments divided into indistinct annulets. Prolegs with crochets in circles. Anal comb often present.

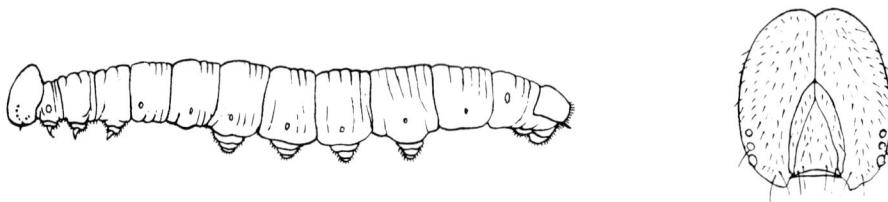

LASIOCAMPIDAE

Medium sized to large larvae feeding on foliage, some living in webs. Body covered with long and short secondary setae but without hair tufts or pencils. Head with secondary setae. Prolegs with crochets in biordinal mesoseries.

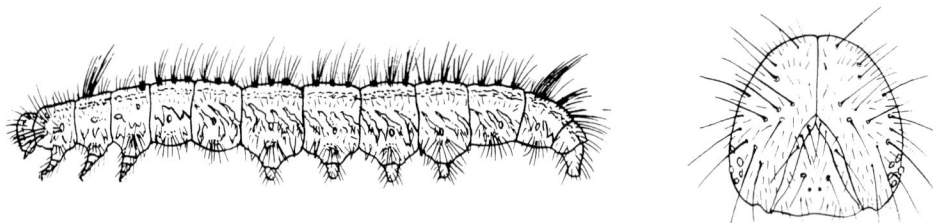

LIMACODODAE

Small to medium sized, slug-like larvae feeding mainly on the foliage of trees and shrubs. Head small, deeply retracted into thorax. Body often with scoli or other protuberances bearing poisonous spines. Prolegs absent but sometimes replaced by sucker-like discs.

216

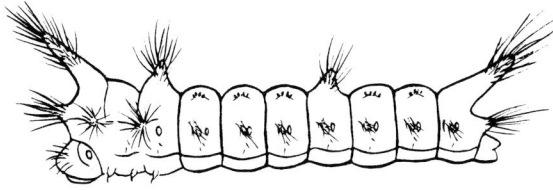

LYCAENIDAE

Small larvae feeding on the foliage, flowers and fruit of various plants but particularly on members of the family Leguminosae. Body somewhat stout and fusiform. Head usually very small and retractile. Body covered with small secondary setae, often with stellate bases. Dorsal honey gland on abdomen indicated by a transverse depression or slit. Prolegs with biordinal or triordinal crochets arranged in interrupted mesoseries with a median fleshy lobe.

LYMANTRIIDAE

Medium to large size larvae feeding mainly on the foliage of trees and shrubs, sometimes living in a communal nest. Body with many plumose secondary setae arising from verrucae. Dorsal tufts or pencils of hairs often present on thorax and abdomen. Mid-dorsal glands present on abdominal segments 6 and 7 although sometimes absent from 6. Head with secondary setae. Prolegs with uniordinal homoideous setae arranged in mesoseries. Some species have poisonous setae.

217

LYONETIDAE

Very small larvae mining leaves, often making blotches or blisters. Head with triangular frons; adfrontals reaching the vertical triangle. Primary setae only present. Crochets in uniordinal circles or transverse bands.

NEPTICULIDAE

Very small, somewhat flattened larvae making linear mines in leaves, bark or fruits of a wide range of plants. Head with frons rectangular and only one ocellus on each side. Segmented thoracic legs absent and if prolegs present, they do not bear crochets. Primary setae only present and those often difficult to see or absent.

NOCTUIDAE

Medium sized to large larvae feeding mainly on foliage although others are flower or fruit feeders, stem borers, root feeders or cutworms. Most have only primary setae although some have well developed secondary setae which are sometimes arranged on verrucae. Head always without secondary setae. Prespiracular (L) group of prothorax normally bisetose. Crochets usually uniordinal, arranged in homoideous mesoseries. Some groups have reduced prolegs but at least those on abdominal segments 5,6, and 10 are developed.

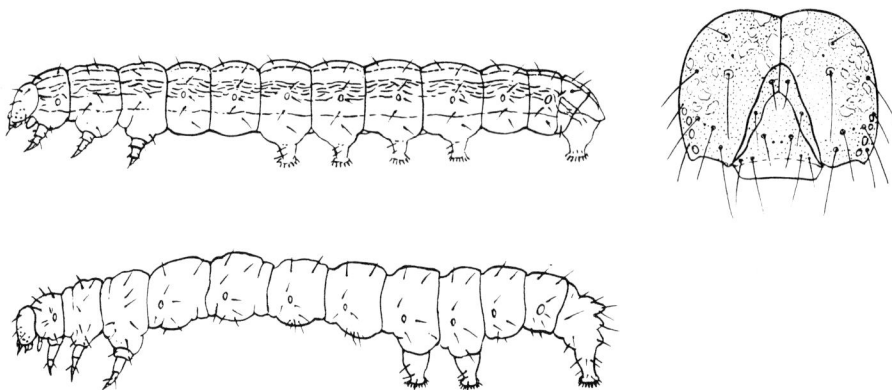

NOTODONTIDAE

Medium sized to large larvae feeding primarily on the foliage of trees and shrubs and plants of the family Leguminosae. Most are without secondary setae except on the prolegs but in some groups numerous secondary setae are scattered over the body. These secondary setae are not grouped on verrucae. In some cases, the anal prolegs are rudimentary or without crochets and sometimes they are greatly elongated to form tails with eversible flagellae. Many species possess dorsal protuberances which may be paired or single.

218

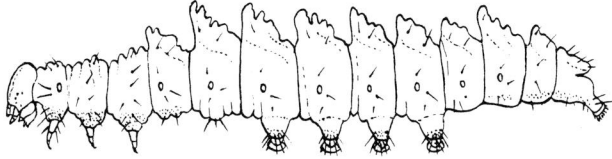

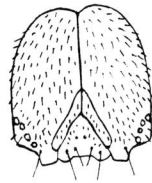

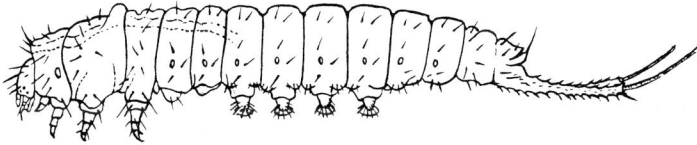

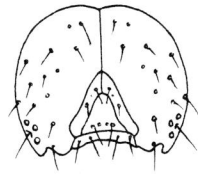

NYMPHALIDAE

Medium sized larvae feeding on the foliage of a wide range of plants. Satyrine larvae feed primarily on cereals and grasses. Head or body with scoli, except in the Satyrinae which have the body segments divided into annuli and the anal plate bifurcate. Prolegs with uniordinal, biordinal or triordinal crochets arranged in homoideous mesoseries.

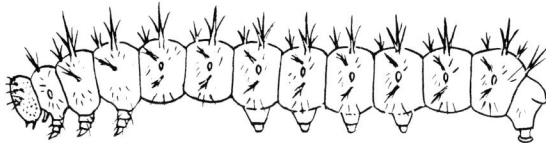

219

OPOSTEGIDAE

Small, extremely slender larvae mining under the bark of stems and stalks. Head strongly flattened and wedge-shaped, with one ocellus on each side or without ocelli. Frons trapezium-shaped with posterior margins not joined by a transverse bridge. A pair of blade-like structures extend posteriorly from the head into the thorax. Thoracic legs and prolegs absent, setae reduced or absent.

OECOPHORIDAE

Small larvae rolling or webbing foliage, flowers or fruits of various plants or feeding in decayed wood or other vegetable material. Primary setae only present, often on strongly pigmented pinacula. Head with ocelli sometimes reduced to one pair. Prolegs short and stout, bearing circles of biordinal crochets. Family characters of larvae difficult to define but various groups can be distinguished by differing setal arrangement.

PAPILIONIDAE

Medium sized to large larvae on the foliage of various plants. Head smaller than prothorax and often slightly retracted into it. Body appearing smooth but with numerous minute secondary setae. Prothorax with a dorsal Y or V-shaped eversible osmeterium (gland). Crochets of prolegs biordinal or triordinal, arranged in mesoseries, often with a biordinal lateroseries of crochets forming a pseudocircle.

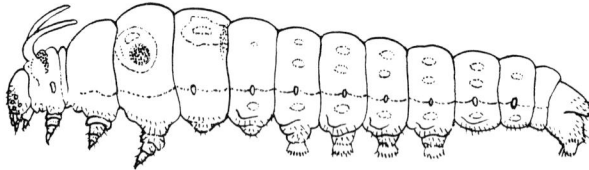

PIERIDAE

Medium sized larvae feeding on the foliage of a wide range of plants. Body segments divided into annuli. Head and body with numerous small secondary setae situated on small to large raised papillae. Prolegs with crochets biordinal or triordinal, arranged in mesoseries.

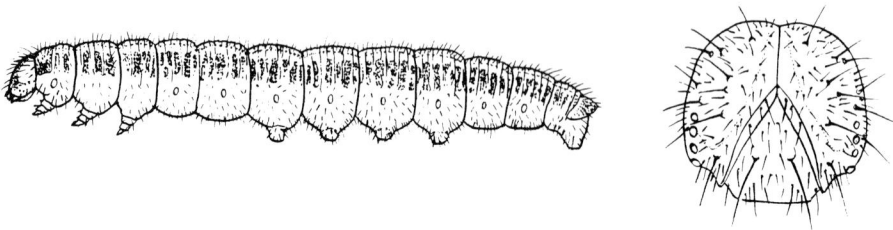

PSYCHIDAE

Small to medium sized larvae living in portable silken bags often covered with fragments of the foodplant. Feeding openly on a wide range of plants. Larvae with well developed thoracic legs. Prothoracic shield strongly developed, fused with prespiracular (L) pinaculum. Primary setae only present. Spiracle of prothorax large and elliptical, situated in a horizontal position. Prolegs with crochets in a uniordinal penellipse.

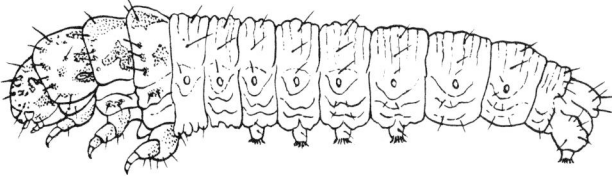

PYRALIDAE

Small to medium sized larvae with diverse feeding habits, although many are leaf rollers. Some species are pests of stored foodstuffs and others are important stem borers of Gramineae. Primary setae only present, setae often on strongly pigmented pinacula. Prespiracular (L) group of prothorax bisetose. Prolegs with crochets in a complete circle or a penellipse.

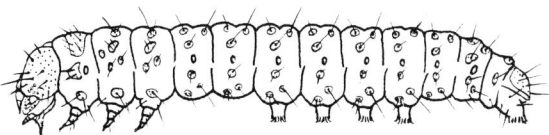

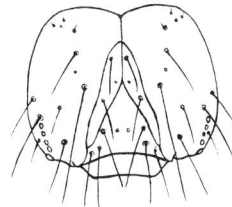

SATURNIIDAE

Large larvae, many feeding openly on the foliage of trees and shrubs. Body often with numerous spines or branched scoli. Prolegs strongly developed, with numerous small secondary setae; crochets biordinal, arranged in mesoseries. Anal prolegs somewhat flattened externally to form a large, triangular plate-like structure.

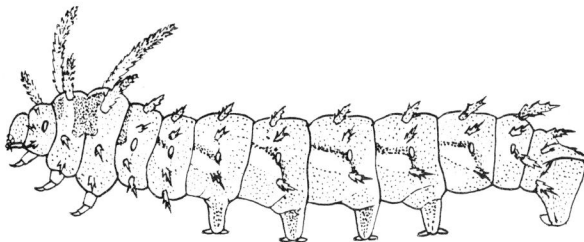

221

SESIIDAE

Small to medium sized larvae boring in wood, stems and roots of various plants. Some larvae form galls and others are inquiline in galls. Head smaller than thorax, which is often somewhat swollen in appearance. Body pale, without pattern; secondary setae absent. Head with ocelli 1-4 arranged in a trapezoid, remote from ocelli 5 and 6. Prolegs with uniordinal crochets in two transverse bands.

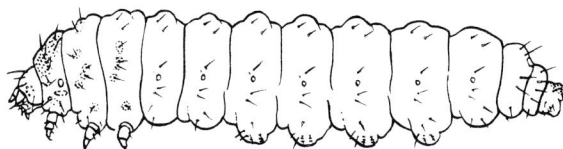

SPHINGIDAE

Large larvae feeding openly on foliage. Body with secondary setae often microscopic. Abdominal segments divided into annulets dorsally. Abdominal segment 8 with a mid-dorsal horn or at least with a sclerotised patch. Prolegs with biordinal crochets in mesoseries. Anal prolegs flattened, forming a triangular pyramid with the anal plate.

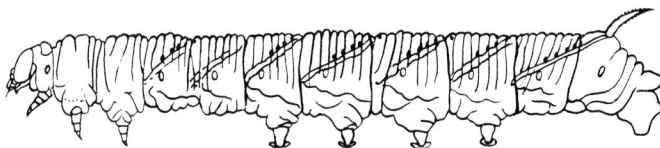

THAUMETOPOEIDAE

Medium sized to large larvae feeding on the foliage of trees and shrubs, usually in a communal silken nest. The larvae of many species move about in columns or processions. Body with numerous plumose setae, some on verrucae. Head and mandibles with secondary setae. Prolegs with uniordinal crochets in homoideous mesoseries.

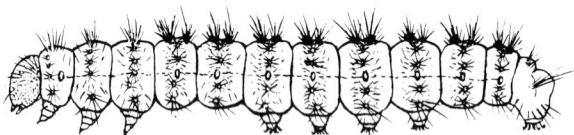

TINEIDAE

Small larvae, many of which are case-bearers, feeding largely on organic matter of animal origin although some feed on seeds, fungi and other plant material. Head with adfrontals reaching the vertical triangle. Ocelli sometimes reduced in number or absent. Primary setae only present, sometimes inconspicuous. Prolegs with crochets in uniordinal uniserial circles.

TORTRICIDAE

Small larvae usually rolling, folding or webbing leaves together, although some feed in buds, fruits, and nuts and others in twigs, stems and rootstocks. Primary setae only are present and these are often situated on strongly pigmented pinacula. Seta SD1 is usually anterior to the spiracle on abdominal segment 8. An anal comb is often present. Abdominal prolegs with uniordinal crochets arranged in circles.

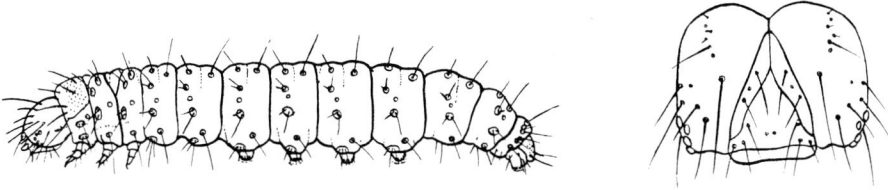

YPONOMEUTIDAE

A heterogeneous group of larvae mining, boring, skeletonizing or webbing foliage, twigs, buds, flowers or fruit. Small larvae with primary setae only. Many species have well developed prolegs with more than one ring of crochets. Distinguished from other small larvae by differences in setal arrangement. Some live in communal webs.

REFERENCES TO WORKS ON LARVAE
(DJC)

BOMBYCIDAE
Miyata, 1970.

DREPANIDAE
Nakajima, 1970.

GENERAL
Capps, 1963; Chu, 1949; Dekle, 1965; Fracker, 1915; Gerasimov, 1952; Hinton, 1943, 1946; Isaac & Rao, 1941; Okumura, 1961; Pant & Chatterjee, 1950, 1951; Peterson, 1962; Stehr, 1987; Williams, 1953; Werner, 1958; Zimmerman, 1978.

GEOMETRIDAE
McGuffin, 1958-1981; Singh, 1953, 1957.

LYMANTRIIDAE
Gardner, 1938.

LIMACODIDAE
Smith, 1965; Cock, Godfray & Holloway, in press.

NOCTUIDAE
Beck, 1960; Brown & Dewhurst, 1975; Crumb, 1929, 1956; Gardner, 1941-1948; Godfrey, 1972.

PSYCHIDAE
Davis, 1975.

PYRALIDAE
Aitken, 1963; Hasenfuss, 1960; Mathur, 1954, 1959; Mathur & Singh, 1963; Neunzig, 1979.

SESIIDAE
Mackay, 1968

TINEIDAE
Hinton, 1956

TORTRICIDAE
Chapman & Lienk, 1971; Mackay, 1959, 1962; Swatschek, 1958.

ZYGAENIDAE
Gardner, 1942.

INTRODUCTION TO PUPAE
(JDH)

The study of lepidopteran pupae and the relevance of their characterisitics to higher classification has not advanced considerably beyond the pioneering studies of Mosher (1916), except perhaps with regard to the subordinal classification. Yet pupal features, and those of the cocoon (if present), promise to be of great value in many instances.

Characters of the cremaster, the arrangement of hooks and spines at the apex of the abdomen with which the pupa anchors itself to the cocoon or to a silken pad, provide strong justification for the naturalness of several noctuid subfamilies (especially the Stictopterinae). Possession of an unusual plate associated with the eye covering indicates a sister-relationship between the Limacodidae and Megalopygidae.

The form of the cocoon, or its equivalent, is also of taxonomic value. A boat shaped cocoon is seen in the noctuid subfamilies Nolinae, Chloephorinae and Sarrothripinae, supporting the hypothesis that, together, they form a natural group. A thoracic silken girdle characterises the Papilionoidea *sensu lato* (see p.181), and an abdominal one is a feature common to nenbers of a tribe of the Sterrhinae (p.177).

It is therefore essential that both pupae and cocoon are retained with eclosed adults when reared.

In the diagrams following, some of the features of 'primitive' (decticous, exarate) and 'advanced' pupae are illustrated, as are male and female genital scars, a means of sexing pupae.

Literature on pupae: Mosher (1916); Gardner (1948).

decticous, exarate
(with mandibles, appendages free)

1+2

3

4

5

6

7

8+9+10

♀

7+8+9+10

♂

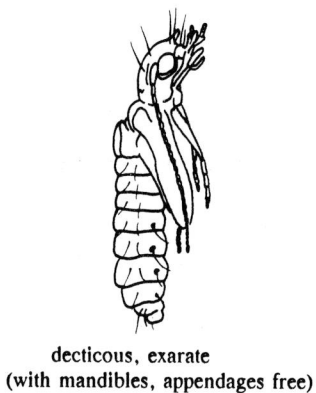

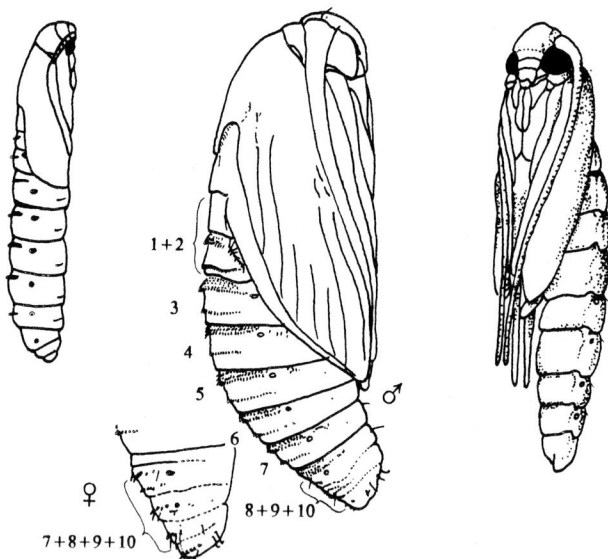

adecticous, obtect
(without mandibles, appendages loosely
or firmly fused to body)

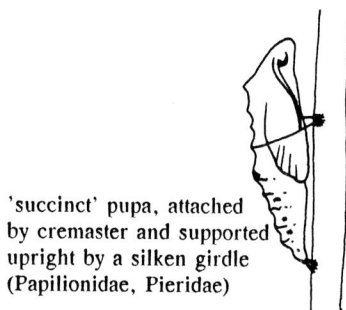

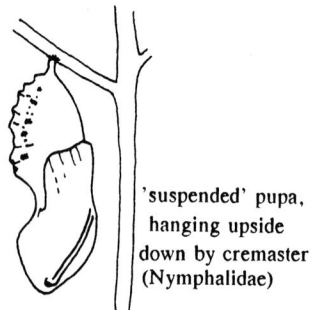

'succinct' pupa, attached
by cremaster and supported
upright by a silken girdle
(Papilionidae, Pieridae)

'suspended' pupa,
hanging upside
down by cremaster
(Nymphalidae)

Papilionoid (butterfly) pupa or chrysalis

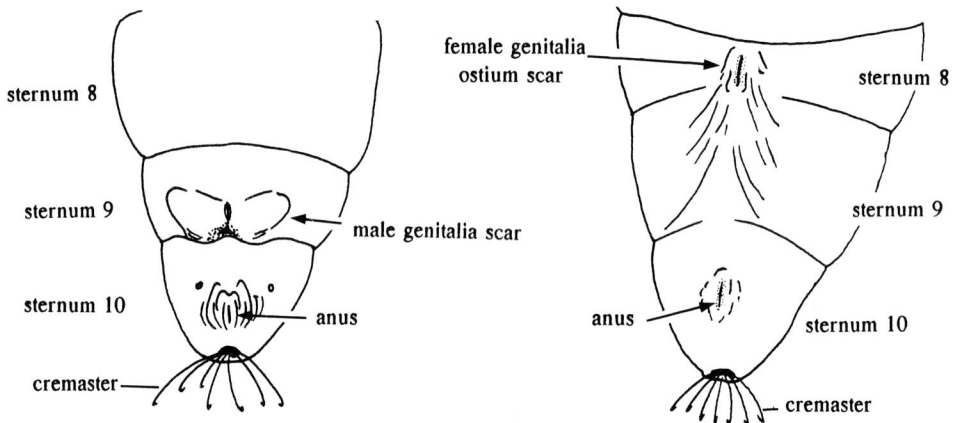

sternum 8

sternum 9

male genitalia scar

sternum 10

anus

cremaster

female genitalia
ostium scar

sternum 8

sternum 9

anus

sternum 10

cremaster

Sexing pupae

GLOSSARY OF TERMS

GENERAL

acuminate	tapering to a long point
adecticous	without mandibles (pupa)
amplexiform	mode of wing coupling through extensive overlap
anal	**angle** - posterior angle of wing (see tornus); **vein** - (loosely) all veins posterior to the cell and (strictly) veins A1, A2, A3, posterior to CuP
anastomosis	convergence of veins with fusion
anchylosed	elbowed, bent or crooked
anteriad	to the front of; the suffix **-ad** indicates "towards"
apomorphic	(character) derived (not primitve) character unique to a taxonomic unit
approximate	(of veins) approaching or close to each other *without* fusion
appressed	pressed closely down, or closely adjacent to
apterous	wingless
areole	secondary cell formed by anastomosis of the veins
bifid	forked, divided into two
brachypterous	with vestigial wings
caudal	'tail' region of insect
cell	area of wing membrane enclosed by radial and cubital veins
chaetosema	(pl. **-ata**) setose blister or zone above eye, posterior to ocellus and antenna
connate	(of veins) arising from the same point on cell margin
coronate	crown-like
costal margin	the 'leading edge' of the wing
counter-tympanal hood	ear-like structure on abdomen opposed to thoracic tympanum (Noctuoidea)
cremaster	cluster of hooks at posterior tip of pupa used to grip silk pad and support pupa
crown	top of head
cruciform	in the shape of a crucifix
decticous	with mandibles (pupa)
dentate	toothed
digitate	finger-like
discal	(of wing) about middle of wing, at end of cell
dorsum	(of wing) the posterior margin or 'trailing edge' of a wing
eclosion	'hatching' or emerging (from pupa/egg)
emarginate	with margin excavated or notched
exarate	(pupa) in which all appendages are free
explanate	splayed out and flattened
falcate	sickle-shaped, 'hook-tipped', or convexly curved (e.g. wing tip)
fascia	(pl. **-ae**) band or transverse line of wing markings (see p.156 for nomenclature)
fasciculate	with small groups of tufts
filiform	thread-like
floricomous	flower-like
frenulum	(frenate = with frenulum) spine or spines arising from base of hindwing costa which coupl with retinaculum on forewing
fringes	marginal scales of wings; row of projecting scales elsewhere
frons	front of head
furcate	forked (bifurcate, trifurcate etc.)
fusiform	spindle shaped

227

girdle	silken thread used to support pupa either thoracic (butterflies) or abdominal (some Sterrhinae)
haustellum	'tongue' or proboscis
humeral	(of hind wing) zone anterior to subcosta, expanded in amplexiform coupling
hyaline	clear or colourless, usually *without* scales, i.e. transparent
jugum	lobe of trailing edge of wing membrane proximal to thorax, separated from outer part of wing by a dorsal (convex) fold (jugal fold)
jugate	method of wing coupling where jugum forms part of the couple
irrorated	minutely speckled, dotted with pale coloured scales
lanceolate	lance or spear shaped; **sub-** less acute
lenticular	lens-shaped, double convex
maculation	spotting
neural, neuration	(of wings) on veins, venation
obtect	(of pupa) where appendages firmly fastened down to rest of body
ocellus	(of head) simple eye on head of adult or larva; (elsewhere) eye-like marking on adult wing or larval body (e.g. Sphingidae)
oligophagous	narrow range of diet
orbicular stigma	a round spot or ring, usually within the forewing cell (Noctuidae)
palpus/palp	(labial or maxillary) a segmented structure associated with the mouthparts
pecten	comb of bristles on labial palp or antennal scape (microlepidoptera) or on hindwing cubital vein (some Tortricidae)
pectinate	feather-like (also **plumose**)
pharate	intermediate between pupa and adult (imago)
plesiomorphic	primitive or ancestral (applied to taxonomic characters)
polyphagous	wide range of diet
porrect	directed straight forward
pubescent	finely hairy
retinaculum	(male only) posteriorly directed scaled lobe below base of subbasally on forewing radial vein stem, which engages the frenulum
rugose	roughened
scape	largest, subbasal segment of the antenna
sclerites	chitinous plates which comprise an insect's exoskeleton
sclerophyllous	(botanical) leaves modified for arid conditions
scobinate	finely and abundantly spined
scolus	(of larva) knob or process bearing secondary setae
serrate	saw-edged
sinuate	slightly curved inwards; **sinaceous** - waves
spiculose	with spicules
stellate	star-shaped
sternum, sternite	ventral sclerite of body segment
stigma	(pl. -ata) a spot or spots usually black and elongated, usually on forewing e.g. triad of stigmata in many Gelechiidae; see p.156 for nomenclature in Noctuidae
subbasal	almost at the base
subtruncate	terminating rather abruptly
suture	groove marking junction between skeletal sclerites
synanthropic	in association with man
tergum, tergite	dorsal sclerite of body segment
termen	outer margin of the wing, between apex and tornus
tornus	posterior angle of wing formed by termen and dorsum
tymbal organ	sound producing sclerite
tympanal organ	sound-sensitive membrane, often set in pocket on thorax or abdomen

GENITALIA FEATURES

aedeagus	phallus, penis, or intromittent organ
anellus	(male) membrane supporting and surrounding aedeagus, sometimes sclerotised
apophysis	(pl. -**es**) see apodeme
apodeme	rod-like structure for muscle attachment in abdominal segments
appendix bursae	(female) diverticulum of bursa copulatrix
bursa copulatrix	(female) large sac for storage of sperm
corema	(pl. -**ata**) eversible, membranous structure usually with scent scales, on male abdomen or genitalia
cornutus	(pl. -**i**)(male) spine or spines on aedeagus vesica
corpus bursae	(female) see bursa copulatrix
ductus bursae	(female) tube leading from ostium to bursa copulatrix
epiptygma	(pl. **ata**) lobes associated with the valves of male genitalia
gnathos	(male) in opposition to uncus, sternum of segment 10
harpe	(male) spine or spur or more complex structure on valve, usually associated with the sacculus
juxta	(male) sclerotised plate between bases of valves, a supporting structure for the aedeagus
lamella (ae) antevaginalis	(female) anterior (cephalic) portion of sterigma
lamella (ae) postvaginalis	(female) posterior (caudal) portion of sterigma
lamellae vaginalis	(female) ventral sclerites of segment 8 surrounding ostium burase
ostium bursae	female genital opening
sacculus	(male) folded ventral part of valve
saccus	(male) sac-like antero-ventral extension of vinculum
signum	(pl. -**a**) spiny or otherwise sclerotised process in bursa copulatrix of female
socius	(pl. -**i**) (male) lateral process associated with the uncus
sterigma	(female) structure, often sclerotised, surrounding ostium
superuncus	(male) dorsal secondary process of uncus
tegumen	dorsal segment of male genital ring from which arises the uncus etc.
transtilla	modification of dorsal part of valve costa
uncus	dorsal part of the male genitalia, tergum of segment 10
valve, valva	(pl. -**ae**) (male) internal clasping plates
vesica	(male) eversible sac, often complex, of the aedeagus
vinculum	(male) typically 'u'-shaped ring forming sternum 9

229

REFERENCES

Ackery, P.R. (1975). A guide to the genera and species of Parnassiinae. *Bull. Br. Mus. nat. Hist. (Ent.)* **31**:73-105.

Ackery, P.R. (1984). Systematic and evolutionary studies on butterflies. *In* Vane-Wright & Ackery (1984), pp. 9-21.

Ackery, P.R. & Smiles, R.L. (1976). An illustrated list of the type-specimens of the Heliconiinae (Lepidoptera: Nymphalidae) in the British Museum (Natural History) *Bull. Br. Mus. nat. Hist. (Ent.)* **32**:173-214.

Ackery, P.R. & Vane-Wright, R.I. (1984). *Milkweed Butterflies: their cladistics and biology.* London: British Museum (Natural History), and Ithaca: Cornell University Press.

Adamczweski, S. (1951). On the systematics and origin of the generic group *Oxyptilus* Zeller (Lep. Alucitidae). *Bull. Br. Mus. nat. Hist. (Ent.)* **1**: 303-387.

Aina, J.O. (1983). *Phycita melongenae* sp. n. (Lepidoptera: Pyralidae) associated with eggplant in West Africa. *Bull. ent. Res.* **73**: 427-429.

Aitken, A.D., (1963). A key to the larvae of some species of Phycitinae (Lepidoptera, Pyralidae) associated with stored products, and some related species. *Bull. ent. Res.* **54**: 175-188.

Allen, M.G. (1981). The saturniid moths of Borneo, with special reference to Brunei. *Brunei Mus. J.* **5(1)**: 100-126.

Altena, C.O. van Regteren, (1953). A revision of the genus *Nyctalemon* Dalman (Lepidoptera, Uraniidae) with notes on the biology, distribution and evolution of its species. *Zoologische Verhandelingen* **19**: 1-56.

Amsel, H.G. (1967). Die afghanistan arten des *Anarsia*-komplexes. *Beitr. naturk. Forsch. SW-Deutschl.* **26**: 17-31.

Arora, G.S. (1976). A taxonomic revision of the Indian species of the family Cossidae (Lepidoptera). *Rec. zool. Surv. India* **69**: 1-160.

Arora, G.S. & Gupta, I.J. (1979). Taxonomic studies on some of the Indian non-mulberry silkmoths (Lepidoptera: Saturniidae: Saturniinae). *Mem. zool. Surv. India* **16**: 1-63.

Austin, G.D. (1931-2). The "nettle grub" pest of tea in Ceylon. *Tea Quarterly* **4**: 74-87; **5**: 4-16, 47-53.

Balachowsky, A.S. et al (1966 & 1972). *Entomologie appliquée a l'agriculture. Lépidoptères,* vols **1** & **2**.

Banziger, H. (1982). Fruit-piercing moths in Thailand: a general survey and some new perspectives. *Mitt. schweiz. ent. Ges.* **55**: 213-240.

Banziger, H. (1983). A taxonomic revision of the fruit-piercing and blood-sucking moth genus Calyptra Ochsenheimer (= Calpe Treitschke) (Lepidoptera: Noctuidae). *Ent. Scand.* **14**: 467-491.

Barcant, M. (1970). *Butterflies of Trinidad and Tobago.* London: Collins.

Barlow, H.S. (1982). *An Introduction to the Moths of South East Asia.* Kuala Lumpur: the author.

Bartiloni, P. (1951). La Phthorimaea operculella Zeller (Lep. Gelechiidae) in Italia. *Redia* **36**: 301-379.

Beck, H., (1960). Die larvalsystematik der Eulen. *Abh. Larvalsyst. Insekten* **4**: 1-406.

Becker, V.O. (1981). Identities and provenance of the gelechiid moths originally described by Francis Walker from "unknown countries". *Syst. Ent.* **6**: 137-141.

Beeson, C.F.C. (1919). The life history of the toon shoot and fruit borer, *Hypsipyla robusta* Moore (Lepidoptera; Pyralidae) with suggestions for its control. *Indian Forest Rec.* **7**: 146-216.

Beirne, B.P. (1943). *Argyresthia conjugella* Zell., and other lepidopterous pests in Ireland during 1942. *Econ. Proc. R. Dublin Soc.* **3**: 163-171.

Beirne, B.P. (1943). The biology and control of the small ermine moths (*Hyponomeuta* spp.)

in Ireland. *Econ. Proc. R. Dublin Soc.* **3:** 191-220.

Bell, T.R.D. & Scott, F.B. (1937). *The fauna of British India including Ceylon and Burma.* Moths, vol.5. Sphingidae. London: Taylor & Francis.

Bender, R. (1985). Notodontidae. *Heterocera Sumatrana* **5**:

Berio, E. (1985). *Fauna d'Italia. Lepidoptera, Noctuidae. I. Generalita Hadeninae Cuculliinae.* Bologna: Calderini.

Berio, E. & Fletcher, D.S. (1958). Monografio dell'antico genera *Sypna* Guen. (Lepidoptera Noctuidae). *Ann. Mus. civ. Stor. nat. Giacomo Doria* **70:** 323-402.

Betts, C.R. & Wootton, R.J. (in prep.). Wing shape & flight behaviour in butterflies (Lepidoptera: Papilionoidea & Hesperioidea); a preliminary analysis. *J. Exp. Biol.*

Birket-Smith, J. (1965). A revision of the West African eilemic moths based on the male genitalia. *Papers fac. Sci. Haile Selassie I. Univ. (Zool.)* **1**.

Bleszynski, S. (1961). Revision of world species of the family Crambidae (Lepidoptera). Part 1. Genus *Calamatropha* Zell. *Acta zool. cracov.* **6**: 137-272.

Bleszynski, S. (1965). Crambinae. *Microlepidoptera Palaearctica* **1**: 533 pp. Karlsruhe: G. Brown.

Bleszynski, S. (1970). A revision of the world species of *Chilo* Zincken (Lepidoptera: Pyralidae). *Bull. Br. Mus. nat. Hist. (Ent.)* **25**: 101-195.

Bleszynski, S. & Collins, R.J. (1962). A short catalogue of the world species of the family Crambidae (Lepidoptera). *Acta zool. cracov.* **7**: 197-389.

Boorman, J. (1961). The Hawk-moths of Nigeria. *Niger. Field* 25(4), 26(1).

Bourgogne, J. (1955). Revision des éspeces africaines du genre *Eumeta* Walker (*Clania* auct.) (Lepidoptera: Psychidae). *Trans. R. ent. Soc. Lond.* **107**: 125-138.

Bourgogne, J. (1959). Mise au point relative a deux Psychides ethiopiennes, *Acanthopsyche brunnescens* Gaede et *Kotochalia junodi* Heylaerts (Lepidoptera Psychidae). *Bull. Inst. fond. Afr. noire* **21**: 1227-6.

Bourgogne, J. (1976). Revision des espèces africaines du genre *Monda* F. Walker (Lep. Psychidae). *Bull. Soc. ent. Fr.* **81**: 163-177.

Boursin, C. (1954-1963). Die *"Agrotis"* - Arten aus Dr.H.Hones China-Ausbeten (Beitra zur fauna Sinica). 7 parts: *Bonner zool. Beitr*, **5**: 213-309 (1-2); *Zeitschr. Wien. ent. Ges.* **40**: 216-223, 232-238 (3-4); *Forsch Landes Nordrhein-Westfalen*, **1170** (5-7).

Box, H.E. (1923). The bionomics of the white coffee-leaf miner, *Leucoptera coffeella* Guer., in Kenya Colony (Lepidoptera, Lyonetiidae). *Bull. ent. Res.* **14**: 133-145.

Box, H.E. (1931). The crambine genera *Diatraea* and *Xanthopherne* (Lep., Pyral.). *Bull. ent. Res.* **22**: 1-50.

Box, H.E. (1935). New records and three new species of American *Diatraea* (Lep., Pyral.). *Bull. ent. Res.* **26:** 323-333.

Box, H.E. (1952). New species and records of *Diatraea* Guild. from Northern Venezuela (Lepid., Pyral.). *Bull. ent. Res.* **42**: 379-398.

Bradley, J.D. (1952a). *Adoxophyes orana* (F.v.R., 1934). (Lep., Tortricidae). *Entomologist* **85**: 1-4.

Bradley, J.D. (1952b). Some important species of the genus *Cryptophlebia* Walsingham, 1899, with descriptions of three new species (Lepidoptera: Olethreutidae). *Bull. ent. Res.* **43**: 659-689.

Bradley, J.D. (1956a). A new clearwing moth from West Africa predaceous on scale-insects (Lep., Aegeriidae). *Entomologist* **89**: 203-205.

Bradley, J.D. (1956b). A new genus for *Tortrix postvittana* (Walker) and certain other Australian and New Zealand species (Lepidoptera: Tortricidae). *Bull. ent. Res.* **47**: 101-105.

Bradley, J.D. (1957). A new species of *Conopia* from Malaya (Lepidoptera: Aegeriidae). *Entomologist* **90**: 67-70.

Bradley, J.D. (1958). Taxonomic notes on *Leucoptera meyricki* Ghesquière and *Leucoptera coffeella* (Guérin-Méneville) (Lepidoptera, Lyonetiidae). *Bull, ent. Res.* **49**: 417-419.

232

Bradley, J.D. (1959). A xyloryctid moth attacking the bark of the rubber tree in Malaya. *Bull. ent. Res.* **50**: 9-10.

Bradley, J.D. (1965) A new species belonging to the genus *Anonaepestis* Ragonot (Lepidoptera, Phycitinae) attacking black pepper (*Piper nigrum*) in West Africa. *Bull. ent. Res.* **56**: 299-302.

Bradley, J.D. (1966). A comparative study of the coconut flat moth (*Agonoxena argaula* Meyr.) and its allies, including a new species (Lepidoptera, Agonoxenidae). *Bull. ent. Res.* **56**: 453-472.

Bradley, J.D. (1968a). Descriptions of two new genera and species of Phycitinae associated with *Hypsipyla robusta* (Moore) on Meliaceae in Nigeria (Lepidoptera, Pyralidae). *Bull. ent. Res.* **57**: 605-613.

Bradley, J.D. (1968b). Two new species of clearwing moths (Lepidoptera, Sesiidae) associated with sweet potato (*Ipomoea batatas*) in East Africa. *Bull. ent. Res.* **58**: 47-53.

Bradley, J.D. et al. (1973). A new species of *Mompha* Hübner (Lepidoptera, Momphidae) from Assam, N.E. India, a potential agent for the biological control of *Ludwigia adscendens*. *Bull. ent. Res.* **63**: 57-63.

Bradley, J.D. (1974). A new species of elachistid moth (Lepidoptera, Elachistidae) reared from sugar cane in Papua New Guinea. *Bull. ent. Res.* **64**: 73-79.

Bradley, J.D. (1981). *Marasmia patnalis* sp. n. (Lepidoptera: Pyralidae) on rice in S.E. Asia. *Bull. ent. Res.* **71**: 323-327.

Bradley, J.D. (1982). New species of Microlepidoptera from Norfolk Island. *J. nat. Hist.* **16**: 367-379.

Bradley, J.D. & Carter, D.J. (1982). A new lyonetiid moth, a pest of winged bean. *Syst. Ent.* **7**: 1-9.

Bradley, J.D. & Povolny, D. (1965). The identity of the pepper flower-bud moth (Lepidoptera, Gelechiidae). *Bull. ent. Res.* **56**: 57-63.

Bradley, J.D., Tremewan, W.G. & Smith, A. (1973 & 1979). *British Tortricoid Moths*, vols 1 & 2. London: The Ray Society.

Braun, A.F. (1963). The genus *Bucculatrix* in America North of Mexico. *Mem. Am. ent. Soc.* **18**: 1-208.

Braun, A.F. (1972). Tischeriidae of America north of Mexico (Microlepidoptera). *Mem. Am. ent. Soc.* **28**: 1-148.

British Museum (Natural History) (1980). Common insect pests of stored food products. *Econ. ser.* **15**; 69 pp.

Brock, J.P. (1971). A contribution towards an understanding of the morphology and phylogeny of the Ditrysian Lepidoptera. *J. nat. Hist.* **5**: 29-102.

Brown, E.S. & Dewhurst, C.F. (1975). The genus *Spodoptera* (Lepidoptera, Noctuidae) in Africa and the Near East. *Bull. ent. Res.* **65**: 221-262.

Brown, F.M. & Heinemann, B. (1972). *Jamaica and its Butterflies*. London: Classey.

Busck, A. (1910). Notes on a horn-feeding lepidopterous larva from Africa. *Smithson. misc. Coll.* **56**: 1-2.

Busck, A. (1914). Two microlepidoptera on *Thurberia thespesioides*. *Proc. ent. Soc. Wash.* **16**: 30-31.

Busck, A. (1917). The pink bollworm, *Pectinophora gossypiella*. *J. agric. Res.* **9**: 343-370.

Busck, A. (1934). Microlepidoptera from Cuba. *Entomologica Am.* **13**: 151-203.

Cai, Rongquan (1979). *Economic Insect Fauna of China*. Fmc. **16**. Lepidoptera: Notodontidae. Beijing: Science Press.

Calora, F.B. (1966). A revision of the species of the *Leucania*-Complex occurring in the Philippines (Lepidoptera, Noctuidae, Hadeninae). *Philipp. Agric.* **50**: 633-728.

Capps, H.W. (1953). A correction in the synonymy of the cabbage webworm *Hellula undalis* (F.) (Lepidoptera: Pyraustidae). *Bull. So. Calif. Acad. Sc.* **52**: 46-47.

Capps, H.W. (1963). Keys for the identification of some lepidopterous larvae frequently intercepted at quarantine. *United States Department of Agriculture Agricultural Research Service* **ARS-33-20-1**: 1-37.

Carcasson, R.H. (1968). Revised catalogue of the African Sphingidae (Lepidoptera) with descriptions of the East African species. *J. East Africa. nat. Hist. Soc. natn. Mus.* **26**(3): 1-148.

Carcasson, R.H. (1981). *Collins hand guide to the Butterflies of Africa.* London: Collins.

Carey, P., Scott, L., Cerda, M. & Garda, R.I. (1978). Ciclo estacional de un minador de coigue (*Nothofagus dombeyi*). *Turrialba* **28**: 151-153.

Carter, D.J. (1984). *Pest Lepidoptera of Europe with special reference to the British Isles.* Series Entomologia, **31**. The Hague: Dr. W. Junk.

Carter, D.J. & Deeming, J.C. (1980). *Azygophleps albovittata* Bethune-Baker (Lepidoptera: Cossidae) attacking groundnuts in northern Nigeria, with descriptions of the immature and imaginal stages. *Bull. ent. Res.* **70**: 399-405.

Chapman, P.J. & Lienk, S.E. (1971). *Tortricid fauna of apple in New York.* New York State Agricultural Experiment Station. Geneva.

Chu, H.F. (1949). *How to know the immature insects.* Wm. C. Brown Co., Dubuque.

Chu, H.F. & Wang, L.Y. (1985). The phylogeny of three subfamilies of the Drepanidae (Lepidoptera). *Acta. entomologica Sinica* **28**: 406-411.

Clarke, J.F.G. (1951). New species of Olethreutinae from Argentina (Lepidoptera). *J. Wash. Acad. Sci.* **41**: 296-299.

Clarke, J.F.G. (1955-70). *Catalogue of the type specimens of Microlepidoptera in the British Museum (Natural History) described by Edward Meyrick,* vols. **1-8**. London: BM(NH)

Clarke, J.F.G. (1971). The Lepidoptera of Rapa Island. *Smithson. Contr. Zool.* **56**.

Clarke, J.F.G. (1976). Microlepidoptera: Tortricoidea. *Insects of Micronesia* **9**: 1-144.

Cock, M.J.W., Godfray, H.C.J. & Holloway, J.D. (Eds.) (in press). *Nettle Caterpillars: A monograph of the South-east Asian limacodid moths of economic importance.* Walingford: C.A.B. International.

Collenette, C.L. (1932). The Lymantriidae of the Malay Peninsula. *Novit. zool.* **38**: 49-102.

Collenette, C.L. (1947). The Lymantriidae of Celebes. *Ann. Mag. nat. Hist.* (2) **14**: 1-60.

Collenette, C.L. (1949a). The Lymantriidae of Java. *Ann. Mag. nat. Hist.* (12) **1**: 685-744.

Collenette, C.L. (1949b). The Lymantriidae of Bali. *Entomologist* **82**: 169-175.

Collenette, C.L. (1955). A key to African genera of the Lymantriidae (Lep.). *Trans R. ent. Soc. Lond.* **107**: 187-197.

Collins, N.M. & Morris M.G. (1985). *Threatened swallowtail Butterflies of the world. The IUCN Red Data Book.* Gland and Cambridge: IUCN.

Common, I.F.B. (1957). The occurrence of *Epinotia lantana* (Busck) (Lepidoptera: Olethreutinae) in Australia. *Proc. Linn. Soc. N.S.W.* **82**: 230-232.

Common, I.F.B. (1958). A revision of the bollworms of cotton (*Pectinophora* Busck (Lepidoptera: Gelechiidae)) and related genera in Australia. *Aust. J. Zool.* **6**: 268-306.

Common, I.F.B., (1958). The Australian cutworms of the genus *Agrotis* (Lepidoptera: Noctuidae). *Aust. J. Zool.* **6**: 69-88.

Common, I.F.B. (1960). A revision of the Australian stem borers hitherto referred to as *Schoenobius* and *Scirpophaga* (Lepidoptera: Pyralidae). *Aust. J. Zool.* **8**: 307-347.

Common, I.F.B. (1963). *Australian Moths.* Brisbane: Jacaranda Press.

Common, I.F.B. (1970). Lepidoptera. *In* CSIRO (D.F. Waterhouse, ed.), *Insects of Australia,* pp. 765-865. Melbourne: Melbourne University Press.

Common, I.F.B. (1973). A new family of Dacnonypha (Lepidoptera) based on three new species from Southern Australia, with notes on the Agathiphagidae. *J. Aust. ent. Soc.* **12**: 11-23.

Common, I.F.B. (1974). Lepidoptera. I. *Insects of Australia. Supplement* **1974,** pp. 98-107. Melbourne: Melbourne University Press.

Common, I.F.B. (1975). Evolution and classification of the Lepidoptera. *Ann. Rev. Ent., Palo Alto* **20**: 183-203.

Common, I.F.B. (1979). The larvae and pupae of *Imma acosma* (Turner) and *Ivaticina* Meyrick (Lepidoptera: Immidae), and the taxonomic relationships of the family. *J.*

Aust. ent. Soc. **18**: 33-38.

Common, I.F.B. & Edwards, E.D. (1981). The life-history and early stages of *Synemon magnifica* Strand (Lepidoptera: Castniidae). *J. Aust. ent. Soc.* **20**: 295-302.

Common, I.F.B. & Waterhouse, D.F. (1972). *Butterflies of Australia.* Sydney: Angus and Robertson.

Conway, G.R. (1971). *Pests of cocoa in Sabah.* pp.125. Sabah: Kementerian Pertanian dan Perikananan.

Corbet, A.S. & Pendlebury, H.M. (1978). *The Butterflies of the Malay Peninsula.* 3rd edn., revised by J.N. Eliot. Kuala Lumpur: Malayan Nature Society.

Corbet, A.S. & Tams, W.H.T. (1943). Keys for the identification of the Lepidoptera infesting stored food products. *Proc. zool. Soc. Lond.*, ser. B **113**: 55-148.

Corbett, G.H. & Dover, C. (1927). Notes on miscellaneous insects of 1926. *Malayan Agric. Journal* **15**: 408-419.

Costa Lima, A.M.Da (1945). *Insetos do Brasil, V. Lepidoteros.* Rio de Janiero: Escola Nacionale de Agronomia.

Cottrell, C.B. (1983). Aphytophagy in butterflies: its relationship to myrmecophily. *Zool. J. Linn. Soc.* **80**: 1-57.

Covell, C.V. (1984). *A Field Guide to the Moths of Eastern North America.* (Peterson series) Boston: Houghton Mifflin.

Cowan, C.F. (1966). Indo-Oriental Horagini. *Bull. Br. Mus. Nat. Hist., (Ent.)* **18**: 103-141.

Cowan, C.F. (1967). The Indo-Oriental tribe Cheritrini. *Bull. Br. Mus. Nat. Hist., (Ent.)* **20**: 77-103.

Crumb, S.E. (1929). Tobacco cutworms. *Tech. Bull. U.S. Dep. Agric.* **88**: 1-179.

Crumb, S.E. (1956). The larvae of the Phalaenidae. *Tech. Bull. U.S. Dep. Agric.* **1135**: 1-356.

D'Abrera, B. (1974). *Moths of Australia.* Melbourne: Lansdowne.

D'Abrera, B. (1975). *Birdwing Butterflies of the World.* Melbourne: Lansdowne.

D'Abrera, B. (1978). *Butterflies of the Australian Region* (2nd edition). Melbourne: Lansdowne.

D'Abrera, B. (1980). *Butterflies of the Afrotropical Region.* Melbourne: Lansdowne.

D'Abrera, B. (1981). *Butterflies of the Neotropical Region.* Part 1. Papilionidae and Pieridae. Melbourne: Lansdowne.

D'Abrera, B. (1982). *Butterflies of the Oriental Region.* Part 1. Papilionidae, Pieridae, Danaidae. Victoria: Hill House.

D'Abrera, B. (1984). *Butterflies of the Neotropical Region.* Part 2. Danaidae, Ithomiidae, Heliconiidae and Morphidae. Victoria: Hill house.

D'Abrera, B. (1985). *Butterflies of the Oriental Region.* Part 2. Nymphalidae, Satyridae and Amathusidae. Victoria: Hill House.

D'Abrera, B. (1986). *Butterflies of the Oriental Region.* Part 3. Lycaenidae and Riodiniidae. Victoria: Hill House.

D'Abrera, B. (1986) *Sphingidae Mundi, Hawkmoths of the World.* Faringdon: E.W. Classey.

D'Abrera, B. (1987). *Butterflies of the Neotropical Region.* Part 3. Brassolidae, Acraeidae, Nymphalidae (partim). Victoria: Hill House.

Dall'Asta, U. (1981a). Revision des espèces du genre *Rahona* Griveaud de l'Afrique Centrale (Lepidoptera Lymantriidae). *Rev. Zool. Afr.* **95**: 254-282.

Dall'Asta, U. (1981b). Nouveaux genres et nouvelles espèces de Lymantriidae africains (Lepidoptera). *Rev. Zool. Afr.* **95**: 733-763.

Dall'Asta, U. (1982). Les espèces du genre *Lemuriana* Griveand de l'Afrique Centrale (Lymantriidae, Lepidoptera). *Rev. Zool. Afr.* **96**: 11-30.

Danilevsky, A.S. & Kuznetsov, V.I. (1968). Tortricidae, Laspeyresiini. *Fauna U.S.S.R.* **5**: 1-635.

Davis, D.R. (1964). Bagworm moths of the western hemisphere. *Bull. U.S. natn. Mus.* **244**: 1-233.

Davis, D.R. (1967). A revision of the moths of the subfamily Prodoxinae (Lepidoptera: Incurvariidae). *Proc. U.S. natn. Mus.* **255**: 1-170.

Davis, D.R. (1968). A revision of the American moths of the family Carposinidae (Lepidoptera: Carposinoidea). *Bull. U.S. natn. Mus.* **289**: 1-105.

Davis, D.R. (1975a). Systematics and zoogeography of the family Neopseustidae with the proposal of a new superfamily (Lepidoptera: Neopseustoidea). *Smithson. Contr. Zool.* **210**: 1-45.

Davis, D.R. (1975b). A revision of the West Indian moths of the family Psychidae with descriptions of new taxa and immature stages. *Smithson. Contr. Zool.* **188**: 1-66.

Davis, D.R. (1975c). A review of Ochsenheimeriidae and the introduction of the cereal stem moth *Ochsenheimeria vacculella* into the United States (Lepidoptera: Tineoidea). *Smithson. Contr. Zool.* **192**: 1-20.

Davis, D.R. (1978). A revision of the North American moths of the superfamily Eriocranioidea with the proposal of a new family Acanthopteroctetidae (Lepidoptera). *Smithson. Contr. Zool.* **251**: 1-131.

Davis, D.R. & Nielsen, E.S. (1980). Description of a new genus and two new species of Neopseustidae from South America, with discussion of phylogeny and biogeographical observations (Lepidoptera: Neopseustoidea). *Steenstrupia* **6**: 253-289.

Dekle, G.W. (1965). Illustrated key to caterpillars on corn. *Bull. Fla. Dep. Agric.* **4**: 1-14.

DeVries, P.J. (in press). *The Butterflies of Costa Rica and Their Natural History.* Princeton Univ. Press.

DeVries, P.J., Kitching, I.J. & Vane-Wright, R.I. (1985). The systematic position of *Antirrhea* and *Caerois*, with comments on the classification of the Nymphalidae (Lepidoptera). *Syst. Ent.* **10**: 11-32.

Diakonoff, A. (1956). Records and descriptions of Microlepidoptera (8). *Zool. Verh. Leiden* **29**: 1-60.

Diakonoff, A. (1963). African species of the genus *Bactra* Stephens (Lep., Tortricidae). *Tijdschr. Ent.* **106**: 285-357.

Diakonoff, A. (1964). Further records and descriptions of *Bactra* Stephens (Lep., Tortricidae). *Zool. Verh. Leiden* **70**: 1-61.

Diakonoff, A. (1967). Microlepidoptera of the Philippine Islands. *Bull. U.S. natn. Mus.* **257**: 1-484.

Diakonoff, A. (1973). The South Asiatic Olethreutini (Lepidoptera, Tortricidae). *Zool. Monogrm Rijkmus. nat. Hist.* **1**: 1-700.

Diakonoff, A. (1976). Tortricidae from Nepal, 2. *Zool. Verh. Leiden* **144**: 3-145.

Diakonoff, A. (1981). Tortricidae from Madagascar, Part 2, Olethreutinae, 1. *Annls. Soc. ent. Fr.* **17**: 7-32.

Diakonoff, A. (1982). On a collection of families of Microlepidoptera from Sri Lanka (Ceylon). *Zool. Verh. Leiden* **193**: 1-124.

Diehl, E.W. (1980). Sphingidae. *Heterocera Sumatrana*, **1**.

Dierl, W. (1965). Festlegung eines lectotypes von *Eumeta variegata* (Snellen) (Lepidoptera, Psychidae). *Zool. Meded. Leiden* **40**: 277-279.

Dierl, W. (1966). Psychidae (Lep.) aus Nepal. *Khumbu Himal.* **1**: 322-342.

Dierl, W. (1971). Biologie und systematik einiger asiatischer Psychidae-arten (Lepidoptera). *Khumbu Himal.* **4**: 58-79.

Dierl, W. (1978). Revision der orientalischen Bombycidae (Lepidoptera). Teil 1: Die *Ocinara* Gruppe. *Spixiana* **1**: 225-268.

Dierl, W. (1979). Revision der orientalischen Bombycidae (Lepidoptera). Teil 2: Erganzungen zur *Ocinara* Gruppe. *Spixiana* **3**: 253-258.

Duckworth, W.D. (1969). A new species of Aegeriidae from Venezuela predaceous on scale insects (Lepidoptera: Yponomeutoidea). *Proc. ent. Soc. Wash.* **71**: 487-490.

Duckworth, W.D. & Eichlin, T.D. (1974). Clearwing moths of Australia and New Zealand (Lepidoptera: Sesiidae). *Smithson. Contr. Zool.* **180**: 1-45.

Dufay, C. (1958). Descriptions de nouvelles *Abrostola* espèces (Lep. Noctuidae). *Bull. Inst. Fr. Afr.* **20**: 199-216.

Dufay, C. (1970). Insects Lépidoptères, Noctuidae Plusiinae. *Faune Ent. Madag. Lep.* **31**: 1-198.

Dufay, C. (1974). Descriptions de nouveaux Plusiinae indo-australiens et neotropicaux (Lep. Noctuidae). *Bull. mens. Soc. Linn. Lyon* **43**: 102-111.

Dugdale, J.S. (1971). Entomology of the Aucklands and other islands south of New Zealand: Lepidoptera, excluding non-crambine Pyralidae. *Pacific insects Monograph* **27**: 55-172.

Dugdale, J.S. (1978). Notes on Lepidoptera of Lakeba and Moce Is., Lau group, Fiji, with an annotated list of the Macrolepidoptera. *Lau-Tonga 1977 Bull.* **17**: 63-89.

Dugdale, J.S. (1980). Australian Trichopterygini (Lepidoptera: Geometridae) with descriptions of eight new taxa. *Aust. J. Zool.* **28**: 301-340.

Dumbleton, L.J. (1952). Coleophoridae (Lep.) as pests of clovers. *N.Z. J. Sci. Technol.* **33**: 109-112.

Dumbleton, L.J. (1963). The biology and control of *Coleophora* spp. (Lepidoptera - Coleophoridae) on white clover. *N.Z. J. agric. Res.* **6**: 277-292.

Dusuzeau, J. Sonthonnax, L., (1897-1904) & Conte, M.A. (1906-1911). Essai de classification des Lepidoptéres producteurs de soie. 7 parts in *Rapports du Laboratorie d'Etudes de la Soie.*

Eichlin, T.D. & Cunningham, H.B. (1978). The Plusiinae (Lepidoptera: Noctuidae) of America north of Mexico, emphasising genitalic and larval morphology. *USDA agric. Res. Service Tech. Bull.* **1567**.

Ehrlich, P.R. (1958). The comparative morphology, phylogeny and higher classification of the butterflies (Lepidoptera: Papilionoidea). *Univ. Kansas Sci. Bull.* **39**: 305-370.

Eliot, J.N. (1969). An analysis of the Eurasian and Australian Neptini. *Bull. Br. Mus. nat. Hist. (Ent.)* Suppl. **15**.

Eliot, J.N. (1973). The higher classification of the Lycaenidae (Lepidoptera): a tentative arrangement. *Bull. Br. Mus. nat. Hist. (Ent.)* **28**: 373-505.

Eliot, J.N. (1986). A review of the Miletini (Lepidoptera: Lycaenidae). *Bull. Br. Mus. nat. Hist. (Ent.)* **53**: 1-105.

Eliot, J.N. & Kawazoe, A. (1983). *Blue Butterflies of the Lycaenopsis group.* London: BMNH.

Eltringham, H. (1912). A monograph of the African species of the genus *Acraea* Fab., with a supplement on those of the Oriental Region. *Trans. R. ent. Soc. Lond.* **1912**: 1-374.

Emmet, A.M. in J. Heath et al (eds). (1976 et seq.). *The moths and butterflies of Great Britain and Ireland.* Vol. **1** (1976), vol. **2** (1985). Great Horkesley: Harley Books.

Emmet, A.M. (1979). *A field guide to the smaller British Lepidoptera.* 271 pp. London: Brit. ent. nat. Hist. Soc.

Engelhardt, G.P. (1946). The North American clearwing moths of the family Aegeriidae. *U.S. natn. Mus. Bull.* **190**: 1-222.

Entwistle, P.F. (1963). Observations on the biology of four species of Psychidae (Lepidoptera) on *Theobroma cacao* L. in Western Nigeria. *Proc. R. ent. Soc. Lond.* (A) **38**: 145-148.

Evans, W.H. 1937. *A catalogue of the African Hesperiidae, indicating the classification and nomenclature adopted in the British Museum.* London: BM(NH).

Evans, W.H. 1949. *A catalogue of the Hesperiidae from Europe, Asia and Australia in the British Museum (Natural History).* London: BM(NH).

Evans, W.H. 1951-1955. *A catalogue of the American Hesperiidae indicating the classification and nomenclature adopted in the British Museum (Natural History).* Parts 1-4. London: BM(NH).

Evans, W.H. 1957. A revision of the *Arhopala* group of Oriental Lycaenidae. *Bull. Br. Mus. nat. Hist. (Ent.)* **5**: 85-141.

Falkovitsh, M.I. 1981. *In* Medvedev (ed). (1981).

Ferguson, D.C. (1971-1972). *The moths of America North of Mexico.* Vol. **20-2A, B.** Bombycoidea, Saturniidae. London: E.W. Classey.

Ferguson, D.C. (1978). *The Moths of America North of Mexico.* Vol. **22-2.** Noctuoidea, Lymantriidae. London: E.W.Classey.

Fibiger, M. & Kristensen, N.P. (1974). The Sesiidae (Lepidoptera) of Fennoscandia and Denmark. *Fauna ent. Scand.* **2**: 1-91.

Fleming, W.A. (1975). *Butterflies of West Malaysia and Singapore.* 2 Vols. Faringdon: Classey.

Fletcher, D.S. (1953). A revision of the genus *Carecomotis* (Lep. Geometridae). *Ann. Mag. nat. Hist. (12)* **6**: 99-142.

Fletcher, D.S. (1958-1968). [Macrolepidoptera]. *Ruwenzori Expedition* 1: parts **4,6,7,8** London: BM(NH).

Fletcher, D.S. (1963). 1 Geometridae, 2 Noctuidae. *Inst. Parc. nat. Congo Rwanda* (2nd series) **15**: 1-155.

Fletcher, D.S. (1967). A revision of the Ethiopian species and a checklist of the World species of *Cleora* (Lep., Geom.). *Bull. Br. Mus. nat. Hist. (Ent.).* Suppl. **8**

Fletcher, D.S. (1974). A revision of the old World genus *Zamarada* (Lepidoptera, Geometridae). *Bull. Br. Mus. nat. Hist. (Ent.).* Suppl. **22**.

Fletcher, D.S. (1979). *The Generic Names of Moths of the World.* Volume **3**. Geometroidea. London: BM(NH).

Fletcher, D.S. & Nye I.W.B. (1982). *The Generic Names of Moths of the World.* Volume **4**. Bombycoidea, Castnioidea, Cossoidea, Mimallonoidea, Sesioidea, Sphingoidea, Zygaenoidea. London: BM(NH).

Fletcher, D.S. & Nye I.W.B. (1984). *The Generic Names of Moths of the World.* Volume **5**. Pyraloidea. BMNH.

Fletcher, T.B. (1909). The plume-moths of Ceylon. *Spolia Zeylandica* **6**: 1-39.

Fletcher, T.B. (1914). *Some South Indian Insects and other Animals of Importance, considered from an Economic point of view.* pp.565, Madras.

Fletcher, T.B. (1919). Life-histories of Indian insects. Microlepidoptera. *Mem. Dept. Agric. India* **6**: 1-217.

Fletcher, T.B. (1933). Life-histories of Indian Cosmopterigidae to Neopseustidae. *Imperial Counc. Agric. Res. Sci. Monogr.* no.**4**, 85 pp.

Forster, W. & Wohlfahrt, T.A. (1954-1974). *Die Schmetterlinge Europas* 5 Vols. Stuttgart: Frankh'sche Verlags handlung.

Fox, R.M., Lindsey Jr. A.W., Clench, H.K. & Miller L.D. (1965). The Butterflies of Liberia. *Mem. Amer. ent. Soc.* **19**: 1-438.

Fracker, S.B. (1915). The classification of lepidopterous larvae. *Illinois biol. Monogr.* **2** (1): 1-169.

Franclemont, J.G. (1951). The species of the *Leucania unipuncta* group, with a discussion of the generic names for the various segregates of *Leucania* in North America (Lepidoptera, Phalaenidae, Hadeninae). *Proc. ent. Soc. Wash.* **53**: 57-85.

Franclemont, J.G. (1973). *The Moths of America North of Mexico.* 20.1. Mimallonoidea and Bombycoidea: Apatelodidae, Bombycidae, Lasiocampidae. London, E.W Classey & R.B.D. Publishers.

Freeman, H.A. (1969). Systematic review of the Megathymidae. *J. Lepid. Soc.* **23**: Suppl. 1.

Freise, G. (1960). Revision der paläearktischen Yponomeutidae. *Beitr. Ent.* **10**: 1-131.

Freise, G. (1969). Beitrage zur inseketen-fauna der DDR: Lepidoptera - Argyresthiidae. *Beitr. Ent.* **19**: 693-752.

Fujioka, T. (1975). *Butterflies of Japan* (in Japanese). Tokyo.

Gaedike, R. (1966). Die genitalien der europaischen Epermeniidae. *Beitr. Ent.* **16**: 633-692.

Gaedike, R. (1970). Revision der paläearktischen Acrolepiidae (Lepidoptera). *Ent. Abh. Mus. Tierk. Dresden* **38**: 1-54.

Gaedike, R. (1979). Katalog der Epermeniidae der Welt (Lepidoptera). *Beitr. Ent.* **29**: 271-288.

Gardner, J.C.M. (1938). Immature stages of Indian Lepidoptera (1) Lymantriidae. *Indian Forest Rec.* N.S. **3** (10): 187-212.

Gardner, J.C.M. (1941). Immature Stages of Indian Lepidoptera (2) Noctuidae, Hypsidae. *Indian Forest Rec.* N.S. **6** (8): 253-298.

Gardner, J.C.M. (1941). Immature stages of Indian Lepidoptera (3). *Indian Forest Rec.* N.S. **6**(9): 299-314.

Gardner, J.C.M. (1942). Immature stages of Indian Lepidoptera (4). Zygaenidae. *Indian Forest Rec. N.S.* **7** (4): 155-162.

Gardner, J.C.M. 1946-1948. On larvae of the Noctuidae (Lepidoptera). I-IV. *Trans. R. ent. Soc. Lond.* **96** (4): 61-72; **97** (10): 237-252; **98** (4): 59-90; **99** (8): 291-318.

Gardner, J.C.M. (1948). Notes on the pupae of the Noctuidae. *Proc. R. ent. Soc. Lond.* (B), **17**: 84-92.

Gerasimov, A.M. (1935). Zur Frage der Homodynamie der Borsten von Schmetterlingsraupen. *Zool. Anz.* **112**: 177-194.

Gerasimov, A.M. (1952). Lepidoptera 1 (2). Larvae. *Fauna SSSR (N.S.)* **56**: 1-338 [In Russian; keys to families translated into English by E.L. Martin. Coridon ser.A, 5].

Gibbs, G.W. (1979). Some notes on the biology and status of the Mnesarchaeidae (Lepidoptera). *N.Z. Ent.* **7**: 2-9

Gibbs, G.W. (1980). *New Zealand Butterflies: Identification and Natural History.* London: Collins.

Godfrey, G.L. (1972). A review and reclassification of larvae of the subfamily Hadeninae (Lepidoptera, Noctuidae) of America north of Mexico. *Tech. Bull. U.S. Dep. Agric.* **1450**: 1-265.

Gozmàny, L.A. & Vari, L. (1973). The Tineidae of the Ethiopian region. *Transv. Mus. Mem.* **18**: 1-238.

Griveaud, P. (1959) Lep: Sphingidae. *Faune Madag. Lep.* **8**: 1-53.

Griveaud, P. (1961) Faune de Madagascar. Eupterotidae and Attacidae. *Inst. Sci. Madag.* **14**: 1-64.

Guillermet, C.W.W. & Guillermet, C.W.W. (1986). *Contribution a l'etude des papillons heteroceres de l'Ile de la Reunion.* Reunion: Conseil General.

Hampson, G.F. (1893-1896). *The Fauna of British India, including Ceylon and Burma. Moths.* 4 volumes. London. Reprinted by W. Junk, The Hague.

Hancock, D.L. (1983). Classification of the Papilionidae (Lepidoptera): a phylogenetic approach. *Smithersia* **2**: 1-48.

Hannemann, H.J. (1977a). Kleinschmetterlinge oder Microlepidoptera. II. Cochylidae, Carposinidae, Pyraloidea. *Die Tierweldt Deutschlands* **50**: 1-401.

Hannemann, H.J. (1977b). Kleinschmetterlinge oder Microlepidoptera. III. Pterophoridae, Yponomeutidae, Tineidae. *Die Tierweldt Deutschlands* **63**: 1-273.

Hardwick, D.F. (1965). The corn earworm complex. *Mem. ent. Soc. Canada* **40.**

Hardwick, D.F. (1970a). The genus *Euxoa* (Lepidoptera: Noctuidae) in North America, 1. Subgenera *Orosagrotis, Longivesica, Chorizagrotis, Pleonectopoda* and *Crassivesica. Mem. ent. Soc. Can* **67**.

Hardwick, D.F. (1970b). A generic revision of the North American Heliothidinae (Lepidoptera: Noctuidae). *Mem. ent. Soc. Canada* **73**.

Harris, K.M. (1962). Lepidopterous stem borers of cereals in Nigeria. *Bull. ent. Res.* **53**: 139-171.

Hasenfuss, I. (1960). Die Larvalsystematik der Zünsler (Pyralidae) aus dem Zoologische Institut der Universitat Erlangen. *Abh. Larvalsyst. Insekten* **5**: 1-263.

Haugum, J. & Low, A.M. (1978-1983). *A monograph of the Birdwing Butterflies.* 2 Vols. Klampenborg: Scandinavia.

Hayes, A.H. (1975). The larger moths of the Galapagos islands (Geometroidea, Sphingoidea & Noctuoidea). *Proc. Calif. Acad. Sci.* **15**: 145-208.

Heath, J. et al. (1976 et seq.) *The moths and butterflies of Great Britain and Ireland,* vol. 1 onwards. Great Horkesley: Harley Books.

Heinrich, C. (1916). On the taxonomic value of some larval characters in the Lepidoptera. *Proc. ent. Soc. Wash.* **18** (3): 154-164.

Heinrich, C. (1919). Notes on the European corn borer (*Pyrausta nubilalis* Hb.) and its nearest American allies, with description of larvae, pupae and one new species. *J. agric. Res.* **18**: 171-178.

Heinrich, C. (1923). Revision of the North American moths of the subfamily Eucosminae of the family Olethreutinae. *Bull. U.S. natn. Mus.* **123**: 1-298.

Heinrich, C. (1926). Revision of the North American moths of the subfamilies Laspeyresiinae and Olethreutinae. *Bull. U.S. natn. Mus.* **132**: 1-216.

Heinrich, C. (1937). Moths of the genus *Rupela* (Pyralidae: Schoenobiinae). *Proc. U.S. natn. Mus.* **84**: 355-388.

Heinrich, C. (1956). American moths of the subfamily Phycitinae. *Bull. U.S. natn. Mus.* **207**: 1-581.

Hemming, F. (1969). The generic names of butterflies and their type-species. *Bull. Br. Mus. nat. Hist. (Ent.)* Suppl. **9**.

Heppner, J.B. (1977). The status of the Glyphipterigidae and a reassessment of relationships in yponomeutoid families and ditrysian superfamilies. *J. Lepid. Soc.* **31**: 124-134.

Heppner, J.B. (1982a). Review of the family Immidae, with a world checklist. *Entomograph* **1**: 257-279.

Heppner, J.B. (1982b). Millieriinae, a new subfamily of the Choreutidae, with new taxa from the United States (Lepidoptera: Sesioidea). *Smithson. Contr. Zool.* **370**: 1-27.

Heppner, J.B. (ed.) (1984). Checklist: Part 1, Micropterigoidea - Immoidea. *Atlas of Neotropical Lepidoptera.* The Hague: Dr. W. Junk.

Heppner, J.B. & Duckworth, W.D. (1981). Classification of the superfamily Sesioidea (Lepidoptera: Ditrysia). *Smithson. Contr. Zool.* **314**: 1-144.

Hering, E.M. (1925). Die Familie der Ratardidae (Lep.). Mit beschreibung 5 neuer Arten. *Mitt. Zool. Mus. Berl.* **12**: 151-158.

Hering, E.M. (1937). Revision der Chrysopolomidae (Lep.) *Ann. Transv. Mus.* **17**: 233-257.

Hering, E.M. (1951). *Biology of the leaf miners.* 420 pp. Gravenhage: Dr. W. Junk.

Hering, E.M. (1953). *Bestimmungstabellen der blattminen von Europa,* 3 vols. Gravenhage: Dr. W. Junk.

Hering, E.M. (1955). Synopsis der Afrikanischen gattungen der Cochlidiidae (Lepidoptera). *Trans. R. ent. Soc. Lond.* **107**: 209-225.

Higgins, L.G. & Riley, N.D. (1980). *A Field Guide to the Butterflies of Britain and Europe.* 4th ed. revised. London: Collins.

Higgins, L.G. (1981). A revision of *Phyciodes* Hubner and related genera, with a review of the classification of the Melitaeinae (Lepidoptera: Nymphalidae). *Bull. Br. Mus. nat. Hist. (Ent.)* **43**: 77-243.

Hinton, H.E. (1943). The larvae of Lepidoptera associated with stored products. *Bull. ent. Res.* **34**: 163-212.

Hinton, H.E. (1946). On the homology and nomenclature of the setae of lepidopterous larvae, with some notes on the phylogeny of the Lepidoptera. *Trans. R. ent. Soc. Lond.* **97**: 1-37.

Hinton, H.E. (1956). The larvae of the species of Tineidae of economic importance. *Bull. ent. Res.* **47**: 251-346.

Hinton, H.E. & Bradley, J.D. (1956): Observations on species of Lepidoptera infesting stored products. XVI. Two genera of clothes moths. *Entomologist* **89**: 42-47.

Hodges, R.W. (1962). A revision of the Cosmopterigidae of America north of Mexico, with a definition of the Momphidae and Walshiidae (Lepidoptera: Gelechioidea). *Entomologica Am.* **42**: 1-171.

Hodges, R.W. (1966). Review of the New World species of *Batrachedra*, with description of three genera (Lepidoptera: Gelechioidea). *Trans. Am. ent. Soc.* **92**: 585-651.

Hodges, R.W. (1969) Nearctic Walshiidae notes and new taxa (Lepidoptera: Gelechioidea). *Smithson. Contr. Zool.* **18**: 1-30.

Hodges, R.W. (1971). *The Moths of America North of Mexico.* Vol. **21**. Sphingoidea, hawkmoths. London: E.W.Classey.

Hodges, R.W. (1978). *The Moths of America North of Mexico.* Vol. **6**. Cosmopterigidae.

Hodges, R.W. et al. (1983). *Check list of the Lepidoptera of America North of Mexico.* London: E.W. Classey.

Holdaway, F.G. (1926). The pink bollworm of Queensland. *Bull. ent. Res.* **17**: 67-83.

Holdaway, F.G. (1929). Confirmatory evidence of the validity of the species *Pectinophora scutigera*, Holdaway (Queensland pink bollworm), from a study of the genitalia. *Bull. ent. Res.* **20**: 179-185.

Holloway, J.D. (1976). *Moths of Borneo with special reference to Mt. Kinabalu.* Kuala Lumpur: Malayan Nature Society.

Holloway, J.D. (1977). *The Lepidoptera of Norfolk Island, their Biogeography and Ecology.* Series Entomologica **13**. The Hague: W. Junk.

Holloway, J.D. (1979a). *A survey of the Lepidoptera, Biogeography and Ecology of New Caledonia.* Series Entomologica **15**. The Hague: W. Junk.

Holloway, J.D. (1982). Taxonomic Appendix. *In* H.S. Barlow, *An Introduction to the Moths of South East Asia,* pp. 174-271. Kuala Lumpur: the author.

Holloway, J.D. (1983a). The moths of Borneo; family Notodontidae. *Malayan Nature Journal* **37**: 1-107

Holloway, J.D. (1983b). The biogeography of the macrolepidoptera of south-eastern Polynesia. *GeoJournal* **7**: 517-525.

Holloway, J.D. (1984). Lepidoptera and the Melanesian Arcs. pp. 129-169. *In* Radovsky, F.J., Raven, P.H., & Sohmer, S.H. (Eds.) *Biogeography of the Tropical Pacific.* Bishop Museum Special Publication **72**.

Holloway, J.D. (1985). The moths of Borneo: family Noctuidae: Subfamilies Euteliinae, Stictopterinae, Plusiinae, Pantheinae. *Malayan Nature Journal* **38**: 157-317.

Holloway, J.D. (1986a). The moths of Borneo: Key to families; Families Cossidae, Metarbelidae, Ratardidae, Dudgeoneidae, Epipyropidae and Limicodidae. *Malayan Nature Journal* **40**: 1-165.

Holloway, J.D. (1986b). Lepidoptera faunas of high mountains in the Indo-Australian tropics. pp. 533-556. *In* F. Vuilleumier & M. Monasterio (eds.) *High Altitude Tropical Biogeography.* New York; Oxford University Press.

Holloway, J.D. (in press). *The Moths of Borneo*: superfamily Bombycoidea; families Lasiocampidae, Eupterotidae, Bombycidae, Brahmaeidae, Saturniidae, Sphingidae. Kuala Lumpur: Southdene.

Holloway, J.D. & Peters, J.V. (1976). The butterflies of New Caledonia and the Loyalty islands. *J. nat. Hist.* **10**: 273-318.

Hopkins, G.H.E. (1927) *Insects of Samoa.* Part III Lepidoptera (1). London: BM(NH).

Hopp, W. 1927. Die Megalopygiden - Unterfamilie der Trosiinae (Lep. Megalopyg.). *Mitt. Zool. Mus. Berlin* **13**: 208-336.

Horak, M. (1984). Assessment of taxonomically significant structures in Tortricinae (Lep., Tortricidae). *Mitt. Schweiz. ent. Ges.* **57**: 3-64.

Houlbert, C. (1918). Revision monographique de la sous familie des Castniinae. *Et. Lepid. comp.* **15**.

Howe, W.H. (1975). *The Butterflies of North America.* Doubleday.

Hubei People's Press (1978). *Illustrations of pests of rice and their natural enemies.* 181 pp., 88 pls.

Hudson, G.V. (1928) *The Butterflies and Moths of New Zealand.* Wellington: Ferguson & Osborn.

Hudson, G.V. (1939) *A Supplement to the Butterflies and Moths of New Zealand.* Wellington: Ferguson & Osborn.

Igarashi, S. (1979). *Papilionidae and their Early Stages.* 2 vols (in Japanese). Tokyo: Kodansha.

Inoue, H. (1956-1957). A revision of the Japanese Lymantriidae. 2 parts. *Japan. J. Med. Sci. Biol.* **9**: 133-163; **10**: 187-219.

Inoue, H. (1970). Geometridae of eastern Nepal. Part 1. Genera *Abraxas* Leech and *Arichanna* Moore. *Spec. Bull. Lepid. Soc. Japan* **4**: 203-239.

Inoue, H. (1971). The Geometridae of the Ryu Kyu Islands. *Bull. Fac. dom. Sci. Otsuma Wom. Univ.* **7**: 141-179.

Inoue, H. (1972). The genus *Abraxas* of Japan, Korea, Saghalien and Manchurin. *Bull. Fac. dom. Sci. Otsuma. Wom. Univ.* **8**: 141-163.

Inoue, H. (1973). An annotated and illustrated catalogue of the Sphingidae of Taiwan (Lepidoptera). *Bull. Fac. dom. Sci. Otsuma Wom. Univ.* **9**: 103-139.

Inoue, H. (1979-1980). Revision of the genus *Eupithecia* of Japan (Lepidoptera, Geometridae). 2 parts. *Bull. Fac. dom. Sci. Otsuma Wom. Univ.* **15**: 157-224, **16**: 153-213.

Inoue, H. Sugi, S. Kuroko, H. Moriuti, S. & Kawabe, A. (1982). *Moths of Japan.* 2 vols. Tokyo: Kodansha.

Isaac, P.V. & Rao, K.V. (1941). A key for the identification of the larvae of the known lepidopterous borers of sugarcane in India based on morphological characters. *Indian J. agric. Sci.* **11**(5): 795-803.

Isaac, P.V. & Venkatraman, T.V. (1941). A key for the identification of pupae of the known lepidopterous borers of sugarcane in India based on morphological characters. *Indian J. Agric. Sci.* **11**(5): 804-815.

Janse, A.J.T. (1932-1964). *The Moths of South Africa.* 7 volumes (incomplete). Durban.

Johnston, G. & Johnston, B. (1980). *This is Hong Kong: Butterflies.* Hong Kong: Crown Copyright.

Jordan, K. (1922). A monograph of the Saturnian subfamily Ludiinae. *Novit. zool.* **29**: 247-326.

Jordan, K. (1924). On the saturnoidean families Oxytenidae and Cercophanidae. *Novit. zool.* **31**: 135-193.

Kalshoven, L.G.E. (1950). *De plagen van de cultuurgewassen in Indonesia* **1**: 360-635.

Kalshoven, L.G.E. (1981). *The Pests of Crops in Indonesia.* Revised and translated by P.A. van der Laan with G.H.L.Rothschild. Jakarta: P.T.Ichtiar Baru - van Hoeve.

Kalsholt, O. & Nielsen, E.S. (1984). A taxonomic review of the stem moths of northern Europe (Lepidoptera: Ochsenheimeriidae). *Ent. Scand.* **15**: 233-247.

Kasy, von F. (1969). Vorlaufige revision de gattung *Ascalenia* Wocke (Lepidoptera, Walshiidae). *Annln naturh. Mus. Wien* **73**: 339-735.

Kato, M. (1940). A monograph of Epipyropidae (Lepidoptera). *Ent. World* **8**: 67-94.

Kawazoe, A. & Wakabayashi, M. (1976). *Coloured Illustrations of the Butterflies of Japan.* Osaka: Hoikusha.

Kimball, C.P. (1965). *The Lepidoptera of Florida, an annotated checklist.* Gainesville: Division of Plant Industry.

Kiriakoff, S.G. (1958). Les Notodontidae africains, *Desmeocraera* et genres voisins (Lep.: Notodontoidea). *Annls Mus. r. Congo. Belge 8 Ser. Zool.*, **66**: 1-88.

Kiriakoff, S.G. (1959). Entomological results of the Swedish expedition 1934 to Burma and the British India, Lepidoptera: family Notodontidae. *Arkiv für Zoologi* (2) **12**: 313-333.

Kiriakoff, S.G. (1960). Thyretidae. *Genera Insectorum* **39** (214e).

Kiriakoff, S.G. (1962). Les Notodontidae africains *Phalera* et genres voisins (Lep.: Notodontoidea). *Revue Zool. Bot. afr.* **66**: 1-44.

Kiriakoff, S.G. (1962). Notes sur les Notodontidae (Lepidoptera): *Pydna* Walker et genres voisins (II). *Bull. ann. Soc. r. Ent. Belg.* **98**: 149-214.

Kiriakoff, S.G. (1963). Die Notodontidae der ausbeten H. Hones aus Ostasien (Lepidoptera). *Bonner. zool. Beitr.* **14**:248-293.

Kiriakoff, S.G. (1963). Les Notodontidae africains. Le 'groupe de *Cerura*' et quelques autres genres (Lep. Notodontoidea). *Bull. Annls. Soc. r. ent. Belg.* **99**: 205-228.

Kiriakoff, S.G. (1964). Notodontidae. *Genera Insectorum* **217a**.

Kiriakoff, S.G. (1967). Notodontidae. *Genera Insectorum* **217b**.

Kiriakoff, S.G. (1968). Lepidoptera, Familia Notodontidae, pars tertia, Genera Indo-Australica. *Genera Insectorum* **217c**.

Kiriakoff, S.G. (1970). Lepidoptera. Familia Thaumetopoeidae. *Genera Insectorum* **219e**.

Kiriakoff, S.G. (1974). Neue und wening bekannte asiatische Notodontidae (Lepidoptera). *Veroft. zool. Staatssamml. München* **17**: 371-421.

Kiriakoff, S.G. (1977). Lepidoptera, Noctuiformes, Agaristidae. 3 vols. *Das Tierreich* **97-99**

Kitching, I.J. (1984). An historical review of the higher classification of the Noctuidae (Lepidoptera). *Bull. Br. Mus. nat. Hist. (Ent.)* **49**: 153-234.

Kitching, I.J. (1987). Spectacles and silver Ys: a synthesis of the systematics, cladisitcs and biology of the Plusiinae (Lepidoptera: Noctuidea). *Bull. Br. Mus. nat.*

Hist. (Ent.) **54**: 75-261

Kobes, L.W.R. (1985). The Thyatiridae, Agaristidae, and Noctuidae (part 1: Pantheinae and Catocalinae) of Sumatra. *Heterocera Sumatrana* **4**.

Kostrowicki, A.S. (1956, 1959) and subsequent authors. *Klucze do Oznaczania owadow Polski.* Series 27-Lepidoptera, part **53** - Noctuidae. Warsaw.

Kozanchikov, I.V. (1969-1956). *Fauna of the U.S.S.R.* **3**. Psychidae 525pp. English translation.

Krampl, F. & Dlabola, J. (1983). A new genus and species of epipyropid moth from Iran ecto-parasitic on a new *Mesophantia* species, with a revision of the host genus (Lepidoptera, Epipyropidae; Homoptera, Flatidae). *Acta ent. bohemoslov.* **80**: 451-472.

Kristensen, N.P. (1967). Erection of a new family in the lepidopterous suborder Dacnonypha. *Ent. Meddr.* **35**: 341-345.

Kristensen, N.P. (1976). Remarks on the family-level phylogeny of the butterflies. *Z. Zool. Syst. Evol. Forsch* **14**: 25-33.

Kristensen, N.P. (1968). The skeletal anatomy of the heads of the adult Mnesarchaeidae and Neopseustidae (Lep., Dacnonypha). *Ent. Meddr.* **36**: 137-151.

Kristensen, N.P. (1978). A new family of Hepialoidea from South America, with remarks on the phylogeny of the suborder Exoporia (Lepidoptera). *Ent. Germ.* **4**: 272-294.

Kristensen, N.P. (1984a). The male genitalia of *Agathiphaga* (Lepidoptera: Agathiphagidae) and the Lepidoptera ground plan. *Ent. Scand.* **15**: 151-178.

Kristensen, N.P. (1984b). Studies on the morphology and systematics of primitive Lepid-optera (Insecta). *Steenstrupia* **10**: 141-191.

Kuroko, H. (1964). Revisional studies on the family Lyonetiidae of Japan (Lepidoptera). *Esakia* **4**: 1-61.

Kuznetsov, V.I. (1978). *In* Medvedev, G.S. (ed), (1978).

Kuznetsov, V.I. & Stekolnokov, A.A. (1985). Comparative and functional morphology of the male genitalia of the Bombycoid moths (Lepidoptera, Papilionomorpha: Lasio-campoidea, Sphingoidea, Bombycoidea) and their systematic position (in Russian with English summary). *Proc. zool. Inst. USSR Acad. Sci.* **134**: 3-48.

Kyrki, J. (1984). The Yponomeutoidea: a reassessment of the superfamily and its supra-generic groups (Lepidoptera). *Ent. Scand.* **15**: 71-84.

Lafontaine, J.D., Mikkola, K. & Kononenko, V.S. (1983). A revision of the genus *Xestia* subg. *Schoyenia* Auriv. (Lepidoptera: Noctuidae) with descriptions of four new species and a new subspecies. *Ent. Scand.* **14**: 337-369.

Lajonquière, Y. de (1968). Description d'un genre nouveau de Lasiocampidae et de son espece type (Lep. Lasioc. Gonometinae). *Bull. Soc. ent. Fr.* **73**: 68-74.

Lajonquière, Y. de (1973). Genres *Dendrolimus* Germar, *Hoenimnema* n. gen., *Cyclophragma* Turner. *Ann. Soc. ent. Fr. (N.S.)* **9**: 509-592.

Lajonquière, Y. de (1974). Formes asiatiques du genre *Cosmotriche* Hübner (= *Selenephera* Rambur = *Selenepherides* Daniel = *Wilemaniella* Matsumura). *Bull. Soc. ent. Fr.* **79**: 132-146.

Lajonquière, Y. de (1976). Le genre *Gastropacha* Ochsenheimer en Asie et le genre *Paradox-opla* nov. gen. *Anns. Soc. ent. Fr. (N.S.)* **12**: 151-177.

Lajonquière, Y. de (1977). Les genres *Arguda* Moore et *Radhica* Moore. *Bull. Soc. ent. Fr.* **82**: 174-191.

Lajonquière, Y. de (1978). Le genre *Philudoria* Kirby, 1892. *Annls. Soc. ent. Fr. (N.S.)* **14**: 381-413.

Lampe, R.E. (1985). *Malayan Saturniidae from the Cameron and Genting Highlands. A Guide for Collectors.* Faringdon: E.W.Classey.

Lange, W.H. (1950). Biology and systematics of plume moths of the genus *Platyptilia* in California. *Hilgardia* **19**: 561-668.

Larsen, T.B. (1974). *Butterflies of Lebanon.* Lebanon: National Council for Scientific Research.

Larsen, T.B. (1984). *Butterflies of Saudi Arabia and its Neighbours.* London: Stacey International.

Lekagul, B. Askins, K. Nabhitabhata, J. & Samwadkit, A. (1977). *Field Guide to the Butterflies of Thailand.* Bangkok: Assoc. for Conservation of Wildlife.

Lemaire, C. (1971-1974). Revision du genre *Automeris* Hübner et des genres voisins, biogeographie, ethologie morphologie, taxonomie (Lep. Attacidae). *Mem. Mus. Nat. Hist. nat. (N.S.) Ser A.*, **68, 79, 92**.

Lemaire, C. (1978). *The Attacidae of America (=Saturniidae). Attacinae.* Neuilly-sur-seine: Edition Lemaire.

Lemaire, C. (1980). *The Attacidae of America (=Saturniidae). Arsenurinae.* Neuilly-sur-seine: Edition Lemaire.

Lever, R.A. (1969). *Pests of the Coconut Palm.* 190 pp. Rome: FAO.

Lewis, H.L. (1973). *Butterflies of the World.* London: Harrap.

Lewvanich, A. (1981). A revision of the Old World species of *Scirpophaga* (Lepidoptera: Pyralidae). *Bull. Br. Mus. nat. Hist. (Ent.)* **42**: 185-298.

Mackay, M.R. (1959). Larvae of the North American Olethreutidae (Lepidoptera). *Can. Ent.*, Suppl. **10**: 1-338.

Mackay, M.R. (1962). Larvae of the North American Tortricinae (Lepidoptera: Tortricidae). *Can. Ent.*, Suppl. **28**: 1-182.

Mackay, M.R. (1963). Problems in naming the setae of lepidopterous larvae. *Can. Ent.* **95**: 996-999.

Mackay, M.R. (1964). The relationship of form and function of minute characters of lepidopterous larvae, and its importance in life-history studies. *Can. Ent.* **96**: 91-104.

Mackay, M.R. (1968). The North American Aegeriidae (Lepidoptera): a revision based on late-instar larvae. *Mem. Ent. Soc. Can.* **58**: 1-112.

Mackay, M.R. (1972). Larval sketches of some Microlepidoptera, chiefly North American. *Mem. ent. Soc. Can.* **88**: 1-83.

Maes, K. (1984). A comparative study of the genitalia of some Lymantriidae Hampson (1893) of the Palaearctic Region. *Academiae Analecta* **46**: 125-149.

Maes, K. (1985). A comparative study of the abdominal tympanal organs in Pyralidae (Lepidoptera). I. Description, terminology, preparation technique. *Nota lepid.* **8**: 341-350.

Marlatt, C.L. (1905). The giant sugar-cane borer. *Bull. Bur. Ent. U.S. Dept. Agric.* **54**: 71-75.

Mathur, R.N. (1954). Immature stages of Indian Lepidoptera No.9 - Pyralidae, sub-family Pyraustinae. *Indian Forest Rec.* N.S. **8** (11): 241-265.

Mathur, R.N. (1959). Immature stages of Indian Lepidoptera No. 12 - Pyralidae, sub-family Pyraustinae. *Indian Forest Rec.* N.S. **9** (10): 183-210.

Mathur, R.N. & Singh, P., (1963). Immature stages of Indian Lepidoptera No. 13 Pyralidae, sub-family Pyraustinae. *Indian Forest Rec.* N.S. **10** (6): 117-148.

McDunnough, J.H. (1949). Revision of the North American species of the genus *Eupithecia* (Lep., Geom.). *Bull. Am. Mus. nat. Hist.* **93**: 537-728.

McDunnough, J.H. (1954). The species of the genus *Hydriomena* occuring in America north of Mexico. (Geom., Larentiinae). *Bull. Amer. Mus. nat. Hist.* **104**: 241-358.

McGuffin, W.C. (1958). Larvae of the Nearctic Larentiinae (Lepidoptera: Geometridae). *Can. Ent.*, *Suppl.* **8**: 1-104.

McGuffin, W.C. (1967). Guide to the Geometridae of Canada (Lepidoptera) I. Subfamily Sterrhinae. *Mem. ent. Soc. Can.* **50**: 1-67.

McGuffin, W.C. (1972-1981). Guide to the Geometridae of Canada (Lepidoptera) II. Subfamily Ennominae. 1-3. *Mem. ent. Soc. Can.* **86**: 1-159; **101**: 1-191. **117**: 1-153.

McKinley, D.J. (1968). Key to some larvae of Lepidoptera attacking cotton in Central Africa. *Cotton Growing Review* July **1968**: 184-197.

McQuillan, P.B. (1985). A taxonomic revision of the Australian autumn gum moth genus *Mnesampela* Guest (Lepidoptera: Geometridae, Ennominae). *Entomologica Scandinavica*, **16**: 175-202.

McQuillan, P.B. & Forrest, J.A. (1985). *A Guide to Common Moths of the Adelaide Region.* 52

pp. South Australian Museum: Special Educational Bulletin No. **5**.

Medvedev, G.S. (ed.) (1978, 1981, 1986). Lepidoptera. *Key to the insects of the European part of the U.S.S.R.* **4** (1) (1978), 710 pp.; **4** (2) (1981), 786 pp.; **4** (3) (1986), 503pp.

Mell, R. (1943). Beitrage zur Fauna sinica XXIV. Uber Phlogophorinae, Odontodinae, Sarrothripinae, "Westermannianae" und Camptolominae (Noctuidae, Lepid.) von Kuangtung. *Zoologische Jahrbucher* **76**: 171-226.

Michener, C.D. (1952). The Saturniidae (Lepidoptera) of the Western Hemisphere, morphology, phylogeny and classification. *Bull. Amer. Mus. nat. Hist.* **98**: 335-502.

Mikkola, K. & Jalas, I. (1977,1979). *Suomen Perhoset*, Yokkoser 1 & 2. Helsinki: Otava.

Miller, J.Y. (1972). Review of the Central American *Castnia inca* complex (Castniidae). *Bull. Allyn Mus.* **6**: 1-13.

Miller, L.D. (1968). The higher classification, phylogeny and zoogeography of the Satyridae (Lepidoptera). *Mem. Amer. ent. Soc.* **24**.

Minet, J.M. (1980). Creation d'une sous-famille particulière pour *Noorda* Walker, 1859, et definition d'un nouveau genre parmi les Odontiinae. *Bull. Soc. ent. Fr.* **85**: 79-87.

Minet, J. (1982). Les Pyraloidea et leurs principales divisions systématiques (Lep. Ditrysia). *Bull. Soc. ent. Fr.* **86**: 262-280.

Minet, J. (1982). Éléments sur la systématique des Notodontidae et nouvelles données concernant leur étude faunistique à Madagascar. *Bull. Soc. ent. Fr.* **87**: 354-370.

Minet, J. (1983). Étude morphologique et phylogénétique des organes tympanique des Pyraloidea. 1. Généralites et homologies (Lep. Glossata). *Annls Soc. ent. Fr.* (N.S.) **19**: 175-207.

Minet, J. (1986). Ébauche d'une classification moderne de l'ordre des Lépidoptères. *Alexanor* **14**: 291-313.

Miyata, T. (1970). A generic revision of the Japanese Bombycidae, with description of a new genus. *Tinea*, **8**: 190-195.

Mohammad, A. (1981). The groundnut leafminer, *Aproaerema modicella* Deventer (*Stomopteryx subsecivella* Zeller) (Lepidoptera: Gelechiidae). *Groundnut Improvement program Occasional Paper* **3**: 1-33.

Moore, F. (1894). A new enemy of the custard apple. *Indian Mus. Notes* **3**: 106-107.

Moriuti, S. (1969). Aryresthiidae (Lepidoptera) of Japan. *Bull. Univ. Osaka Prefect.* **21**: 1-50.

Morishita, K. (1981). Danaidae. *Butterflies of the south-east Asian islands.* **2**: 441-598. Tokyo.

Mosher, E. (1916). A classification of the Lepidoptera based on characters of the pupa. *Bull. Illinois St. Lab. nat. Hist.* **12**: 17-159.

Moulds, M.S. (1981). Larval food plants of hawk moths (Lepidoptera: Sphingidae) affecting commercial crops in Australia. *Gen. appl. Ent.* **13**: 69-80.

Moulds, M.S. (1984). Larval food plants of hawk moths (Lepidoptera: Sphingidae) affecting garden ornaments in Australia. *Gen. appl. Ent.* **16**: 57-64.

Munroe, E. (1961). Synopsis of the North American Odontiinae, with descriptions of New genera and species (Lepidoptera: Pyralidae). *Can. Ent., Suppl.* **24**: 1-93.

Munroe, E. (1972 et seq.). Pyralidae. *The moths of America North of Mexico* fasc. **13**: et seq. Faringdon: E.W.Classey.

Mutuura, A. et al, (1965, 1969). *Early stages of Japanese moths in colour.* Vols. I & II. Hoikusha, Osaka.

Mutuura, A. & Munroe, E. (1970). Taxonomy and distribution of the European corn borer and allied species; genus *Ostrinia* (Lepidoptera: Pyralidae). *Mem. ent. Soc. Can.* **71**: 1-112.

Nakajima, H. (1970). A contribution to the knowledge of the immature stages of Drepanidae occuring in Japan. *Tinea*, **8**: 167-184.

Neunzig, H.H. (1979). Systematics of immature phycitines (Lepidoptera: Pyralidae) associated with leguminous plants in the southern United States. *Tech. Bull.*

U.S. Dep. Agric. **1598**: 1-119.

Nielsen, E.S. (1985). Primitive (non-ditrysian) Lepidoptera of the Andes: diversity, distribution, biology and phylogenetic relationships. *J. Res. Lepid.*, *Suppl.* **1**: 1-16.

Nielsen, E.S. (1987). The recently discovered primitive (non-ditrysian) family Palaephatidae (Lepidoptera) in Australia. *Invertbr. Taxon.* **1**: 201-229.

Nielsen, E.S. & Davis, D.R. (1985). The first Southern Hemisphere prodoxid and the phylogeny of the Incurvarioidea (Lepidoptera). *Syst. Ent.* **10**: 307-322.

Nielsen, E.S. & Robinson, G.S. (1983). Ghost moths of southern S. America (Lepidoptera: Hepialidae). *Steenstrupia* **7**: 25-57.

Nye, I.W.B. (1960). The insect pests of graminaceous crops in East Africa. *Colonial Research Studies* **31**. London: HMSO.

Nye, I.W.B. (1975). *The Generic Names of Moths of the World.* Volume **1**, Noctuoidea: Noctuidae. London: BMNH.

Okumura, G.T. (1961). Identification of lepidopterous larvae attacking cotton. *Spec. Publs. California Department of Agriculture* **282**: 1-80.

Orfila, R.N. (1961). Les Dalceridae (Lep. Zygaenoidea) Argentinas. *Revist. Invest. Agric.* **40**: 249-264.

PANS, (1970). *Pans manual no. 3, Pest control in rice.* 270 pp. London: Ministry for Overseas Development (PANS).

Pant, G.D. & Chatterjee, N.C., (1950). A list of described immature stages of Indian Lepidoptera. Part I Rhopalocera. *Indian Forest Rec.* N.S. **7** (7): 213-255.

Pant, G.D. & Chatterjee, N.C., (1951). A list of described immature stages of Indian Lepidoptera. Part II Heterocera. *Indian Forest Rec.* N.S. **7** (9): 267-333.

Parsons, M. (1986a). A revision of the genus *Callictita* Bethune-Baker (Lepidoptera: Lycaenidae) from the mountains of New Guinea. *Bull. Allyn Mus.* **103**: 1-27.

Parsons, M. (1986b). A new genus and twenty-six new species of butterflies (Lepidoptera: Hesperiidae, Lycaenidae, Nymphalidae) from Papua New Guinea and Irian Jaya. *Tyo To Ga* **37**: 103-177.

Patzak, H. (1974). Beitrage zur insekten fauna der DDR: Lepidoptera - Coleophoridae *Beitr. Ent..Berlin* **24**: 153-278.

Pennington, K.M. (1978). *Pennington's Butterflies of South Africa.* (Ed. C.G.C. Dickson). Johannesburg. A.D. Douke.

Peterson, A., (1962). *Larvae of insects. Part 1, Lepidoptera and plant infesting Hymenoptera.* Columbus, Ohio.

Pierre, J. (1983). *Systématique Evolutive, Cladistique et Mimetisme chez les Lépidoptères du genre* Acraea *(Nymphalides).* Doctoral thesis, Universite Paris VI.

Pinhey, E.C.G. (1962). *Hawk Moths of Central and Southern Africa.* Cape Town: Longman.

Pinhey, E.C.G. (1965). *The Butterflies of South Africa.* Johannesburg: Nelson.

Pinhey, E.C.G. (1972). *Emperor Moths of South and South Central Africa.* Cape Town: Struik.

Pinhey, E.C.G. (1975). *Moths of Southern Africa.* Capetown: Tafelberg.

Pinratana, A. (1981-1983). *Butterflies in Thailand.* 4 Volumes. Bangkok: Viratham Press.

Pittaway, A.R. (in prep.) *Hawkmoths of the Western Palaearctic.* Harley Books.

Pottinger, R.P. & Le Roux, E.J. (1971). The biology and dynamics of *Lithocolletis blancardella* (Lepidoptera: Gracillariidae) on apple in Quebec. *Mem. ent. Soc. Can.* **77**: 1-437.

Povolný, D. (1964). Gnorimoschemini trib. nov. - eine neue tribus der familie Gelechiidae nebst bemerkungen zu ihrer taxonomic (Lepidoptera). *Cas. ceske Spol. ent.* **61**: 330-359.

Povolný, D. (1966). A type revision of some Old-World species of the tribe Gnorimoschemini with special regard to pests (Lepidoptera). *Acta ent. bohemoslovaca* **63**: 128-148.

Povolný, D. (1967). Ein kritischer beitrag zur taxonomischen klarung einiger palaearktischer arten der gattung *Scrobipalpa* (Lepidoptera: Gelechiidae) *Acta sc. nat. Brno* **1**: 209-250.

Povolný, D. (1967). Die stammesgeschichtlichen beziehungen der tribus Gnorimoschemini in weltrahmen (Lepidoptera, Gelechiidae). *Acta ent. Mus. natn. Pragae* **37**: 161-232.

Povolný, D. (1967). Genitalia of some nearctic and neotropical members of Gnorimoschemini (Lepidoptera, Gelechiidae). *Acta ent. Mus. natn. Pragae* **37**: 51-127.

Povolný, D. (1973a). *Scrobipalpopsis solanivora* sp. n. - a new pest of potato (*Solanum tuberosum*) from Central America. *Sb. vys. Sk. zemed. Brně* **21**: 133-146.

Povolný, D. (1973b). *Keiferia brunnea* sp. n., taxonomic studies of the neotropical genera *Keiferia* Busck and *Tildenia* Povolný, and their economic importance. *Sb. vys. Sk. zemed. Brně* **21**: 603-615.

Povolný, D. (1975a). *Keiferia colombiana* sp. n., a new species of Gelechiidae from South America. *Acta Univ. Agric. Brno* **23**: 109-113.

Povolný, D. (1975b). On three neotropical species of Gnorimoschemini (Lepidoptera, Gelechiidae) mining Solanaceae. *Sb. vys. Sk. zemed. Brně* **23**: 379-393.

Povolný, D. (1977). Notes on Gnorimoschemini of Australia and New Zealand (Lepidoptera, Gelechiidae). *Acta ent. Mus. natn. Pragae* **39**: 403-443.

Povolný, D. (1979). On some little known moths of the family Gelechiidae (Lepidoptera) as pests of crops. *Sb. vys. Sk. zemed. Brně* **27**: 139-165.

Povolný, D. & Bradley, J.D. (1980). *Megalocypha melongenae* sp. n. reared from *Solanum melongena* in southern India (Lepidoptera, Gelechiidae). *Acta ent. bohemoslovaca* **77**: 112-117.

Powell, J.A. (1964). Occurrence in California of *Oinophila v-flava*, a moth probably introduced from Europe. *Pan-Pacif. Entom.* **40**: 155-157.

Powell, J.A. (1976). The giant blastobasid moths of yucca (Gelechioidea). *J. Lepid. Soc.* **30**: 219-229.

Prout, L.B. (1929). A revision of the Indo-Australian *Cleora* of the *alienaria* group. *Bull. Hill Mus.*, **3**: 179-222.

Prout, L.B. (1937). A revision of the *decisaria* group of *Cleora*. *Novit. zool.*, **40**: 190-198.

Pyle, R.M. (1981). *The Audubon Society Field Guide to North American Butterflies.* New York: Alfred Knopf.

Razowski, J. (1976). Phylogeny and system of Tortricidae (Lepidoptera). *Acta zool. cracov.* **21**: 73-118.

Razowski, J. (1977). Monograph of the genus *Archips* Hübner (Lepidoptera, Tortricidae). *Acta. zool. cracov.* **22**: 55-205.

Reidl, T. (1969). Materiaux por la connaissance de Momphidae palearctiques (Lepidoptera). *Polskie Pismo ent.* **39**: 645-923.

Riley, N.D. (1975). *A Field Guide to the Butterflies of the West Indies.* London: Collins.

Rindge, F.H. (1959). A revision of *Glaucina*, *Synglochis* and *Eubarnesia* (Lepidoptera, Geometridae) *Bull. Am. Mus. nat. Hist.* **118**: 259-366.

Rindge, F.H. (1961). A revision of the Nacophorini (Lepidoptera, Geometridae). *Bull. Am. Mus. nat. Hist.* **123**: 87-154.

Rindge, F.H. (1964). A revision of the genera *Melanolophia*, *Pheroteria* and *Melanotesia* (Lep., Geometridae). *Bull. Am. Mus. nat. Hist.* **126**: 241-434.

Rindge, F.H. (1965). A revision of the Nearctic species of the genus *Glena* (Lepidoptera, Geometridae). *Bull. Am. Mus. nat. Hist.* **129**: 265-306.

Rindge, F.H. (1966). A revision of the moth genus *Anacamptodes* (Lepidoptera, Geometridae). *Bull. Am. Mus. nat. Hist.* **132**: 175-244.

Rindge, F.H. (1967). A revision of the Neotropical species of the moth genus *Glena* (Lepidoptera, Geometridae). *Bull. Am. Mus. nat. Hist.* **135**: 109-171.

Rindge, F.H., (1968). A revision of the moth genus *Stenoporpia* (Lepidoptera, Geometridae) *Bull. Am. Mus. nat. Hist.* **140**: 65-134.

Rindge, F.H., (1970). A revision of the moth genera *Hulstina* and *Pterotaea* (Lepidoptera, Geometridae). *Bull. Am. Mus. nat. Hist.* **142**: 255-342.

Rindge, F.H. (1971). A revision of the Nacophorini from cool and cold temperate southern South America (Lepidoptera, Geometridae). *Bull. Am. Mus. nat. Hist.* **145**: 303-392.

Rindge, F.H. (1972). A revision of the moth genus *Mericisca* (Lepidoptera, Geometridae).

Bull. Am. Mus. nat. Hist. **149**: 341-406.

Rindge, F.H. (1973). A revision of the moth genera *Nepterotaea* and *Chesiadodes* (Lepidoptera, Geometridae). *Bull. Am. Mus. nat. Hist.* **152**: 205-252.

Rindge, F.H. (1975). A revision of the New World Bistonini (Lepidoptera, Geometridae). *Bull. Am. Mus. nat. Hist.* **156**: 69-156.

Rindge, F.H. (1978). A revision of the genus *Sabulodes* (Lepidoptera, Geometridae). *Bull. Am. Mus. nat. Hist.* **160**: 193-292.

Riotte, J.C.E., & Peigler, R.S. (1980). A revision of the American genus *Anisota* (Saturniidae). *J. Res. Lepid.* **19**: 101-180.

Roberts, H. (1966). A survey of the important shoot, stem, wood, flower and fruit boring insects of the Meliaceae in Nigeria. *Niger. For. Inf. Bull.* (N.S.) **15**: 1-38.

Robinson, G.S., (1975). *Macrolepidoptera of Fiji and Rotuma, a taxonomic and geographic study.* Faringdon: Classey.

Robinson, G.S. (1979). Clothes-moths of the *Tinea pellionella* complex: a revision of the world's species (Lepidoptera; Tineidae). *Bull. Br. Mus. nat. Hist. (Ent.)* **38**: 57-128.

Rockburne, E.W., & Lafontaine, J.D. (1973). The Cutworm Moths of Ontario and Quebec. *Research Branch, Canada Dept. Agriculture Publications* **1593**.

Roepke, W. (1951). The genus *Trabala* Walk. in the Far East (Lep. Het. fam. Lasiocampidae). *Meded. landbhoogesch. Wageningen* **50**: 104-133.

Roepke, W. (1955). Notes and descriptions of Cossidae from New Guinea (Lepidoptera: Heterocera). *Trans. R. ent. Soc. Lond.* **107**: 281-288.

Roepke, W. (1957a). The cossids of the Malay Region (Lepidoptera: Heterocera). *Verh. Akad. Wet. Amst. (Afd. Natuurk.)* (2) **52**: 1-60.

Roepke, W. (1957b). The genus *Nyctemera* Huebner. II. *Tijdschr. ent.* **100**: 147-178.

Roesler, U. (1983). Die Phycitinae von Sumatra (Lepidoptera: Pyralidae). *Heterocera Sumatrana* **3**.

Rothschild, W. & Jordan, K. (1903). A revision of the lepidopterous family Sphingidae. *Novit. zool.* **9**, Suppl. (2 vols).

Rougeot, P.-C. (1955). Les Attacides (Saturniidae) de l'Equateur Africain Francais. *Encycl. Ent.* **34**: 1-116.

Rougeot, P.-C. (1962). Les Lépidoptères de l'Afrique noire. Attacides. 4. *Bull. Inst. fr. Afrique noire*, **14**: 1-205.

Rougeot, P.-C. & Viette, P., (1978). *Guide des Papillons nocturnes d'Europe et' d'Afrique du Nord.* Paris: Delachauxe & Niestle.

Rungs, C. (1953). Le complexe de *Leucania loreyi* auct. nec Dup. (Lep. Phalaenidae). *Bull. Soc. ent. Fr.* **58**: 138-141.

Sakai, S. (1981). *Butterflies of Afghanistan.* Tokyo.

Samson, C. (1983). Butterflies (Lepidoptera: Rhopalocera) of Vanuatu. *Naika* (J. Vanuatu nat. Hist. Soc.). **June 1983:** 2-6.

Sato, R. (1984). Taxonomic study of the genus *Hypomecis* Hübner and its allied genera from Japan (Lepidoptera: Geometridae: Ennominae). *Special Bulletin of Essa Entomological Society.*

Sattler, K. & Tremewan, W.G. (1974 & 1978). Catalogue of the family-group and genus names of the Coleophoridae (Lepidoptera). *Bull. Br. Mus. Hist. (Ent.)* **30** (1974): 185-214; **37** (1978): 73-96.

Sbordoni, V. & Forestiero, S. (1985). *The World of Butterflies.* London: Guild Publishing.

Schreiber, H. (1978). Dispersal centres of Sphingidae (Lepidoptera) in the Neotropical region. *Biogeographica* **10**.

Scoble, M.J. (1986). The structure and affinities of the Hedyloidea: a new concept of the butterflies. *Bull. Br. Mus. nat. Hist. (Ent.)* **53**: 251-286.

Scoble, M.J. & Edwards, E.D. (in press). *Hypsidia* Rothschild: a review and a reassessment (Lepidoptera: Drepanoidea, Drepanidae). *Ent. Scand.*

Scott, J.A. (1984). The phylogeny of butterflies (Papilionoidea and Hesperioidea). *J. Res. Lepid.* **23**: 241-281.

248

Scott, J.A. (1986). On the monophyly of the Macrolepidoptera, including a reassessment of their relationship to Cossoidea and Castnioidea, and a reassignment of Mimallonidae to Pyraloidea. *J. Res. Lepid.* **25**: 30-38.

Sekskyaeva, S. (1983). *In* Medvedev (ed). (1983).

Shaffer, J.C. (1968). A revision of the Peoriinae and Anerastiinae (auctorum) of America north of Mexico (Lepidoptera: Pyralidae). *U.S. nat. Mus. Bull.* **280**: 1-124.

Shields, O. (1984). A revised annotated checklist of World Libytheidae. *J. Res. Lepid.* **22**: 264-266.

Shields, O. (1985). Zoogeography of the Libytheidae (Snouts or beaks). *Tokurana* 9: 1-58.

Shirozu, T. (1960). *Butterflies of Formosa in Colour.* Osaka: Hoikusha.

Shirozu, T. & Hara, A. (1960-1962). *Early Stages of Japanese Butterflies in Colour.* 2 Vols. Osaka: Hoikusha.

Sick, H. (1939). Family Dalceridae. In A. Seitz (ed). *Macrolepidoptera of the World* 6: 1303-1312.

Singh, B. (1953). Immature stages of Indian Lepidoptera No. 8 - Geometridae. *Indian Forest Rec.* N.S. **8** (7): 67-158.

Singh, B. (1955). Description and systematic position of the larva and pupa of the teak defoliator, *Hyblaea puera* Cramer (Lepidoptera, Hyblaeidae). *Indian Forest Rec.* **9**: 1-16.

Singh, B. (1957). Some more Indian geometrid larvae (Lepidoptera) with a note on the identity of components of various groups of setae. *Indian Forest Rec.* N.S. 9(6): 131-162.

Skou, P. (1984). *Nordens Malere, Handbog over de danske og fennoskandiske arter af Drepanidae og Geometridae (Lepidoptera).* Kobenhavn & Suenborg: Fauna Boger & Apollo Boger.

Smart, P. (1975). *The Illustrated Encyclopaedia of the Butterfly World.* London: Hamlyn.

Smiles, R.L. (1982). The taxonomy and phylogeny of the genus *Polyura* Billberg (Lepidoptera; Nymphalidae). *Bull. Br. Mus. nat. Hist. (Ent.).* **44**: 115-237.

Smith, M.R. (1965). A list of Lepidoptera associated with cocoa in West Africa, with notes on identification and biology of species in Ghana. *Cocoa Res. Inst. Tech. Bull.* **9.**

Speidel, W. (1977). *Ein Versuch zur Unterteilung der Lepidopteren in Unterordnungen.* Atalanta, Wurzburg 8(2): 119-121.

Speidel, W. (1984). Revision der Acentropinae der palaearktischen faunengebeites (Lepidoptera, Crambidae). *Neue Entomologische Nachrichten* **12**: 1-157.

Stehr, F.W. (ed.) (1987). *Immature Insects.* [Lepidoptera, pp. 288-596]. Dubuque, Iowa: Kendall/Hunt.

Stempffer, H. (1967). The genera of the African Lycaenidae (Lep. Rhopal.). *Bull. Br. Mus. nat. Hist. (Ent.)* Suppl. **10**: 1-322.

Sukhareva, M.I. (1978). *In* Medvedev, G.S. (ed). (1978).

Swatschek, B., (1958). Die Larvalsystematik der Wickler. *Abh. Larvalsyst. Insekton* 3: 1-269.

Talbot, G. (1928-1937). *A monograph of the Pierine genus* Delias. London: John Bale, Danielsson and British Museum (Natural History).

Talbot, G. (1939). *The fauna of British India, including Ceylon and Burma. Butterflies.* London: Taylor & Francis., reprint New Delhi, 1975.

Tams, W.H.T. (1935). Heterocera (exclusive of the Geometridae and the microlepidoptera). *Insects of Samoa* **3**: 169-290.

Tams, W.H.T. & Bowden, J. (1953). A revision of the African species of *Sesamia* Guenee and related genera (Agrotidae - Lepidoptera). *Bull. ent. Res.* **43**: 645-678.

Tarmann, G. (1984). Generische Revision der amerikanischen Zygaenidae mit beschreibung neuer guttungen und Arten (Insecta: Lepidoptera). *Entomofauna, Zeitschrift fur Entomologie* Suppl. 2 (2 vols.).

Tite, G.E. (1963). A synonymic list of the genus *Nacaduba* and allied genera. *Bull. Br. Mus. nat. Hist. (Ent.)* **13**: 67-116.

Tite, G.E. (1963b). A revision of the genus *Candalides* and allied genera. *Bull. Bri. nat.*

Hist. (Ent.). **14**: 197-259.

Todd, E.L. (1959). The fruit piercing moths of the genus *Gonodonta* Hübner (Lepidoptera, Noctuidae). *U.S. Dept. Agri. Agr. Res. Ser.* **1201**: 1-52.

Todd, E.L. & Poole, R.W. (1980). Keys and illustrations for the armyworm moths of the noctuid genus *Spodoptera* Guenée from the Western Hemisphere. *Annals. ent. Soc. Amer.* **73**: 722-738.

Toll, S. (1953). Eupistidae of Poland. *Mater. Fizjogr. Kraju* **32**: 1-292.

Tothill, J.D. Taylor, T.H.C. & Pavie, R.W. (1930). *The Coconut Moth in Fiji.* London: Imperial Bureau of Entomology.

Toulgoet, H. de (1984). Liste recapitulative des lépidoptères Arctiidae et Nolidae de Madagascar et de l'archipel des Comores. *Miscellanea entomologica,* **50**: 69-108.

Traugott-Olsen, E. & Nielsen, E.S. (1977). The Elachistidae (Lepidoptera) of Fennoscandia and Denmark. *Fauna ent. Scand.* **6**: 1-299.

Tsukada, E. & Nishiyama, Y., (1982). Papilionidae. *Butterflies of the South East Asian Islands.* **1**. (transl. K. Morishita). Tokyo: Plapac.

Tsukada, E. (1985). Nymphalidae (1). *Butterflies of the south-east Asian islands.* **4**. Tokyo: Plapac.

Ueda, K. (1984). A revision of the genus *Deltote* R.L. and its allied genera from Japan and Taiwan. (Lepidoptera: Noctuidae: Acontinae). Part 1. A generic classification of the genus *Deltote* R.L. and allied genera. *Bull. Kitakyushu. Mus. nat. Hist.* **5**: 91-133.

Ueda, K. (1987). A revision of the genus *Deltote* R.L. and its allied genera from Japan and Taiwan (Lepidoptera: Noctuidae; Acontiinae). Part 2. Systematics of the genus *Deltote* R.L. and its allied genera. *Bull. Kitakyushu Mus. nat. Hist.* **6**: 1-117.

Vane-Wright, R.I. & Ackery, P.R. (Eds.) (1984). *The Biology of Butterflies.* Symposium of the Royal Entomologcal Society of London, **11**. London: Academic Press.

Van Someren, V.G.L. (1963-1975). Revisional notes on African *Charaxes* (Lepidoptera: Nymphalidae). 10 parts. *Bull. Br. Mus. nat. Hist. (Ent.).*

Vari, L. (1961). Lithocolletidae. *South African Lepidoptera,* **1**: 1-238.

Viette, P. (1967). Noctuidae (Amphipyrinae, Melicleptrinae). *Faune Madag. Lep.* **20**.

Washbourn, R. (1940). On the distribution of *Leucoptera daricella* (Meyr.); with the description of a new leaf-miner from coffee. *Bull. ent. Res.* **30**: 455-462.

Watson, A., (1965). A revison of the Ethiopian Drepanidae (Lepidoptera). *Bull. Br. Mus. nat. Hist. (Ent.)* Suppl. **3**: 1-177.

Watson, A. (1967). A survey of the extra-Ethiopian Oretinae (Lepidoptera: Drepanidae). *Bull. Br. Mus. nat. Hist. (Ent.)* **13**: 152-221.

Watson, A. (1968). The taxonomy of the Drepaninae represented in China, with account of their world distribution (Lepidoptera: Drepanidae). *Bull. Br. Mus. nat. Hist. (Ent.)* **12**: 1-151.

Watson, A. (1971). An illustrated catalogue of the Neotropical Arctiinae types in the United States National Museum. I. *Smiths. Contr. Zool.* **50**.

Watson, A. (1973). An illustrated catalogue of the Neotropical Arctiinae types in the United States National Museum. II. *Smiths. Contr. Zool.* **128**.

Watson, A. (1975). A reclassification of the Arctiidae and Ctenuchidae formerly placed in the thyretid genus *Automolis* Hibner (Lepidoptera). *Bull. Br. Mus. nat. Hist. (Ent.)* Suppl. **25**.

Watson, A. (1980). A revision of the *Halysidota tesselaris* species group. *Bull. Br. Mus. nat. Hist. (Ent.)* **40**: 1-65.

Watson, A., Fletcher, D.S. & Nye, I.W.B. (1980). *The Generic Names of Moths of the World.* Volume 2. Noctuoidea (part). London: British Museum (Natural History).

Watson, A., & Goodger, D.T. (1986). Catalogue of the Neotropical tiger-moths. *Occasional papers on Systematic Entomology.* **1**.

Watson, A. & Whalley, P.E.S. (1975). *The Dictionary of Butterflies and Moths in Colour.* London: McGraw-Hill. [Reprinted with minor revision in 1983]

250

Werner, K. (1958). Die Larvalsystematik einiger Kleinschmetterlingsfamilien. *Abh. Larval-syst. Insekten* **2**: 1-145.

Werny, K. (1966). *Untersuchen uber die Systematik der Tribus Thyatirini, Macrothyatirini, Habrosynini, und Tetheini (Lepidoptera: Thyatiridae)*. Innaugural Dissertation, Universitat des Saarlandes.

Whalley, P.E.S. (1963). A revision of the World species of the genus *Endotricha* Zeller (Lepidoptera: Pyralidae). *Bull. Br. Mus. nat. Hist (Ent.)* **13**: 395-454.

Whalley, P.E.S. (1964a). Catalogue of the World genera of the Thyrididae (Lepidoptera) with type selection and synonymy. *Ann. Mag. nat. Hist.* **7**: 115-127.

Whalley, P.E.S. (1964b). Catalogue of the Galleriinae (Lepidoptera, Pyralidae) with descriptions of new genera and species. *Acta zool. cracov.* **9**: 561-618.

Whalley, P.E.S. (1971). The Thyrididae of Africa and its islands, a taxonomic and zoogeographic study. *Bull. Br. Mus. nat. Hist. (Ent.)* **17**: 1-198.

Whalley, P.E.S. (1973). The genus *Etiella* Zeller (Lepidoptera: Pyralidae): a zoogeographic and taxonomic study. *Bull. Br. Mus. nat. Hist. (Ent.)* **28**: 1-21.

Whalley, P.E.S. (1974). Scent dispersal mechanisms in the genus *Striglina* Guenee (Lepid. Thyrididae), with a description of a new species. *J. Ent.* **43**: 121-128.

Whalley, P.E.S. (1976). *Tropical leaf moths*. London: BM(NH).

Wilkinson, C. (1967). A taxonomic revision of the genus *Teldenia* Moore (Lepidoptera: Drepanidae, Drepaninae). *Trans. R. ent. Soc. Lond.* **119**: 303-362.

Wilkinson, C. & Scoble, M.J. 1979. The Nepticulidae of Canada. *Mem. ent. Soc. Can.* **107**: 1-129.

Williams, J.G. (1969). *A Field Guide to the Butterflies of Africa*. London: Collins.

Williams, J.R. (1953). The larvae and pupae of some important Lepidoptera. *Bull. ent. Res.* **43**: 691-701.

Wiltshire, E.P. (1948-49). The Lepidoptera of the Kingdom of Egypt. In 2 parts. *Bull. Soc. Fouad Ent.* **32**: 203-296; **33**: 381-460.

Wiltshire, E.P. (1957). *The Lepidoptera of Iraq*. London: Kaye.

Wiltshire, E.P. (1977). Lepidoptera, Parts 1 & 2. *J. Oman Stud.* Spec. Rep. No. **1**, 1977: 155-176.

Wiltshire, E.P. (1979). A revision of the Armadini (Lep., Noctuidae). *Entomonograph* **2**.

Wiltshire, E.P. (1980). The larger moths of Dhofar and their zoogeographic compositions. *J. Oman Stud.* Spec. Rep. No. **2**, 1980: 187-216.

Wiltshire, E.P. (1980-1983). Lepidoptera. *In*: *Fauna of Saudi Arabia* **2**: 179-240; **4**: 271-332; **5**: 293-332.

Wiltshire, E.P. (1984). Insects of Saudi Arabia. Lepidoptera: Fam. Noctuidae (part 4). *Fauna of Saudi Arabia* **6**: 388-412.

Wiltshire, E.P. (1986). Lepidoptera of Saudi Arabia: Families Cossidae, Sesiidae, Metarbelidae, Lasiocampidae, Sphingidae, Geometridae, Lymantriidae, Arctiidae, Nolidae, Noctuidae (Heterocera); Family Satyridae (Rhopalocera) (part 5). *Fauna of Saudi Arabia* **8**: 262-323.

Wood, B.J. (1968). *Pests of oil palms in Malaysia and their control*. Kuala Lumpur: Incorporated Society of Planters.

Woodhouse, L.G.O. (1950). *The Butterfly Fauna of Ceylon*. 2nd Edition. Colombo: Colombo Apothecaries Co.

Wynter-Blyth, M.A. (1957). *Butterflies of the Indian Region*. Bombay: Bombay Natural History Society.

Yano, K. (1963). Taxonomic and biological studies of Pterophoridae of Japan (Lepidoptera). *Pacif. Insects* **5**: 65-209.

Yasuda, T. (1972 & 1975). The Tortricinae and Sparganothinae of Japan (Lepidoptera: Tortricidae). *Bull. Univ. Osaka Prefect.* **24** (1972): 53-134; **27** (1975): 79-251.

Yata, O. (1981). Pieridae. *Butterflies of the south-east Asian islands*, **2**: 205-438. Tokyo.

Zagulyaev, A.K. (1960). Tineinae. *Fauna U.S.S.R.* **4** (3), 266 pp. In Russian.

Zagulyaev, A.K. (1964). Nemapogoninae. *Fauna U.S.S.R.* **4** (2), 435 pp. In Russian.

Zagulyaev, A.K. (1975). Myrmecozelinae. *Fauna U.S.S.R.* **4** (5), 426 pp. In Russian.

Zagulyaev, A.K. (1979) Meessiinae. *Fauna U.S.S.R.* **4** (6), 408pp. In Russian.

Zagulyaev, A.K. (1981). *In* Medvedev. G.S. (ed). (1981)

Zimmerman, E.C. (1958). Lepidoptera: Pyraloidea. *Insects of Hawaii* **8:** 465 pp.

Zimmerman, E.C. (1978). Microlepidoptera. *Insects of Hawaii* **9** (parts 1 & 2): 1903 pp.

SPECIES COMBINATIONS USED IN THIS MANUAL

Species	Genus	Species	Genus
acetosae	Stigmella	cramerella	Conopomorpha
africana	Psara	crocicapitella	Monopis
agrippina	Thysania	derogata	Syllepte
albipunctella	Raghuva	demodocus	Papilio
albovittata	Azygophleps	depressella	Emmalocera
alcyonipennella	Coleophora	diffusa	Naranga
alternus	Neostauropus	dimidiatella	Opogona
anonella	Cerconota	disciplaga	Squamura
aphidivora	Cryptoblabes	doleus	Heteropan
aphidivora	Isauria	entella	Oeonistis
arenosella	Batrachedra	eriosoma	Chrysodeixis
argaula	Agonoxena	euadrusalis	Orthaga
articollis	Ectoedemia	eustrigata	Calyptra
atomosa	Marasmarcha	farinalis	Pyralis
assectella	Acrolepiopsis	ficipastica	Stathmopoda
atlas	Attacus	flava	Anomis
auricilia	Chilo	fluctuosalis	Parapoynx
badia	Pyroderces	frischella	Coleophora
basalis	Milionia	frugalis	Mocis
belina	Gonimbrasia	fullonia	Othreis
berberidella	Carposina	fusca	Busseola
bhurmitra	Ectropis	gemmatalis	Anticarsia
bimacula	Aegocera	gnidiella	Cryptoblabes
binotalis	Crocidolomia	gossypiella	Pectinophora
biselliella	Tineola	gossypii	Bucculatrix
blancardiella	Phyllonorycter	granella	Nemapogon
cactorum	Cactoblastis	hecabe	Eurema
capitella	Lampronia	hercules	Coscinocera
catoxantha	Artona	hosei	Endoclita
cautella	Ephestia	humuli	Hepialus
celerio	Hippotion	inana	Blastobasis
celtis	Selepa	incertulas	Scirpophaga
cephalonica	Corcyra	innotata	Scirpophaga
ceramica	Xyleutes	insulana	Earias
cerealella	Sitotroga	interpunctella	Plodia
chaerophyllella	Epermenia	ipomoeella	Stigmella
chalcites	Chrysodeixis	ipsilon	Agrotis
citrella	Phyllocnistis	iridescens	Levuana
clerkella	Lyonetia	irorata	Anticarsia
coccidivora	Laetilia	jocosatrix	Penicillaria
coccidophaga	Euzophora	julianalis	Dicymolomia
coffeae	Zeuzera	kuehniella	Ephestia
coffearia	Homona	laceratalis	Hypena
coffeella	Leucoptera	leilus	Urania
conflictaria	Epiplema	lepida	Parasa
conjugella	Argyresthia	leucostoma	Cydia
conspersa	Argyresthia	leucotreta	Cryptophlebia
convolvuli	Agrius	libania	Gorgopis
convolvuli	Brachmia	licoides	Castnia
corbetti	Mahasena	lineola	Amsacta
coronata	Ophiusa	littoralis	Spodoptera

loreyi
loxoptila
lupulinus
malella
malinella
mandarina
materna
mathias
medinalis
medosa
micaceana
minor
modicella
molesta
morgani
mori
myrmecias
myrsusalis
nilotica
norax
nubilalis
oecophila
oenochares
ombrodelta
operculella
orichalcea
pariana
partellus
pastinacella
pelodes
pennatula
penniseti
peponis
pilaria
plagiophleps
polychrysa
pomella
pomonella
proleucella
pseudospretella
psophocarpella
puera
pulverulenta
pusillidactyla
queenslandensis
radda
reticulata
ricini
rileyi
rubiella
rubraria
rufivena
rutella

Mythimna
Bucculatrix
Hepialus
Stigmella
Yponomeuta
Bombyx
Othreis
Pelopides
Cnaphalocrocis
Dasychira
Archips
Phycodes
Aproaerema
Cydia
Xanthopan
Bombyx
Copromorpha
Banisia
Characoma
Hibrildes
Ostrinia
Oecia
Phthoropoea
Cryptophlebia
Phthorimaea
Thysanoplusia
Choreutis
Chilo
Depressaria
Autosticha
Psara
Sesamia
Anadevidia
Thalassodes
Pteroma
Chilo
Stigmella
Cydia
Cryptoblabes
Hofmannophila
Leucoptera
Hyblaea
Epiceura
Lantanophaga
Agathiphaga
Eublemma
Anaphe
Pericallia
Pyroderces
Lampronia
Scopula
Tirathaba
Setomorpha

sabulifera
saccharalis
scitula
semibrunnea
separatella
sericeus
simplex
simplicella
siva
sodaliata
solena
somnulentella
sperbius
splendiflorella
strigatum
sundascribens
tamsi
testacea
theae
thrax
thurberiella
tipuliformes
trapezialis
ubaldus
undalis
unipuncta
vaculella
valida
variegata
venosa
venosana
virescens
viridana
vishnou
vitiensis
xanthorrhoea
xylostella
yuccasella
zebrina
zinckenella

Anomis
Diatraea
Eublemma
Odites
Maliarpha
Endoclita
Nisaga
Glyphipterix
Streblote
Xanthorhoe
Eupneusta
Bedellia
Syntomis
Coptodisca
Hypena
Aegilia
Epicampoptera
Luperina
Parametriotes
Erionota
Bucculatrix
Synanthedon
Marasmia
Azanus
Helulla
Mythimna
Ochsenheimeria
Sagalassa
Eumeta
Hibrildes
Bactra
Heliothis
Tortryx
Trabala
Agathiphaga
Euproctis
Plutella
Tegeticula
Erecthias
Etiella

INDEX TO GENERA AND SPECIES
(Italicised numbers represent pages containing relevant illustrations usually in addition to textual references)